PLANTS AND THEIR
ATMOSPHERIC ENVIRONMENT

PLANTS AND THEIR ATMOSPHERIC ENVIRONMENT

The 21st Symposium of
The British Ecological Society
Edinburgh 1979

EDITED BY

J. GRACE
Department of Forestry and Natural Resources
University of Edinburgh

E. D. FORD
Institute of Terrestrial Ecology
Penicuik

AND

P. G. JARVIS
Department of Forestry and Natural Resources
University of Edinburgh

BLACKWELL SCIENTIFIC PUBLICATIONS
OXFORD LONDON EDINBURGH
BOSTON MELBOURNE
1981

Typeset by Enset Ltd.
Midsomer Norton, Bath,
printed in Great Britain at
Billing & Sons Ltd., Guildford and
bound by
Kemp Hall Bindery, Oxford

DISTRIBUTORS

USA and Canada
   Halsted Press,
   a Division of
   John Wiley & Sons Inc.
   New York

Australia
   Blackwell Scientific Book
   Distributors
   214 Berkeley Street, Carlton
   Victoria 3053

British Library
Cataloguing in Publication Data

Plants and their atmospheric environment,
   (British Ecological Society. Symposia;
   21st).
   1. Botany – Ecology – Congresses
   2. Atmosphere – Congresses
   I. Grace, John, b. 1945
   II. Ford, E. D.
   III. Jarvis, Paul Gordon
   IV. Series
   581.5′222   QK754.7.A/

ISBN 0-632-00525-4

CONTENTS

EDITORS' PREFACE

Of the twenty Symposia already published in this series, four have been concerned with aspects of the relationships between vegetation and the atmospheric environment. The first of these, held in 1961, was devoted to the water relations of plants. The second, held in 1967, was on the measurement of the plant environment; and the third and fourth were on the subject of light — in 1965 and 1974. In the broad field of plant-atmosphere relations there have been several developments over this period of nearly two decades. Some of these are the result of closer links between physical and biological scientists, leading to a common language and a better understanding of the fundamental processes which occur at the surface of vegetation. We planned this Symposium to reflect these recent developments, inviting a number of speakers to deliver 'keynote' addresses in those fields which seemed important to us. At the same time, through the *Bulletin*, we invited contributions from others.

The Symposium was scheduled to occur immediately prior to a meeting organised at the School of Agriculture, University of Nottingham, on *Limiting processes in plant productivity*. After our Symposium, a number of participants travelled to Nottingham for the meeting there, an arrangement which was appreciated by visitors from overseas.

The Symposium was held at the University of Edinburgh between the 26th and 30th of March 1979. There were 120 participants, of whom 29 came from abroad. All the papers delivered at the Symposium are printed in this book, though we regret that abstracts from the poster session cannot be published for reasons of space.

Contributors were asked to use S.I. units throughout and to follow a standard set of symbols. Strict conformity turned out to be impracticable as some authors used quantities not in the standard set which, for good reason, required representation by some of the symbols already in the list. Furthermore, in the matter of choice of symbols, individual opinions sometimes run high and we thought it imprudent to insist on uniformity. All symbols for quantities are set in italics, except for those which denote flux densities — these are set in bold type. Roman type is used for mathematical operators (like log and exp) and for the symbols for dimensionless numbers in fluid dynamics (otherwise they appear to be two symbols multiplied together).

We are grateful to Drs Ian Rorison and Paul Biscoe who, with us, constituted the original planning committee and to Dr John Lee who provided the benefit of his experience as Meetings Secretary. Special thanks are due to Professor Rutter, who—in his retiral year—provided a summary of the proceedings. The University of Edinburgh invited participants to a sherry reception and offered the use of the Upper Library of the Old College for the Symposium Dinner. The Department of Forestry and Natural Resources of the University of Edinburgh hosted the poster session and provided refreshments. We are grateful to the staff of the Institute of Terrestrial Ecology at Bush, the Northern Research Station of the Forestry Commission and the Royal Botanic Gardens for accommodating our excursions. We would like to thank our staunch friends who gave up their time to help in the task of organising the Symposium, our secretaries who assisted us in our correspondence with the contributors and finally the staff of the Edinburgh office of Blackwell Scientific Publications with whom it has been a pleasure to work.

1. COUPLING OF PLANTS TO THE ATMOSPHERE

J. L. MONTEITH

*Department of Physiology and Environmental Studies, University
of Nottingham School of Agriculture, Sutton Bonington, Loughborough
LE12 5RD*

ECOLOGY AND THE ATMOSPHERE

The theme of plants and their atmospheric environment is particularly appropriate for the last British Ecological Society symposium of the seventies—a decade in which this major aspect of physical ecology has made substantial progress. If, thirty years ago, the Society had tried to hold a meeting exclusively concerned with such a topic, the input from physics and meteorology would have been slender, perhaps confined to controversy about a new evaporation formula developed by a worker from Rothamsted. Following a similar theme, many microclimatic papers published in the fifties dealt with heat balance of vegetation and its relation to evaporation. In the early sixties, the International Biological Programme fostered closer collaboration between ecologists and microclimatologists and by the end of that decade, an IBP symposium at Trebon (Šetlík 1970) reviewed the related topics of light, carbon dioxide exchange and photosynthesis. Atmospheric ecology continued to develop and expand in the 1970s. The volume of work on pollution has grown rapidly; the subject of wind has achieved text-book status, and vapour pressure deficit has emerged from relative obscurity as a significant element in the microclimate of plants. All these topics are discussed in later chapters of this volume.

By 1979, the common ground between ecologists, physiologists and physicists has extended to the point where it is possible to begin a symposium by talking about how plants are 'coupled' to their environment using a physical concept in a biological context. Fig. 1.1 summarizes the nature of this coupling on several related scales. Continentally, the location of major biomes is determined by the general circulation of the atmosphere and, in particular, by the spatial and temporal distribution of rainfall, temperature and solar radiation. There is an element of feedback in the system because air masses become cooler and moister when they pass over extensive areas of transpiring vegetation. This aspect of cooling is difficult to quantify but meteorologists working on general circulation models have begun to

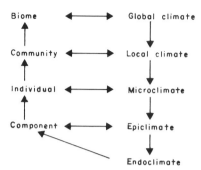

FIG. 1.1.   The coupling of vegetation to the atmosphere: scales and interactions.

recognize the importance of specifying the physical behaviour of vegetation in an appropriate way.

Within the large scale global climate, local climates (sometimes referred to as meso- or topo-climates) determine the survival and rate of growth of plant communities and associations. The interaction of these communities with the climate imposed on them produces a microclimate to which individual members of the community are exposed. In a similar way, the interaction of the individual with its microclimate establishes a climate on an even smaller scale confined to the layer of air in immediate contact with components such as leaves, stems and inflorescences. Some workers refer to this as the teleoclimate (Gates 1968) but the Greek $\tau\epsilon\lambda\epsilon o\varsigma$ means 'end' in a philosophical rather than a physical sense and a more appropriate term would be epiclimate from $\epsilon\pi\iota$ 'upon' or 'resting on'. Finally, the nature of the epiclimate determines the temperature of internal tissue, the gas concentration of inter-cellular spaces etc and these conditions could be called the endoclimate from $\epsilon\nu\delta o\nu$ 'within'.

It is the endoclimate which determines rates of growth and development of individual organs, thus establishing another system of feedback from the individual plant through the behaviour of communities to the biome.

This chapter is mainly concerned with the nature of epiclimates and with the coupling between plants and the atmosphere which occurs across the boundary layer of individual organs, a system which Grace describes in greater detail in the following chapter. Most attention will be given to the thermal characteristics of the epiclimate, partly because they illustrate a number of general principles and partly because several later chapters are concerned with the response of plants to temperature. The chapter by Unsworth deals with some aspects of the relation between epiclimate and microclimate.

THE CONCEPT OF COUPLING

Two systems are said to be coupled when they are capable of exchanging force, momentum, energy or mass. In physics, one of the most versatile forms of coupling is the electrical circuit in which energy in the form of charge moves from high to low 'potential' at a rate described as a 'current' (i). The simplest component of an electrical circuit is a resistor through which a current will flow when a difference of potential δV is maintained across it, and the 'resistance' of a resistor is defined as the potential difference needed to sustain unit current or $\delta V/i$.

The concept of resistance was implicit in the work of Brown & Escombe (1900) who measured the diffusion of water vapour and carbon dioxide in simple physical systems, relating their analysis to the processes of transpiration and assimilation by leaves. Maskell (1928) clearly identified the chain of resistances which govern the diffusion of CO_2 into a leaf. These early studies on single leaves were extended theoretically by Penman & Schofield (1951) and experimentally by Gaastra (1959). Comparable electrical analogues were then applied to the gas exchange of uniform stands of plants (Monteith 1963) and applications both to plant and to animal ecology are reviewed in several recent texts (Monteith 1975; Gates & Schmerl 1975; Campbell 1977). The flow of water through plants has also been analysed in terms of the drop in potential across a chain of resistances and according to Richter (1973), this line of work was initiated by Huber.

In analogues of heat and mass transfer, the potential of any entity Z is usually expressed as the amount of Z per unit volume of air and the rate of transport of Z is the amount moving through unit area of a system in unit time. Resistance then has the dimensions of time per unit length. The fact that these dimensions are independent of the nature of the entity is one of the main reasons for preferring resistances (or their reciprocals—conductances) to transfer coefficients of the type used by physicists and engineers.

The coupling of an organ to its environment can be described by a set of analogous electrical circuits, each describing the transport of a specific entity. There is no absolute scale of coupling; but if the resistance between A and B in a circuit is much smaller than the resistance between B and C, then A and B are said to be 'tightly coupled', and the comparison with B/C is implicit. A well known physiological example is the dependence of evaporation on stomatal resistance, r_s, and boundary layer resistance, r. Bange (1953) showed that when r_s is much larger than r, evaporation is almost independent of windspeed implying that the vapour pressure at the leaf surface is tightly coupled to the vapour pressure in the surrounding air.

The resistance analogues which have been widely used in physiology and ecology are appropriate for systems in a steady state, whereas the environment of most organisms is continuously changing. Records of temperature, vapour pressure and wind near the ground reveal a complex spectrum of erratic fluctuations associated with turbulent eddies whose higher frequencies extend beyond 10 Hz. At the low-frequency end of the spectrum, random fluctuations with periods of several minutes merge into the systematic 24 hour cycle imposed by solar radiation. To explore the significance of these fluctuations in physical ecology, the concept of coupling can be extended to include components such as capacitances and inductances which determine the distribution of current through a circuit in response to changes of potential. The concept of capacitance has a number of applications in physical ecology including thermal problems discussed later. Inductance will not be considered further here because its relevance is more limited.

A capacitor (or condenser) is a device for accumulating and storing charge. When current i flows into a capacitor whose capacitance is C, the voltage across the capacitor increases at a rate which is proportional to the current and inversely proportional to C, i.e. $\partial V/\partial t = i/C$ so that $V = \int i \, dt/C$. As the capacitor accumulates charge, V increases, opposing the flow of current which decreases and tends to zero when V is equal and opposite to the voltage in the rest of the circuit.

BOUNDARY LAYER RESISTANCE

Principles

In the steady state, the coupling of a plant organ to air passing over it depends on the resistance of its boundary layer, the epiclimate within which air movement is slowed by friction. Depending on the nature of the air-stream and of the surface, flow in the boundary layer may be either laminar, i.e. following smooth streamlines, or turbulent. Moreover, the pattern of flow can be determined either by an external pressure gradient—forced convection—or by a gradient of density established by a difference of temperature or of vapour pressure between the surface and the surrounding air—free convection. In plant ecology, rates of heat and mass transfer are rarely determined by free convection alone but in a very light wind, and especially over large leaves, forced and free convection may act together. Because hybrid convection is difficult to analyse, nearly all experimental work on heat loss from plant organs has been concerned with forced convection.

The depth of a boundary layer and the pattern of flow within it can be investigated with a hot-wire anemometer (Grace & Wilson 1976). The thermal boundary has been studied by Schlieren photography (Gates & Benedict 1963; Yabuki, Ishibashi & Miyagwa 1970); and the water vapour boundary layer with a microwave refractometer (Gates, Vetter & Thompson 1963).

It is characteristic of forced convection from smooth surfaces that the thickness of the boundary layer increases with distance from the leading edge and leaves exhibit this type of epiclimate when they are held rigid in a wind tunnel (see p. 35). The mean depth of the boundary layer depends primarily on air velocity and on the size and geometry of the object. Because the mean depth is usually two orders of magnitude less than the characteristic dimension of the object, the diffusion of heat, mass or momentum from a surface into the surrounding air occurs across the boundary layer, i.e. at right angles to the surface, rather than in the direction of the flow. If i is a flux per unit area and δV is the corresponding mean difference of potential between the surface and the free stream, i.e. across the boundary layer, the effective mean depth of the layer t is defined by the relation $i = D \, \delta V / t$ where D is a molecular diffusion coefficient. By analogy with Ohm's law, the resistance of the layer is $r = t/D^*$. For a given object at a fixed windspeed, the depth t is weakly dependent on the diffusion coefficient, being approximately proportional to $D^{1/3}$ (Grace, this volume). The depth is therefore different for heat, mass and momentum. Relevant values of diffusion coefficients and relative resistances were given by Jarvis (1971) and Monteith (1975).

Because the appropriate boundary layer depth of an object is rarely known *a priori*, resistances for forced convection are usually derived from standard formulae for heat transfer from objects of simple geometry, as tabulated, for example, by Leyton (1975) and Monteith (1975). From such formulae, resistances can be expressed in the form

$$r = Bd^{1-n}u^{-n} \tag{1.1}$$

where u is windspeed. The quantity d is known as the 'characteristic dimension' of the object, e.g. the downwind length of a flat plate, the diameter of a sphere, or the diameter of a cylinder with its axis at right angles to the flow. Table 1.1 contains values of the constants B and n to give r in units of s m^{-1} when d is in m and u in m s^{-1}. Fig. 1.2 shows how the resistances of plates and spheres decrease with windspeed in the range

TABLE 1.1. Constants of equation 1.1 for calculating the boundary layer resistance (s m^{-1}) for heat transfer from an object with a smooth isothermal surface at an arbitrary temperature of 20 °C.

	d	B	n
Flat plate (uniform width crosswind; resistance for two surfaces in parallel)	downwind length	151	0·5
Cylinder (length effectively infinite; axis at right angles to airflow)	diameter	248	0·6
Sphere	diameter	175	0·6

* This symbol, r, is equivalent to r_a used by other authors (e.g. Chapter 2).

0·1 to 10 m s^{-1} and increase with characteristic dimension in the range 0·3 to 10 cm. Plates with $d \simeq 3$ cm exposed to windspeeds between 0·5 and 5 m s^{-1} have resistances in the range 40 to 12 s m^{-1} corresponding to mean boundary layer thickness in the range 2 to 0·5 mm on both sides of the leaf.

The corresponding resistance for free convection is independent of windspeed and is usually expressed as a function of $d^n/(T_o - T_a)^p$ where T_o is the surface temperature, assumed uniform, and T_a is air temperature. For laminar flow, $n = p = 0.25$. However, when the flow induced by buoyancy is turbulent rather than laminar—and turbulence is an inherent

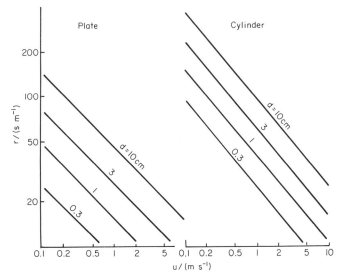

FIG. 1.2. Maximum value of boundary layer resistance for a flat plate (two surfaces in parallel) and a cylinder as function of windspeed u. The characteristic dimension d is the downwind length of the plate or the diameter of the cylinder and convection is forced.

feature of flow in plant communities even at low windspeeds—the index n is zero and $p = 0.33$. In units of s m^{-1}, the resistance for the upper side of a heated horizontal plate is then

$$r = 680/(T_o - T_a)^{1/3} \qquad (1.2)$$

The relevance of equation 1.2 for vegetation has never been tested but in the absence of relevant field measurements, it can be used as a guide to the *maximum* value of resistance likely to be associated with a given value of $(T_o - T_a)$. When the resistance calculated for forced convection (eqn 1.1) exceeds the corresponding value for free convection (eqn 1.2), then eqn 1.2 should be adopted for further analysis. More exact criteria are discussed by Kreith (1973).

Practice

Ecologists are not concerned with the aerodynamic behaviour of smooth, flat rigid plates but with leaves with rough or hairy surfaces, often curved, and prone to flutter. Moreover, leaves (and stems) in plant communities are exposed to turbulent air in the wakes of their neighbours. How do these complications affect the coupling between plant organs and the microclimate of the canopy? The experimental evidence, some of which is contradictory, is briefly reviewed here in terms of boundary layer resistances and is discussed in terms of non-dimensional groups by Grace in the following chapter.

Several recent papers have been concerned with the extent to which the boundary layer resistance of an organ depends on turbulence in the airstream. In these studies, the effectiveness of turbulence is expressed as the ratio (β) of the resistance for a smooth, rigid object of simple and appropriate geometry (Table 1.1) to the smaller, measured resistance of the organ at the same velocity.

In wind-tunnel studies, turbulence has usually been induced in a somewhat arbitrary way by grids of wires or an array of obstacles. When windspeed is measured with a hot-wire anemometer in a circuit which records both instantaneous and mean velocities, u and $\bar{u}$, the intensity of turbulence (I) is defined as $\sqrt{\overline{(u - \bar{u})^2}}/\bar{u}$. In plant communities, I often has a value in the range 0·4 to 0·7 (Legg & Monteith 1975), and it is not difficult to achieve this level in wind tunnels. However, the disturbance of a boundary layer by turbulence depends on the size of the eddies and not simply on their frequency as represented by an index of intensity. Transfer across the boundary layer is not affected either by very small or by very large eddies (Schlichting 1960) but by the spectrum of intermediate frequencies which deserve more attention in this type of study.

Haseba (1973a) attempted to generate turbulence with an appropriate spectrum by working in the airstream behind a canopy of citrus leaves. He estimated boundary layer conductances* ($1/r$) by measuring the rate of evaporation from a section of a model leaf. Fig. 1.3 shows the dependence of β on leaf area density (LAD) (cm²/cm³). For many agricultural crops, LAD is less than 0·1 so Fig. 1.3 implies that a figure of $\beta = 1·2$ should be representative of turbulence within a canopy. A similar result was obtained by Chamberlain (1974) who found $\beta \simeq 1·25$ for the transfer of radioactive lead vapour to bean (*Vicia faba*) leaves and stems within a stand of plants on the floor of a wind-tunnel.

Nobel (1974a) found that the mean boundary layer depth of a wet cylinder, simulating a plant stem, decreased with increasing turbulence to

* In Japanese and Russian literature, conductances are often called transfer coefficients and given the symbol D.

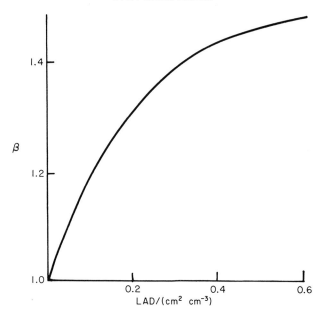

FIG. 1.3.   Ratio of boundary layer resistance for isolated citrus leaf to resistance measured within a canopy of leaves of specified leaf area density (*LAD*); redrawn from Haseba (1973a).

give a maximum value of $\beta \simeq 1\cdot3$ when $I > 0\cdot5$ (Fig. 1.4). In a similar study with wet spheres, simulating fruit or fungi (Nobel 1974b), β was about $1\cdot16$, consistent with the expectation that boundary layers surrounding completely streamlined objects should be more difficult to disturb by turbulence than those which are attached to flat plates.

Both Haseba and Nobel used rigid models and most of Chamberlain's measurements on bean leaves were made at windspeeds below the value of 2 m s^{-1} at which fluttering started. In experiments where model leaves have been forced to flutter at a frequency of 4 Hz, β reached a value of $2\cdot7$ (Parlange, Waggoner & Heichel 1971). However, when leaves flutter naturally in response to changes of windspeed and direction, β is likely to be substantially less than this figure.

Thorpe & Butler (1977) measured the resistance of the thermal boundary layer for leaves attached to apple trees growing in rows. For a set of measurements with considerable scatter, the mean relation between resistance and windspeed, measured between the rows, was close to the value predicted for a smooth flat plate, i.e. $\beta \simeq 1$. By referring to measurements on the relation between windspeed and shelter (Landsberg & Powell 1973), Thorpe & Butler deduced that two effects were complementary: the decrease of windspeed between the alley and the canopy; and the increase of turbulence within the canopy. Since the *LAD* was between 0·04 and 0·05 cm^{-1}, a value of $\beta = 1\cdot1$ would be expected from Fig. 1.2.

Evidence in conflict with these figures was provided by Grace & Wilson (1976) who obtained $\beta = 2\cdot5$ for a rigid metal plate simulating a poplar leaf. Furthermore, Wigley & Clark (1974), working with a model bean leaf in an airstream with $I \simeq 0\cdot35$, found that β increased from about 1·5 at 0·5 m s^{-1} to 3·0 at 4·5 m s^{-1}. Their evidence that β is a function of windspeed is not consistent with the more detailed work of Haseba (1973a).

Several workers have examined the dependence of resistance on the angle of inclination φ between a flat plate and the airstream. In general, the resistance of the windward side of a plate decreases as φ increases and according to measurements of evaporation by Haseba (1973b), β is about 1·2 for normal incidence ($\varphi = \pi/2$), irrespective of windspeed. The resistance of the leeward side increases with φ, and β is about 0·8 behind a plate when $\varphi = \pi/2$. Using radioactive vapour, Chamberlain was able to measure resistance as a function of φ and of distance from the leading edge. For a flat plate wetted on both sides, the resistance of the two surfaces combined in

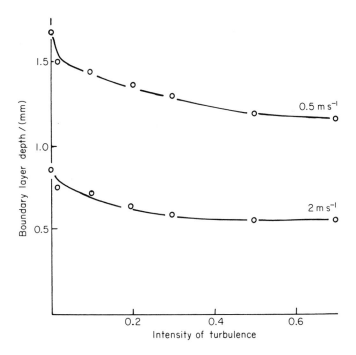

Fig. 1.4. Mean boundary depth for a cylinder of 10 cm diameter exposed in a wind-tunnel to specified windspeeds and turbulent intensities; redrawn from Nobel (1974a).

parallel was almost independent of φ. However, the difference in vapour resistance between windward and leeward surfaces would be significant for a leaf with stomata on one surface only.

The effect of leaf roughness on resistance has not been thoroughly studied, partly because of the problem of specifying or simulating the irregular surface of leaves. In related work on heat loss from rough cylinders, Achenbach (1974) showed that the main effect of increasing roughness was to lower the windspeed at which the boundary layer became turbulent—a point stressed by Grace (this volume) and Grace & Wilson (1976). Above this critical windspeed, there was a range in which resistance was proportional to u^{-1} rather than to $u^{-1/2}$ as in laminar flow.

In summary, it appears that the maximum resistance coupling a plant to its environment cannot exceed the value for a smooth flat surface. In the field, the effects of turbulence, fluttering, roughness and free convection combine to reduce the resistance by a factor $1/\beta$, usually between 0·5 and 1. The same processes combined with local differences of stomatal resistance make the distribution of temperature and other potentials far from uniform as Wigley & Clark demonstrated by thermography and this point is discussed later. At best, wind-tunnel measurements on real or model leaves are a useful guide to the way in which the foliage in a plant community is coupled to its microclimate. However, assuming that appropriate resistances for heat and mass transfer *can* be measured or estimated we now consider how they are used to calculate surface temperature and rates of transpiration.

SIMULTANEOUS HEAT AND WATER VAPOUR TRANSFER

When a leaf is transpiring, the extent to which it is coupled to the surrounding air depends on its stomatal resistance, r_s, as well as on the boundary layer resistances already considered. For convenience, and without significant loss of accuracy, the resistance to heat and vapour transfer by convection will be assumed to have the same value r. Different values can be used in the same analysis if required.

If we define the potential for sensible heat transfer as ρcT, where ρ is the density and c the specific heat of air, the loss of sensible heat from a leaf with a mean surface temperature T_o to air at T_a is

$$C = \rho c(T_o - T_a)/r \qquad (1.3)$$

where r is the boundary layer resistance for the two sides of a leaf coupled in parallel. The corresponding potential for latent heat transfer is $\rho ce/\gamma$ where e is vapour pressure (mbar) and γ ($= 0·66$ mbar K^{-1}) is the psychrometer constant. For an amphistomatous leaf with a stomatal resistance of r_s for two sides in parallel ($2r_s$ on each epidermis), it is convenient to work with a modified psychrometer constant obtained by multiplying γ by the ratio of the total resistances for water vapour and for heat transfer, i.e. $\gamma^* = \gamma(r + r_s)/r$ (Monteith 1965). The corresponding formula for a leaf with stomata on one epidermis only is $\gamma^* = \gamma(r + 2r_s)/r$ and a more complex expression must be used if the two epidermises have different resistances.

For an amphistomatous leaf, the loss of latent heat is

$$\lambda E = \rho c[e_s(T_o) - e_a]/\gamma(r_s + r) = \rho c[e_s(T_o) - e_a]/\gamma^*r \qquad (1.4)$$

where $e_s(T_o)$ is the saturation vapour pressure at leaf temperature and e_a is the value in the surrounding air. A form of expansion first used by Penman (1948) gives

$$\lambda E = \rho c\, \Delta(T_o - T_a)/(\gamma^*r) + \rho c\, \delta/(\gamma^*r) \qquad (1.4a)$$

where Δ is the rate of change of $e_s(T)$ with temperature at a point between T_o and T_a and δ is the saturation deficit (i.e. $e_s(T_a) - e_a$). In practice, Δ is usually evaluated at temperature T_a. Neglecting metabolic exchanges of heat which are usually small and assuming for the time being that leaf temperature is constant, the sum of sensible and latent heat can be equated to the net absorption of radiant energy, i.e.

$$\mathbf{R}_n = \rho c(T_o - T_a)[1 + \Delta/\gamma^*]/r + \rho c\delta/(\gamma^*r) \qquad (1.5)$$

For a non-transpiring leaf γ^* is infinite so that

$$\mathbf{R}_n = \rho c(T_o - T_a)/r \qquad (1.5a)$$

Comparison of eqns 1.5 and 1.5a shows that the effect of transpiration on the heat balance of the leaf is twofold: the thermal resistance is reduced from r to $r' = r/(1 + \Delta/\gamma^*)$ and the heat load is reduced from $\mathbf{R}_n$ to $\mathbf{R}_n - \rho c\, \delta/(\gamma^*r)$.

The value of an electrical resistance can be reduced from x to $x/(1 + \varepsilon)$ by connecting a resistance x/ε in parallel. Eqn 1.5 can therefore be represented by the circuit shown in Fig. 1.5 with r and $r(\gamma^*/\Delta)$ in parallel, similar to the circuit described by Cowan (1972a). To make the role of stomatal resistance explicit, $r(\gamma^*/\Delta)$ is shown as the two resistors $r(\gamma/\Delta)$ and $r_s(\gamma/\Delta)$ in series. The net radiation is represented by a current generator. The opposing current $\rho c\, \delta/(\gamma^*r)$ can be introduced in two ways: as a battery of voltage $\rho c\, \delta/(\Delta + \gamma^*)$ in the main circuit; or more simply as a voltage $\rho c\, \delta/\Delta$ in series with $r(\gamma^*/\Delta)$. In the latter arrangement, the sensible heat fraction of the total current $\mathbf{R}_n$ passes through r and the latent heat fraction through $r(\gamma^*/\Delta)$.

The potential across the pair of resistances is $\rho c(T_o - T_a)$. By evaluating the current flowing through the two branches of the circuit (or by manipulating eqn 1.5), the surface temperature can be expressed as the sum of air temperature and two terms which represent temperature differences, viz.

$$T_o = T_a + [r'\mathbf{R}_n/\rho c] - [\delta/(\Delta + \gamma^*)] \qquad (1.6)$$

The first term in square brackets is an apparent *increase* in air temperature proportional to net radiation and for a non-transpiring leaf ($r' = r$) it is identical to the 'radiation increment' used by Burton & Edholm (1955) and other human physiologists. The second term in brackets is an apparent *decrease* in temperature proportional to saturation deficit.

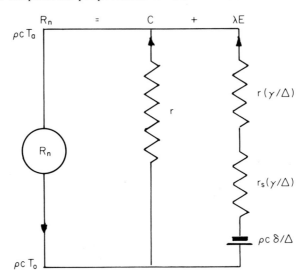

FIG. 1.5. Electrical circuit analogous to heat balance of transpiring leaf (eqn 1.5). The generated current $\mathbf{R}_n$ is partitioned between a current $\mathbf{C}$ passing through the resistor r and a current λE passing through a battery of voltage $\rho c \delta/\Delta$ and two resistors whose sum is $r_t\gamma^*/\Delta$.

The temperature T_o is the equilibrium mean surface temperature of an organ in a fixed microclimate (specified by $\mathbf{R}_n$, δ, etc.). In a changing microclimate, T_o can be regarded as the effective temperature of the environment towards which surface temperature will move in an attempt to reach thermal equilibrium and in the next section, this quantity is given the symbol T_e.

Further analysis of the circuit gives the evaporation rate in the form

$$\lambda E = (\Delta\mathbf{R}_n + \rho c \, \delta/r)/(\Delta + \gamma^*) \qquad (1.7)$$

A full derivation is given by Jarvis (this volume).

Although equations 1.6 and 1.7 are formally exact, a minor modification is needed before they can be used to estimate the surface temperature and evaporation rate of a leaf in a specified microclimate particularly in problems where r is treated as a variable. Then $\mathbf{R}_n$ must also be allowed to vary because it depends on the temperature of the surface over which the flux is measured, i.e. the term T_o is concealed in the right-hand side of equation 1.7. An appropriate procedure is to replace $\mathbf{R}_n$ by a fixed value of the 'isothermal' net radiation $\mathbf{R}_{ni}$ which is the net radiant flux the surface would absorb if it were at air temperature (Monteith 1975). The equation defining $\mathbf{R}_{ni}$ is therefore

$$\mathbf{R}_{ni} = \mathbf{R}_n + \sigma(T_o{}^4 - T_a{}^4) \tag{1.8}$$

where σ is the Stefan Boltzmann constant and the temperatures are now expressed in K. Because the difference between T_o and T_a is rarely more than a few degrees Kelvin, the term $\sigma(T_o{}^4 - T_a{}^4)$ can be replaced by $4\sigma T_a{}^3(T_o - T_a)$ with little error. This expression can be written in the same form as the corresponding term for heat transfer by convection if a radiative resistance r_R is defined by writing

$$4\sigma T_a{}^3(T_o - T_a) = \rho c(T_o - T_a)/r_R \tag{1.9}$$

from which

$$r_R = \rho c/4\sigma T_a{}^3 \; (= 213 \text{ s m}^{-1} \text{ at } 20\,^{\circ}\text{C})$$

Since convection and radiation are parallel modes of heat loss, the combined thermal resistance coupling a surface to its environment is

$$r_t = (r^{-1} + r_R{}^{-1})^{-1} \tag{1.10}$$

Retaining the convention that γ^*/γ is the ratio of resistances for water vapour and heat transfer (p. 11), γ^* is now defined as

$$\gamma^* = \gamma(r + r_s)(r^{-1} + r_R{}^{-1}) \tag{1.11}$$

where the first term in brackets is the total resistance to vapour transfer as before and the second term in brackets is the reciprocal of r_t. The corresponding value of r' to be used in eqn 1.6 is

$$r_t' = r_t/(1 + \Delta/\gamma^*) \tag{1.12}$$

with γ^* defined by eqn 1.11.

In his pioneering analysis of leaf heat balance, Raschke (1956) derived expressions for surface temperature and transpiration rate which are essentially the same as equations 1.6 and 1.7 but with an entirely different set of symbols and without reference to Penman's related work. The control of evaporation rate by stomata was expressed by a '*Wasserbedeckungsfaktor*' proportional to $r/(r + r_s)$ in the nomenclature used here. It is therefore difficult to interpret graphs and tables in which r was treated as a variable whereas $r/(r + r_s)$ was assumed to have a constant value. This anomaly was corrected in a later paper (Raschke 1958) in which stomatal resistance was identified as a separate parameter and assigned appropriate values.

The analysis of the heat balance of the equation presented here depends on two similar approximations: the relation between saturation vapour pressure and temperature is assumed to be linear; and radiative heat loss from a surface at T_o to an environment at T_a is assumed proportional to $T_o - T_a$. With these assumptions, both Δ and r_R can be calculated as functions of the known temperature T_a whereas in exact analysis, they should be evaluated at an appropriate temperature between T_a and T_o. Raschke (1958) used the same approximations to solve the heat balance equation and demonstrated that values of surface temperature so derived were almost identical to exact values obtained by solving the equation graphically. Gates & Papian (1971) and other later workers avoided the use of fixed values of Δ and r_R by determining the heat balance components iteratively with a computer. However, because they used fixed and arbitrary relationships to estimate boundary layer resistance from windspeed and leaf dimensions, all their calculations are subject to uncertainties much larger than the error introduced by the approximate solution of the heat balance equation as represented by equations 1.6 and 1.7. A few of the implications of these equations will now be considered with the help of simple graphs.

GRAPHICAL EXAMPLES

Equation 1.6 was used to calculate the excess surface temperature $T_o - T_a$ of an organ exposed to bright sunshine ($\mathbf{R}_{ni} = 300$ W m^{-2}) with $T_a = 20\,°C$ and $\delta = 10$ mbar, values characteristic of a summer day in a temperate climate (Fig. 1.6). The independent variable was r, the resistance for heat and mass transfer by convection as determined from Table 1.1 or Fig. 1.2, for example.

For a non-transpiring surface ($r_s = \infty$), $T_o = T_a$ when $r_a = 0$ and $(T_o - T_a)$ increases linearly with the total thermal resistance $r_t = (r^{-1} + r_R^{-1})^{-1}$. The increase with r is therefore non-linear as shown and in the limit when r is very large $T_o - T_a$ tends to the value $r_R\mathbf{R}_{ni}/\rho c$ or $52\,°C$ in this example.

For a perfectly wet surface ($r_s = 0$), the minimum value of T_o is the wet-bulb temperature and $T_a - T_o$ is given by $\delta e/(\Delta + \gamma)$.

Curves for three finite values of r_s are also shown. The value $r_s = 50$ s m^{-1} corresponds to a leaf with stomata wide open and in this environment $T_o - T_a$ is almost proportional to r. Since the convective heat flux $\mathbf{C}$ is proportional to $(T_o - T_a)/r$ (eqn 1.3) it follows that $\mathbf{C}$ must be nearly independent of r which implies that $\lambda E \, (= \mathbf{R}_n - \mathbf{C})$ is also nearly independent of r. It is shown elsewhere (Monteith 1965) that the exact condition for evaporation rate to be independent of windspeed is $\lambda E/\mathbf{C} = \Delta/\gamma$ which implies that $r_s = \rho c \, \delta(\Delta + \gamma)/\Delta\gamma\mathbf{R}_n$. When r_s is greater than this critical

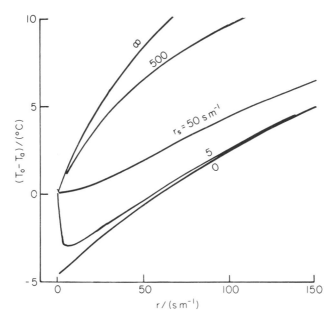

FIG. 1.6. Predicted difference between surface and air temperature for an amphistomous leaf in sunlight with specified boundary layer and stomatal resistances (for both laminae in parallel). Assumed microclimate: $\mathbf{R}_{ni} = 300$ W m^{-2}, $T_a = 20$ °C $\delta = 10$ mbar.

value, λE increases with decreasing windspeed and C decreases, implying that the curve for $T_o - T_a$ should be concave to the axis of r. When r_s is less than the critical value, the curve should be convex to the r axis.

The curves for $r_s = 500$ and 5 s m^{-1} demonstrate these differences. A value of $r_s = 500$ s m^{-1} is appropriate for a leaf with stomata partly closed by water stress, and $r_s = 5$ for a leaf incompletely covered by evaporating rain or dew. For values of r exceeding 30 s m^{-1}, surface temperatures are similar for $r_s = 0$ and 5 s m^{-1}. The theory predicts that if r_s has a finite value, however small, T_o will tend to T_a as r tends to zero. For small values of r_s, there is a regime in which surface temperature would be less than air temperature but would increase towards air temperature if windspeed decreased. For most values of r_s however, including $r_s = 0$, surface temperature decreases with increasing windspeed.

For the range of r values commonly encountered in the field, say 20 to 60 s m^{-1} for leaves, cooling as a consequence of transpiration ranges from about 1 to 2 °C for $r_s = 50$ s m^{-1} and from 4 to 7 °C when $r_s = 500$ s m^{-1}. For a freely evaporating surface ($r_s = 0$) the corresponding range is 7·5 to 9·5 °C.

Figure 1.7 shows the corresponding calculation of $T_o - T_a$ at night when $\mathbf{R}_{ni} = -100$ W m^{-2} and $T = 10\,°C$. The line marked 'dry' corresponds to a system in which there is no condensation so that $\mathbf{R}_n = C$. For a surface wetted by dew, eqn 1.7 is valid with $r_s = 0$. Cooling of the surface below air temperature is least when the atmosphere is saturated (dew 100). Dew may form when the atmosphere is unsaturated but only when the resistance exceeds a limiting value, e.g. about 25 s m^{-1} for a relative humidity of 90% in the example shown. In the very light winds usually associated with dew formation, the boundary layer resistance of leaves is expected to fall between 50 and 100 s m^{-1} implying that the rise in temperature attributable to the release of latent heat is of the order of 1·5 to 3°C. When the air temperature is close to 0°C, this rise will sometimes be sufficient to prevent, or at least to delay, damage to tissue by freezing.

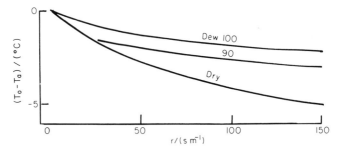

FIG. 1.7.   Predicted difference between surface and air temperature for a leaf with specified boundary layer resistance in the dark ($r_s = \infty$). Assumed microclimate: $\mathbf{R}_n = -100$ W m^{-2}; $T = 10\,°C$. Dew formation occurs when r.h. is 100 or 90%. The 'dry' curve is appropriate when the r.h. is too low to allow condensation.

Salisbury & Spomer (1964) and other workers have noted that the leaf temperature excess in bright sunshine tends to be high when the leaf temperature is low and vice versa. This effect can be predicted from eqn 1.6 on two grounds.

(i) Because Δ increases with temperature, $(T_o - T_a)$, if positive, will tend to decrease with increasing temperature.

(ii) For a fixed relative humidity, the saturation deficit δ increases with temperature. Figure 1.8 shows the relation between $T_o - T_a$ and T_a for a leaf with $r = r_s = 50$ s m^{-1} exposed to radiation at 300 W m^{-2}. The upper line corresponds to a relative humidity of 70%, characteristic of the level often recorded in crop canopies. The lower line corresponds to much drier air at 35% r.h. Fortuitously, both relations are almost linear, and their slopes are about $-2·2$ and $-3·1$. The effect of strong radiative heating at low air temperatures may be particularly significant for arctic and for alpine communities as observed by Warren Wilson

(1957) and others. In the tropics, on the other hand, a leaf adapted to keep stomata open when air temperature is 30–40 °C should be able to maintain its tissue at a similar temperature. Linacre (1964a) suggested that observed leaf temperatures were usually below air temperature when $T_a \simeq 35$ °C, but the calculation presented here shows that the critical temperature for $T_o = T_a$ must be a function of several micro-meteorological variables and of stomatal resistance.

Attention has recently been focused on the tendency for the leaves of some species to reduce stomatal aperture in response to an increase of saturation deficit and therefore of potential transpiration rate. The analysis considered here shows that a modest degree of stomatal closure is unlikely to cause overheating of very small leaves such as coniferous needles, in any climate, or of any leaves exposed to high air temperatures.

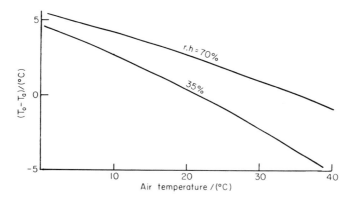

FIG. 1.8. Predicted difference between leaf surface and air temperatures as a function of air temperature at two specified levels of relative humidity. Assumed microclimate $R_{ni} = 300$ W m^{-2}, $r = r_s = 50$ s m^{-1}.

THERMAL CAPACITANCE

In a previous section, an expression was derived for $T_o = T_e$, the mean surface temperature of an organ in thermal equilibrium with its environment. When T_e changes as a result of a change in radiation, windspeed, air temperature or humidity, surface temperature cannot change instantaneously but must lag behind the change in T_e to an extent which depends on the heat capacity of the organ and on other factors now considered.

Suppose the mean surface temperature of an organ is $T \neq T_e$ so that the organ gains heat from the environment at a net rate given by

$$\mathbf{R}_n - \mathbf{C} - \lambda \mathbf{E} = \mathbf{R}_{ni} - \frac{\rho c(T - T_a)}{r_t} - \frac{\rho c[e_s(T) - e_a]}{\gamma^* r_t} \qquad (1.13)$$

$$= \mathbf{R}_{ni} - \frac{\rho c (T - T_a)}{r_t'} - \frac{\rho c \delta}{\gamma^* r_t} \tag{1.13a}$$

where r_t, r_t' and γ^* are defined in equations 1.10–1.12. By definition, the gain of heat from the environment becomes zero when $T = T_e$, i.e.

$$0 = \mathbf{R}_{ni} - \frac{\rho c (T_e - T_a)}{r_t'} - \frac{\rho c \delta}{\gamma^* r_t} \tag{1.13b}$$

Subtracting eqn 1.13b from eqn 1.13a, the gain of heat when $T = T_e$ is

$$R_n - \mathbf{C} - \lambda E = \rho c (T_e - T)/r_t' \tag{1.14}$$

The rate of gain of heat can also be expressed in terms of the rate at which the heat content of the organ increases. With the important proviso that the whole organ is at a uniform temperature T, the rate of change of heat content can be written as $(\rho' c' v/A)\dot{T}$ where $\dot{T} = \partial T/\partial t$, $\rho' c'$ is the volumetric heat capacity and v is the volume of tissue with an external area of A (one side only for a leaf). The heat balance of the organ is therefore given by

$$\rho c (T_e - T)/r_t' = (\rho' c' v/A)\dot{T} \tag{1.15}$$

which can be reduced to the form

$$\dot{T} = (T_e - T)/\tau \tag{1.16}$$

where τ, with dimensions of time is given by

$$\tau = r_t'(\rho' c'/\rho c)(v/A) \tag{1.17}$$

For a leaf, the term v/A is simply a mean thickness. For a stem, treated as an infinite cylinder of diameter d, it is $d/4$ and for a sphere it is $d/6$.

Unfortunately, values for the heat capacity of plant tissue are extremely hard to find in the literature. According to Miller (1938), measurements of intercellular space in leaves range from 3 to 70% of tissue volume, and assuming that the tissue is composed entirely of water with a specific heat of 4.2 J g^{-1} K^{-1}, the corresponding range of volumetric specific heat is 4.1 to 1.3 MJ m^{-3} K^{-1}. For an intermediate value of 3 MJ m^{-3} K^{-1} and with $\rho c = 1.2$ kJ m^{-3} K^{-1} for air at $20\,°C$, $(\rho' c'/\rho c) \simeq 2.5 \times 10^3$. The specific heat of fruit depends mainly on fractional water content (Turrell & Perry 1957) and a range from 2 to 3 MJ m^{-3} K^{-1} is probably appropriate. For wood, the range of specific heat must be substantial because the density of wood ranges from 0.1 t m^{-3} for balsa to 1.33 t m^{-3} for ebony (Weast 1977). Furthermore, large differences of specific heat are associated with the radial differences of water content observed in tree trunks. For example, Herrington (1969) found that $\rho' c'$ for red pine increased from 0.9 MJ m^{-3} K^{-1} in the heartwood to 2.6 MJ m^{-3} K^{-1} in sapwood with a mean value of 2.0 MJ m^{-3} K^{-1} for the whole trunk.

Eqn 1.16 is formally identical to the expression for the rate of change of voltage V across a capacitance C in series with a resistor R. When a voltage E is imposed across the components, the subsequent change of V is given by

$$\dot{V} = (E - V)/RC \qquad (1.18)$$

and $\dot{V}$ decreases as the condenser accumulates charge. The term RC is known as the 'time-constant' of the circuit and this name is appropriate for τ.

The voltage across the condenser increases from $V = 0$ to a maximum value of E. In the same way, the temperature of an organ will always move in the direction of an equilibrium temperature at which $T = T_e$, i.e. the excess of surface over air temperature will tend to the value given in eqn 1.6. Fig. 1.9 is an analogue circuit which is electrically equivalent to the circuit of Fig. 1.5. The battery imposes a voltage of $\rho c(T_e - T)$ across a resistor r_t' and a capacitor C in series. As the time constant of the circuit τ is the product $r_t'C$, the thermal capacity can be identified as $\tau/r_t' = (\rho'c'/\rho c)(v/A)$ which has dimensions of length.

To indicate scale, time constants were calculated for plant organs of different shape and size and for two windspeeds, assuming $\rho'c'/\rho c = 2\cdot5 \times 10^3$. Table 1.2 shows that the value of τ for small leaves is of the order of a few seconds (and is therefore difficult to measure accurately). For larger leaves, τ ranges from the order of a minute for a characteristic dimension of 5 cm to several minutes for very large leaves. Stems, buds and fruits have a larger range of τ because their volume/area factor is larger. Even for small twigs and berries, τ is of the order of 1 min and for the trunks of mature trees and very large fruits τ is a significant fraction of 1 day.

TABLE 1.2. Time constants for leaves, stems and fruits treated as simple geometrical shapes. r_t is calculated from eqns 1.1 and 1.10 and τ from eqn 1.17. Values are for non-transpiring organs except those in brackets which were calculated from leaves with $r_s = 50$ s m^{-1} at 20 °C.

		Arbitrary dimensions (cm)		Time constant (minutes) u(m s^{-1})	
		d	(v/A)	1	4
Leaves	grass	0·6	0·05	0·22 (0·16)	0·11 (0·10)
	beech	6	0·10	1·13 (0·63)	0.66 (0·43)
	giant hogweed	60	0·15	3·50 (1·61)	2·35 (1·21)
Stems, etc.	bramble	0·6	0·15	1·73	0·81
	whin	6	1·5	37	18·8
	beech trunk	60	15	650	390
Fruits	rowan	0·6	0·1	0·85	0·39
	crab apple	6	1	18·7	9·2
	Jack fruit	60	10	360	200

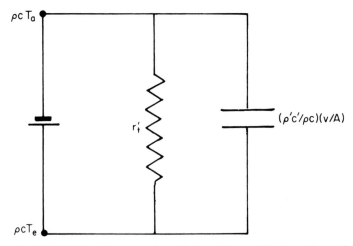

FIG. 1.9.   Electrical circuit analogous to dynamic heat balance of leaf or other object (eqn 1.15).

The ratio of time constants for a transpiring leaf and for the same leaf in the same environment with stomata closed is

$$r_t'/r_t = (1 + \Delta/\gamma^*)^{-1} \qquad (1.19)$$

For a completely wet surface ($r_s = 0$), r_t'/r_t has a value of about 0·3 at 20 °C, and this ratio describes the relative response times of a wet and dry bulb thermometer of the same size. For a leaf with $r_s \simeq r$, a common condition, $r_t'/r_t = 0·5$ at 20 °C.

Solutions of the dynamic heat balance equation

Equation 1.16 can be solved for a number of standard boundary conditions to obtain surface temperature as an explicit function of time. For example, if the equilibrium temperature T_e increases (or decreases) instantaneously from T_1 to T_2 as a result of a change in T_a, R_n or δ, the surface temperature will increase (or decrease) exponentially as shown in Fig. 1.10a. The corresponding solution is

$$T = T_2 - (T_2 - T_1) \exp(-t/\tau) \qquad (1.20)$$

This equation is appropriate for a leaf exposed to sudden changes of irradiance when small clouds obscure the sun. The response time of a leaf which has been shaded can be estimated by measuring the mean surface temperature as a function of time and plotting $\ln[(T_2 - T)/(T_2 - T_1)]$ against t to give $-1/\tau$ as a slope. Linacre (1964b) and others have used this method to estimate heat transfer coefficients which are inversely proportional

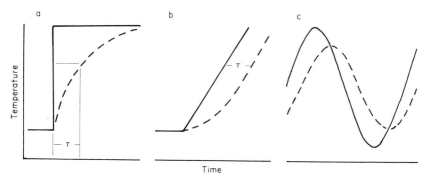

Fᴵɢ. 1.10. Change of surface temperature (dashed line) in response to change of environmental temperature (full line). (a) Step change (eqn 1.20); τ is the time for a fractional change of $1 - e^{-1}$ or 0·63. (b) Ramp change (eqn 1.23); τ is the constant time lag established after the term $\exp(-t/\tau)$ becomes negligible. (c) Harmonic oscillation for the case $\omega\tau = 1$ (eqns 1.25, 1.25a and b).

to τ, but the way in which transpiration reduces the effective time constant, as shown by equation 1.17, has usually been ignored.

By combining equations 1.19 and 1.20, it is possible, at least in principle, to determine both r and r_s for a leaf, using a stopwatch and a radiation thermometer or a carefully installed set of thermocouples. The leaf must first achieve a constant measured temperature in sunshine. It is then shaded and the change of temperature with time recorded so that the 'wet' time constant τ_w can be estimated from equation 1.20. The leaf is then covered with vaseline to stop transpiration and the shading process is repeated to find the 'dry' constant τ_d (Linacre 1972). Since $\tau_w/\tau_d = (1 + \Delta/\gamma^*)^{-1}$, it can be shown from the definition of γ^* that

$$r_s/r \simeq \{(\Delta/\gamma)/[(\tau_d/\tau_w) - 1]\} - 1 \qquad (1.21)$$

when $r_t \simeq r$. If the mean thickness x of the leaf is measured, r can be calculated as $\tau_d/(\rho'c'/\rho c)x$ so that r_s can be estimated from equation 1.21. (In some circumstances the estimate would not be very exact but the procedure is worth noting as an instructive exercise.)

A second case is illustrated in Fig. 1.10b, where the equilibrium temperature changes at a constant rate α from an initial temperature T_1. The boundary condition is

$$T_e = T_1 + \alpha t \qquad (1.22)$$

and the corresponding solution of equation 1.16 is

$$T = T_1 + \alpha(t - \tau) + [\alpha\tau \exp(-t/\tau)] \qquad (1.23)$$

The term in square brackets can be neglected when $t > 2\tau$ and thereafter the temperature of the organ increases at the same rate as air temperature but with a time lag $\alpha\tau$. Monteith & Butler (1979) used this equation to calculate the surface temperature of cocoa pods on which dew may form after dawn when the surface temperature rises more slowly than the dew-point of the surrounding air.

A harmonic oscillation of temperature (Fig. 1.10c) has the form

$$T_e = \bar{T}_e + A \sin \omega t \qquad (1.24)$$

where $\bar{T}_e$ is a mean equilibrium temperature with maximum and minimum values of $\bar{T}_e \pm A$ and ω is the frequency of the oscillation equal to $2\pi/P$ where P is the period. The appropriate solution of eqn 1.16 is

$$T = \bar{T}_e + A' \sin (\omega t - \varphi) \qquad (1.25)$$

where the semi-amplitude of the surface temperature is

$$A' = A \cos \varphi \qquad (1.25a)$$

and the phase lag is

$$\varphi = \tan^{-1} (\omega\tau) \qquad (1.25b)$$

which has a maximum value of $\pi/2$ or 6 h for a 24 h cycle. In principle, this type of analysis can be used to predict how the surface temperature of an organ should respond to a diurnal oscillation of T_e. For organs such as leaves, buds and small fruits, τ is much smaller than ω^{-1} which has a value of $P/2\pi$ or about 4 h when $P = 24$ h. When the product $\omega\tau$ is small, the amplitude of surface temperature A' is close to the imposed value of A and the phase lag φ is negligible. For very bulky organs such as tree trunks, τ and ω^{-1} are of comparable magnitude so that the simple theory represented by eqn 1.25a predicts that A' may be substantially smaller than A. In rigorous analysis, however, the temperature gradients which exist within large organs cannot be neglected as they were in deriving eqn 1.16 and this point is considered in the next section.

The ecological significance of the thermal time constants which are characteristic of plant organs has received little attention but in terms of adaptation the response to rapid or systematic changes of environmental temperature may be just as important as differences in thermal resistance discussed by Lewis (1972) and others.

In an environment where there is a risk of damage to tissue by extremes of heat or cold, large organs are likely to survive stress more readily than small ones. It is possible, for example, that giant cacti maintain internal temperatures which are well below air temperature and surface temperature in bright sunshine. Large organs also avoid the effect of very short periods of high or low temperature often associated with a brief drop in windspeed. Low growing arctic and alpine species may achieve similar protection because they are thermally coupled to the underlying soil rather than to the atmosphere (see Grace, this volume).

When microclimate and species are matched so that rates of physiological processes increase more or less linearly with temperature, the mean rate over a whole day will be independent of the range of tissue temperature. But when the relation between rate and temperature is *not* linear, the mean rate for the day will depend on the details of its thermal history and not simply on the mean temperature. In the tropics, environmental temperature T_e around midday may exceed the optimum for photosynthesis, particularly for plants which are short of water. Thermal inertia as a result of a large mass could be an advantage if it postponed the time of maximum tissue temperature until air temperature started to decline in the middle of the afternoon. This argument is contrary to the widespread view that small leaves are well adapted to survive in high air temperatures and bright sunlight. In fact, many tropical species are notable for the large size of their leaves whereas many temperate and arctic species have relatively small leaves.

When vegetation is exposed to fire, the thermal response time of essential tissue must be an important factor in the survival of species. In a textbook reviewing the effects of fire on ecosystems (Kozlowski & Ahlgren 1974) this aspect of environmental physics is referred to briefly but is not treated quantitatively.

CAVEAT — LACK OF UNIFORMITY

The calculations of heat transfer and rate of temperature change in the last section were based on the assumption that the tissue within a plant organ can be treated as isothermal. In practice, this condition is not always satisfied and the existence of temperature gradients has a number of implications.

In the first place, the formula for the thermal resistance of a flat plate (eqn 1.1 and Table 1.1) is appropriate for a surface with uniform temperature. In laminar flow, the loss of sensible heat per unit area from such a surface decreases with the square root of distance (x) from the leading edge. When heat flux rather than temperature is uniform over the surface, the temperature excess decreases with $x^{1/2}$ as Wigley & Clark (1974) demonstrated for a model leaf.

In practice, for a given mean temperature excess the difference in flux between the two cases is small. The value of A in the constant flux equation is about 10% of the 'isothermal' value in Table 1.1 and the value of n is the same. A real leaf within a canopy will tend to behave like a constant flux surface when the whole lamina is in bright sunshine but when part of the lamina is shaded, the distribution of flux and of temperature is complex. Fortunately, the thermal coupling of leaves to their environment appears to be unaffected by the fact that sensible heat transfer occurs over the whole epidermis whereas latent heat is lost mainly from tissue surrounding the substomatal cavity (Cowan 1972b).

For any curved element of surface on a fruit or tree trunk, for example, the boundary layer resistance depends on the direction of the airflow whereas the irradiance is a function of orientation with respect to the sun and sky. Tanner & Golz (1972) measured the temperature of ovaries attached to the umbel of onions and observed the maximum difference between ovary and air temperature over the segment directly facing the sun. The excess temperature was greatest when this segment was at right angles to the direction of the wind and decreased by a factor of about 5 when the segment was facing into the wind. For a fixed angle between sun and wind, the excess temperature was inversely proportional to $u^{2/3}$ and exceeded 15 °C when u fell below 1 m s^{-1}. It was suggested that ovary temperatures of 50 to 60 °C could be responsible for loss of seed production.

Very few systematic measurements or estimates of internal temperature have been published. Thorpe (1974) calculated the temperature distribution inside a sunlit apple and predicted that when radiant energy was absorbed at 500 W m^{-2}, the excess of tissue temperature above air temperature would be 9 °C just below the skin at a point facing the sun, 4 °C at the centre and 2 °C at the coolest point on the surface. Although the theory took no account of differences in boundary layer thickness associated with wind direction, similar measurements of surface temperature were obtained with a radiation thermometer. The thermal time constant for a change in irradiance was about 10 min at a point facing the sun and this figure is consistent with an estimate from eqn 1.17. The time constant was more than 1 h at the centre of the fruit.

Herrington (1969) solved the equations for the radiant flow of heat in a tree trunk and compared predicted changes of temperature with measurements in the stem of a 48 year old specimen of *Pinus resinosa* in a plantation. He used a mean value of volumetric specific heat in the analysis but the measured value increased by a factor of three between the heartwood and the cambium. In contrast to the system studied by Thorpe, the trunk was not exposed to bright sunlight so there was no significant difference in surface temperature around the trunk and diurnal changes of internal temperature agreed well with theory. The amplitude of surface temperature was 0·75 of the air temperature amplitude compared with a value of 0·63 from eqn 1.25a but in the centre of the trunk the relative amplitude was only 0·25. The measured phase lag was 0·6 h at the surface increasing to 9·3 h at the centre of the trunk compared with a value of about 3 h from eqn 1.25b.

From the comparison of elementary theory assuming uniform tissue temperature (eqn 1.16) and the more exact analyses of Thorpe and Herrington, it appears that eqn 1.25 provides a good estimate of *surface* temperature amplitude even for bulky organs but substantially overestimates the amplitude within the tissue. The phase lag derived from simple theory (eqn 1.25b) overestimates the lag at the surface and underestimates the lag at the centre of a tree trunk or a large fruit.

OTHER FORMS OF COUPLING: BRIEF CRITIQUE

At least two other forms of coupling between plants and the atmosphere have been represented by electrical analogues. The treatment of water flow as a current passing through a resistance was referred to in the introduction. Measurements have shown, however, that the resistance (R) often decreases with increasing flow rate (i) (Weatherley 1976) and in the limit when $R \propto 1/i$ the 'resistance' becomes a constant potential device. Similarly, the water capacity of plants (C) defined as a change in water content per unit change of potential is not a unique property of the tissue but changes with the value of the potential (Milburn 1979). Several workers have analysed diurnal changes of plant water potential using an equation of the form

$$\dot{\psi} = (\psi_o - \psi)/RC$$

where ψ_o is a constant base potential. Wallace (1978) found that the hydraulic time constant (RC) for a stand of winter wheat fluctuated between 30 and 110 min during the summer of 1975 and Jones (1978) reported an even wider range from 17 min to 174 min for three varieties of wheat growing in three seasons. In neither analysis could the variability of RC be related to age, weather, soil water history or morphology and it is doubtful whether the concept of a unique time-constant is valid in a circuit whose components are functions of current and potential. More complete models have been developed in which individual organs are represented by separate R-C circuits (Powell & Thorpe 1977; Meidner & Sherriff 1976) but the validity and usefulness of such analogues remain to be demonstrated.

The awkwardness of non-linear components has also inhibited the development of resistance analogues of photosynthesis and respiration. In the most elementary models of photosynthesis, the resistance to CO_2 diffusion is expressed as the sum of three components: the boundary layer and stomatal resistances already discussed and a 'mesophyll' resistance r_m often evaluated as a residual term (Jarvis 1971; Unsworth, this volume). The mesophyll resistance can be treated as the sum of a physical resistance for CO_2 diffusion in the liquid phase and a chemical quasi-resistance determined by the kinetics of the photochemical process; but neither of these components can be readily measured or estimated. Unlike conventional resistances, the so-called carboxylation component of the chemical resistance is proportional to the local potential. Moreover, the derivation of a carboxylation resistance from the shape of a light response curve (Prioul & Chartier 1977) is based on the implicit but implausible assumption that the physical sector of the CO_2 diffusion pathway is identical for all cells.

Resistance models of respiration have also been proposed (Lake 1967). Ryle, Cobby & Powell (1976) showed that the respiration of labelled carbon from maize and barley plants decreased at a rate which could be represented by two time constants, one of about 7 to 11 hours and the other about 4 to 5 days. Respiration could therefore be represented by the discharge of two RC circuits but as the rate of discharge is effectively independent of the inter-cellular CO_2 concentration, it is difficult to see how an R-C model of respiration could be coupled to a resistance model of photosynthesis.

Despite these limitations, electrical analogues of CO_2 coupling provide a useful complement to similar analogues of water vapour coupling, particularly for estimating the rate of water transpired per unit of carbon assimilated (Penman & Schofield 1951; Rijtema 1968). Cowan & Farquhar (1977) have suggested that this ratio may be optimized by diurnal changes of stomatal resistance—a highly sophisticated form of coupling between plants and the atmosphere. Supporting evidence comes from Goudriaan & van Laar (1978) who found that the stomatal resistance of some species tends to change in a way which stabilizes the intercellular CO_2 concentration.

Electrical analogues have also been useful in establishing the relative importance of the resistances which couple plant leaves to their CO_2 supply. The mesophyll resistance usually dominates the circuit so that the rate of carbon assimilation changes less than transpiration when stomata close and is usually independent of windspeed. In a few experiments, however, the dependence of CO_2 assimilation on boundary layer resistance has been clearly demonstrated (e.g. by Yabuki & Miyagwa 1970) and this aspect of coupling deserves closer attention in field studies.

REFERENCES

Achenbach E. (1974) Heat transfer from smooth and rough surfaced circular cylinders in a cross-flow. *Heat Transfer Conference, Tokyo* (10) *FC* **6.1**, 229–233.

Bange G.G.J (1953) On the quantitative explanation of stomatal transpiration. *Acta Botanica Neerlandica*, **2**, 255–297.

Brown H. & Escombe F. (1900) Static diffusion of gases and liquids in relation to the assimilation of carbon and translocation in plants. *Philosophical Transactions of the Royal Society of London*, **B 193**, 223–291.

Burton A.C. & Edholm O. (1955) *Man in a Cold Environment*. Edward Arnold, London.

Campbell G.S. (1977) *An Introduction to Environmental Biophysics*. Springer-Verlag, New York.

Chamberlain A. (1974) Mass transfer to bean leaves. *Boundary-Layer Meteorology*, **6**, 477–486.

Cowan I.R. (1972a) An e'ectrical analogue of evaporation from and flow of water in plants. *Planta*, **106**, 221–226.

Cowan I.R. (1972b) Mass and heat transfer in laminar boundary layers with particular reference to assimilation and transpiration in leaves. *Agricultural Meteorology*, **10**, 311–329.

Cowan I.R. & Farquhar G.D. (1977) Stomatal function in relation to leaf metabolism and environment. Symposium of the Society for Experimental Biology 31, pp. 471–505. Cambridge University Press.

Gaastra P. (1959) Photosynthesis of crop plants. *Mededelingen van de Landbouwhogeschool te Wageningen*, **59**, 1–68.

Gates D.M. (1968) Energy exchange in the biosphere. *Functioning of Terrestrial Ecosystems*. (Ed. by F.E. Eckardt). UNESCO, Paris.

Gates D.M. & Benedict C.M. (1963) Convection phenomena from plants in still air. *American Journal of Botany*, **50**, 563–573.

Gates D.M. & Schmerl R.B. (1975) *Perspectives of Biophysical Ecology*. Springer-Verlag, New York.

Gates D.M. & Papian L. E. (1971) *Atlas of Energy Budgets of Plant Leaves*. Academic Press, New York.

Gates D.M., Vetter M.J. & Thompson M.G. (1963) Measurements of moisture boundary layers and leaf transpiration with a microwave refractometer. *Nature*, **197**, 1070–1072.

Goudriaan J. & van Laar H.H. (1978) Relations between leaf resistance, CO_2 concentration and CO_2 assimilation. *Photosynthetica*, **12**, 241–249.

Grace J. & Wilson J. (1976) The boundary layer over a *Populus* leaf. *Journal of Experimental Botany*, **27**, 231–241.

Haseba T. (1973a) Water vapour transfer from leaf-like surfaces within canopy models. *Journal of Agricultural Meteorology, Tokyo*, **29**, 25–33.

Haseba T. (1973b) Water vapour transfer from plant-leaf surface by forced convection with special reference to inclination of leaf. *Journal of Agricultural Meteorology, Tokyo*, **29**, 1–9.

Herrington L.P. (1969) *On Temperature and Heat Flow in Tree Stems*. Yale University School of Forestry, Bulletin No. 73.

Jarvis P.G. (1971) The estimation of resistances to carbon dioxide transfer. *Plant Photosynthetic Production* (Ed. by Z. Šesták, J. Čatský and P.G. Jarvis). Junk, The Hague.

Jones H.G. (1978) Modelling diurnal trends of leaf water potential in transpiring wheat. *Journal of Applied Ecology*, **15**, 613–623.

Kozlowski T.T. & Ahlgren C.E. (1974) *Fire and Ecosystems*. Academic Press, New York.

Kreith F. (1973) *Principles of Heat Transfer*, 3rd edn. International Text Book Co., Scranton, Pennsylvania.

Lake J.V. (1967) Respiration of leaves during photosynthesis. *Australian Journal of Biological Sciences*, **20**, 487–493.

Landsberg J.J. & Powell D.B.B. (1973) Surface exchange characteristics of leaves subject to mutual interference. *Agricultural Meteorology*, **13**, 169–179.

Legg B. & Monteith J.L. (1975) Heat and mass transfer within plant canopies. *Heat and Mass Transfer in the Biosphere*. (Ed. by D.A. de Vries and N.H. Afgan). John Wiley, New York.

Leyton L. (1975) *Fluid Behaviour in Biological Systems*. Oxford University Press.

Lewis M.C. (1972) The physiological significance of variation in leaf structure. *Science Progress*, **60**, 25–51.

Linacre E.T. (1964a) A note on a feature of leaf and air temperatures. *Agricultural Meteorology*, **1**, 66–72.

Linacre E.T. (1964b) Heat transfer coefficients of leaves. *Plant Physiology*, **39**, 687–690.

Linacre E.T. (1972) Leaf temperatures, diffusion resistances and transpiration. *Agricultural Meteorology*, **9**, 365–382.

Maskell E.J. (1928) Experimental studies on vegetable assimilation and respiration XVIII. The relation between stomatal opening and assimilation. *Proceedings of the Royal Society of London* B 102, 488–453.

Me.dner H. & Sherriff D.W. (1976) *Water and Plants.* Blackie, Glasgow and London.

Milburn J.A. (1979) *Water Flow in Plants.* Longman, London and New York.

Miller E.C. (1938) *Plant Physiology.* McGraw Hill, New York.

Monteith J.L. (1963) Gas exchange in plant communities. *Environmental Control of Plant Growth.* (Ed. by L.T. Evans). Academic Press, New York.

Monteith J.L. (1965) Evaporation and environment. Symposium of the Society for Experimental Biology 19, pp. 205–234. Cambridge University Press.

Monteith J.L. (1975) *Principles of Environmental Physics.* Edward Arnold, London.

Monteith J.L. & Butler D.R. (1979) Dew and thermal lag: a model for cocoa pods. *Quarterly Journal of the Royal Meteorological Society,* 105, 207–215.

Nobel P.S. (1974a) Boundary layers of air adjacent to cylinders. *Plant Physiology,* 54, 177–181.

Nobel P.S. (1974b) Effective thickness and resistance of the air boundary layer adjacent to spherical plant parts. *Journal of Experimental Botany,* 26, 120–130.

Parlange J-Y, Waggoner P.E. & Heichel G.H. (1971) Boundary layer resistance and temperature distribution on still and flapping leaves. *Plant Physiology,* 48, 437–442.

Penman H.L. (1948) Natural evaporation. *Proceedings of the Royal Society, Series A,* 193, 120–143.

Penman H.L. & Schofield R.K. (1951) Some physical aspects of assimilation and transpiration. Symposium of the Society for Experimental Biology 5, pp. 115–130. Cambridge University Press.

Powell D.B.B. & Thorpe M.R. (1977) Dynamic aspects of plant-water relations. *Environmental Effects on Crop Physiology.* (Ed. by J.J. Landsberg and C.V. Cutting). Academic Press, London.

Prioul J.L. & Chartier P. (1977) Partitioning of transfer and carboxylation components of intracellular resistance. *Annals of Botany,* 41, 789–800.

Raschke K. (1956) Über die physikalischen Beziehungen zwischen Wärmeübergangszahl, Strahlungsaustauch, Temperatur und Transpiration eines Blattes. *Planta,* 48, 200–238.

Raschke K. (1958) Über den Einfluss der Diffusionswiderstande auf die Transpiration und die Temperatur eines Blattes. *Flora,* 146, 546–578.

Richter H. (1973) Frictional potential losses and total water potential in plants. *Journal of Experimental Botany,* 24, 983–999.

Rijtema P.E. (1968) On the relation between transpiration, soil physical properties and crop production as a basis for water supply plans. *Institute for Land and Water Management Research Technical Bulletin* 58, 29–58.

Ryle G.J.A., Cobby J.M. & Powell C.E. (1976) Synthetic and maintenance respiratory losses of $CO_2{}^{14}$ in uniculm barley and maize. *Annals of Botany,* 40, 571–586.

Salisbury F.B. & Spomer G.E. (1964) Leaf temperatures of alpine plants in the field. *Planta,* 60, 497–505.

Schlichting, H. (1960) *Boundary-Layer Theory,* 4th edn. McGraw Hill, New York.

Šetlík I. (Ed.) (1970) *Prediction and Measurement of Photosynthetic Productivity.* Centre for Agricultural Publishing, Wageningen.

Tanner C.B. & Golz S.M. (1972) Excessively high temperatures of seed onion umbels. *Journal of the American Society for Horticultural Science,* 97, 5–9.

Thorpe M. (1974) Radiant heating of apples. *Journal of Applied Ecology,* 11, 755–760.

Thorpe M.R. & Butler D.R. (1977) Heat transfer coefficients for leaves on orchard apple trees. *Boundary-Layer Meteorology,* 12, 61–73.

Turrell F.M. & Perry R.C. (1957) Specific heat and conductivity of citrus fruit. *Proceedings of the American Society of Horticultural Science*, **70**, 261–265.

Wallace J.S. (1978) *Water Transport and Leaf Water Relations in Winter Wheat Crops.* Ph.D. Thesis, University of Nottingham.

Warren Wilson J. (1957) Observations on the temperatures of Arctic plants and their environment. *Journal of Ecology*, **45**, 499–531.

Weast R.C. (Ed.) (1977) *Handbook of Physics and Chemistry*, 58th edn. ORC Press.

Weatherley P.E. (1976) Water movement through plants. *Philosophical Transactions of the Royal Society, Series B*, **273**, 435–444.

Wigley G. & Clark J.A. (1974) Heat transport coefficient for constant energy flux models of broad leaves. *Boundary-Layer Meteorology*, **1**, 123–456.

Yabuki K. & Miyagwa H. (1970) Studies on the effect of wind-speed upon the photosynthesis. *Journal of Agricultural Meteorology, Tokyo*, **26**, 137–142.

Yabuki K., Miyagwa & Ishibashi K. (1970) Studies of the effect of wind-speed upon the photosynthesis. 1. Boundary layer near the surface. *Journal of Agricultural Meteorology, Tokyo*, **26**, 65–70.

2. SOME EFFECTS OF WIND ON PLANTS

JOHN GRACE

*Department of Forestry and Natural Resources,
University of Edinburgh, Edinburgh EH9 3JU*

SUMMARY

1. Field observations suggest that wind affects the shape and size of trees, the physiognomy of the vegetation, and the productivity of crops. Ecologists and foresters use the term 'exposure' to describe this impact of wind on vegetation. The present article explores the physical and biological meaning of this term.

2. Transport of heat, gases and momentum across the boundary layer between the atmosphere and the plant depend on structural features of the vegetation. High transport rates between plant surfaces and the air occur when the leaves, stems and branches are small rather than large, and when the vegetation is tall rather than short. In parts of the world where overall temperatures are low and windspeeds high it is argued that, if all other variables are equal, tall or small-leaved plants will be colder than short or broad-leaved plants, and so be at a disadvantage.

3. The effect of wind on transpiration can be calculated from the Penman-Monteith equation. Wind does not always increase transpiration rate: in many ordinary conditions an increase in wind causes a decrease in transpiration.

4. Leaf surfaces may sustain damage in wind, causing a disruption of proper cuticular control of water loss. This damage has been observed in grasses, *Acer* and *Fragaria* but does not seem to occur in conifers.

5. The effect of wind on exchanges of gases can be calculated if the other (physiological) resistances in the total diffusion path are known. In general, the effect of wind on the rate of photosynthesis is small (except at very low windspeeds), because the stomatal and mesophyll resistances are large in relation to the boundary layer resistance.

6. There are several reports which suggest that motion *per se* is a potent inhibitor of plant growth. The mechanisms involved are unknown.

7. In conclusion, the shaping of trees by the wind and the general effects of 'exposure' already referred to are likely to be caused by the effect of wind on the temperature of leaves and meristems, damage to leaf surfaces, and the direct effect of motion as an inhibitor of growth.

INTRODUCTION

Effects of wind on vegetation may be seen almost everywhere. Trees growing in windy places are often shaped by wind, their branches appearing permanently swept to the leeward, earning the names 'wind-brushed' or 'flag tree', and these shaped trees have even been used to assess the average strength and direction of the wind (e.g. Putnam 1948; Sekiguti 1951; Holroyd 1970; Yoshino 1973; Wardle 1977). The extreme condition occurs at the

31

altitudinal limit of tree growth, where tree species may adopt a prostrate form—the 'krummholz' condition (Wardle 1968).

Effects like these are not confined to trees. At high altitudes, above the treeline, the shrub vegetation is locally dwarfed. Many years ago Raunkiaer (1909, 1934) proposed a functional classification of plants, defining certain 'life forms' on the basis of height above the ground of their perennating organs. He showed that cold and windy areas supported only a restricted spectrum of life forms—those with buds near the ground. Similarly, Tansley (1939) showed that above 1000 m a.s.l. in the Scottish Grampian mountains, 27 per cent of the species are chamaephytes as compared to 9 per cent in the flora of the world as a whole; and Gimingham (1951) demonstrated a similar shift towards dwarf and prostrate growth forms with increasing exposure to the wind in sand dune systems.

TABLE 2.1. The spectrum of growth forms at sites of increasing exposure to the wind on Monte Maiella, Italy (site A is most exposed). Figures refer to % of total vegetation at that site (from Whitehead 1954).

Growth form	Site exposure to the wind			
	D	C	B	A
Medium herb, 100 mm	1·56	—	—	—
Small sub-shrub, 50 mm	0·82	—	—	—
Small herb, 50–100 mm	5·04	4·03	—	—
Large tussock, diam. 500 mm	18·1	3·07	—	—
Large cushion, diam. 50 mm	9·14	20·3	—	—
Woody mat	58·0	22·0	0·30	—
Herbaceous mat	—	5·68	21·1	—
Large rosette, diam. 50 mm	4·55	6·90	30·2	—
Small rosette, diam. 50 mm	0·67	7·40	8·45	6·02
Small tussock, diam. 50 mm	0·01	0·67	8·59	4·30
Small cushion, diam. 50 mm	2·36	2·2	6·03	30·4
Prostrate, suffruticose or woody	—	—	5·73	15·7
Prostrate herb	—	—	19·6	43·5

In two most interesting papers Whitehead (1954, 1959) described the influence of creating shelter around small plots of mountain vegetation with walls of stones (Table 2.1). Over just one year the resulting shelter increased the average plant height from 22 mm to 32–68 mm. In later years new species invaded the sheltered plots and the percentage cover of established species changed markedly. More recently Nägeli (1971) showed a similar dependence of vegetation type on wind speed in an alpine valley; he demonstrated that the natural shelter afforded by topography created physiognomic differences in the vegetation.

Shelter is also significant to agricultural production. Since the early work of Bates (1911) there have been many demonstrations that shelter belts or windbreaks increase the yield of crops. In a recent review covering eighty-six different cases, substantial improvements in the yield ($>10\%$) of the crop, attributable to shelter, were reported in sixty-eight cases (Grace 1977).

How are these effects to be interpreted? Ecologists and foresters have used the word 'exposure' to describe the combined stresses that plants suffer in cold and windy places. The word has been used in a rough and intuitive manner and 'exposure' is rarely defined in a measurable way. Yet our knowledge of microclimates and the aerodynamics of vegetation has advanced considerably in recent years, to the point where we ought now to understand the meaning of this 'exposure'. Electric resistance models have provided a focal point whereby knowledge from the physical and engineering sciences may be brought to bear on biological problems and, in particular, on the estimation of transpiration rates and surface temperatures. This chapter brings together some recent advances in the general field of wind relations so that the concept of 'exposure' may be reassessed.

EXCHANGES IN THE BOUNDARY LAYER

The important role of air movement is to ventilate plant surfaces, mixing air near the leaf with new air from the bulk of the atmosphere. Hence, in conditions of bright sunlight, fresh supplies of carbon dioxide are brought to the leaves, while water vapour and heat are transported away.

The basic equation which expresses the transfer of any entity between a surface and the atmosphere states that the flux is proportional to the concentration gradient.

$$\mathbf{F} = -D \cdot \delta\chi/\delta z \qquad (2.1)$$

where $\mathbf{F}$ is the flux, $\delta\chi/\delta z$ is the vertical concentration gradient above the surface and D is a transfer coefficient, in appropriate units.

If the flow of air is laminar, the appropriate value for D is the molecular diffusivity. In many other circumstances, the flow of air is turbulent and the appropriate transfer coefficient is that of eddy diffusion, K. In general, values of K are highly variable and always larger than the coefficients of molecular diffusion. Within vegetation K may vary from 10^{-5} near leaves to 10^{-1} m² s⁻¹ near the top of the plants, while above the crop K increases linearly with height and may be as high as 10^2 m² s⁻¹. This large variation occurs because the size of the individual eddies, which are the medium for transport, increases with height above the surface. In such a medium the rate of transport is independent of molecular size, and so, according to the principle of Prandtl, the value of K is independent of the entity being transported.

A diffusion resistance r, or its inverse the conductance g, is defined in relation to vertical distance z as

$$r = g^{-1} = \int_{z_1}^{z_2} \frac{dz}{D} \qquad (2.2)$$

Thus if we consider diffusion from a hairy leaf, a layer of still air trapped by dense hairs 1 mm long might be expected to impose a resistance to water vapour diffusion of $1 \times 10^{-3}/(2\cdot4 \times 10^{-5}) \simeq 40$ s m^{-1}. This is equivalent to half a metre or so of turbulent air, assuming $K = 10^{-2}$ m^2 s^{-1} above the leaf.

When is the flow of air over the leaves laminar and when turbulent? Classical studies in fluid dynamics suggest that this question may be answered if the Reynolds number Re is known:

$$\mathrm{Re} = \frac{ul}{\nu} \qquad (2.3)$$

where u is the fluid velocity, l is the characteristic dimension of the test object—for a leaf it would be the average length in the direction of the airflow, and ν is a property of the fluid called the kinematic viscosity. At low values of Re the fluid moves as a coherent mass because of its viscous properties. As Re rises above a critical point Re_{crit} the inertial forces within the fluid predominate and the smooth flow breaks down to give chaotic motion in which individual lumps of fluid move in directions other than parallel to the main flow and mixing occurs.

For smooth flat plates exposed to laminar flow in a wind tunnel the transition from a laminar to a turbulent boundary layer occurs at a Reynolds number of about 2×10^4. As Reynolds numbers for leaves in natural conditions are less than this, the view has arisen that the air flow over leaves must be laminar (Oke 1977), and that the theory of laminar boundary layers may be used in any calculations. However, when the air flow over real leaves has been investigated the boundary layer is seen to be turbulent (Perrier et al. 1973; Grace & Wilson 1976). For example over a *Populus* leaf exposed in a wind tunnel to smooth air flow at 1 m s^{-1} the boundary layer was laminar only for a short distance along the adaxial surface (Fig. 2.1). Elsewhere there was considerable turbulence, which seemed to originate from surface irregularities such as veins standing proud, or the decurrent leading edge (Fig. 2.1). The critical value of Re needed for the transition from laminar to turbulent flow seems to be between 400 and 3000, much lower than those normally quoted in engineering texts which refer to smooth flat plates. Turbulence in the boundary layer is encouraged by turbulence in the air stream incident on the leaf, and accentuated by the uneven topography and roughness of the leaf. Decurrent leading edges, or a serrated margin may serve to 'trip' the airflow and so precipitate turbulence.

These measurements suggest that the boundary layers over leaves are usually turbulent or partly turbulent when the wind blows, except in the very

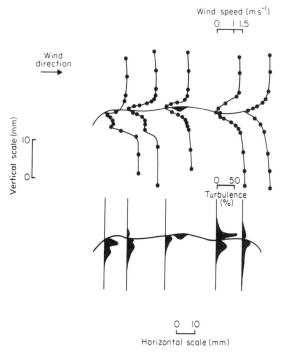

FIG. 2.1. Profiles of mean wind speed (*upper*) and turbulence (*lower*) around a *Populus* leaf shown in transverse section. The free air-stream in the wind tunnel was laminar, but the boundary layer was turbulent over much of the surface. The scale for height is exaggerated X2 compared to the horizontal scale for length (from Grace & Wilson 1976).

low wind speeds and in the case of very small leaves where the dimension is so low as to give a very small Reynolds number. It should be realized that even in a fully-developed boundary layer there is a *viscous sublayer* very close to the surface in which the flow remains laminar. Some idea of the thickness of this sublayer can be obtained by observing the motion of minute objects which happen to occur at the leaf surface, such as fungal conidiophores. From one series of experiments the sublayer would appear to be not more than a few tens of micrometres though in hairy leaves it might be considerably more (Grace & Collins 1976).

EFFECT OF WIND SPEED ON r_a OF LEAVES

Attempts to measure the effect of wind speed on the boundary layer resistance r_a have been made many times. The most common method has been to make a paper model of the leaf, wet it with water, and then measure the evaporation rate of water from it in an airstream of known humidity (Jarvis 1971). Calculation of the resistance requires an estimate of the water vapour pressure at the evaporating surface; this value is very sensitive to any small differences in temperature over the surface. Moreover, it is not always easy to maintain the air flow at a constant and known humidity (for a discussion of errors see Grace & Wilson 1976). These problems have led some workers to use completely different diffusing species (Thom 1968; Macleod & Todd 1973; Chamberlain 1974).

There is perhaps a larger source of error in this procedure; leaf micro-topography cannot be represented in the paper model. In a real leaf much of the resistance to transfer may reside in the sublayer trapped between hairs, grooves or papillae.

Boundary layer resistance has been measured by engineers as well as by plant scientists. Many fluids other than air have been used, and surfaces of diverse size and shape have been employed according to the needs of the investigator. It is useful to be able to bring results from all these experiments together to make 'predictive' formulae: this may be done using *dimensionless groups* to express the transfer rates (see also the papers in this volume by Monteith, Chamberlain and Proctor). The application of dimensionless groups is discussed in textbooks on heat and mass transfer and also by Monteith (1973), Leyton (1975) and Campbell (1977); there follows a definition of the dimensionless numbers which are used universally to describe mass and heat transfer. To express mass transfer we use the Sherwood number Sh

$$\text{Sh} = \frac{\mathbf{F} \, l}{D(\chi_s - \chi_a)} \tag{2.4}$$

where $\mathbf{F}$ is the mass flux density (kg m^{-2} s^{-1}), l is the dimension of the object parallel to the wind speed (m), D is the appropriate coefficient of molecular diffusion (m^2 s^{-1}), χ_s is the concentration at the surface and χ_a is the concentration in the bulk of the fluid (kg m^{-3}). The more familiar boundary layer resistance can be found as $r_a' = 1/D\text{Sh}$ where the suffix ' is used to denote that this is for one side of the leaf only. The Nusselt number Nu is the analogous number for heat transfer, enabling direct comparison of heat transfer rates in any fluid (ecologists' interests are usually, but not always, confined to air and water):

$$\text{Nu} = \frac{\mathbf{C} \, l}{\rho c_p D_H (T_s - T_a)} \tag{2.5}$$

where $\mathbf{C}$ is the heat flux density (J m^{-2} s^{-1}), ρ is the density of the fluid (kg m^{-3}), c_p is the specific heat at constant pressure (J °C^{-1} kg^{-1}), D_H is the molecular diffusion coefficient for heat in the fluid concerned (m^2 s^{-1}), T_s is the surface temperature and T_a is the temperature of the bulk of the fluid (°C). A boundary layer resistance for heat transfer from one side of the leaf is calculated as $r_a' = 1/D_H\text{Nu}$.

Formulae for laminar boundary layers

To estimate the effect of wind on boundary layer resistance of a flat plate the following formulae are widely used because they can be derived from theoretical considerations and are also a good representation of the large body of experimental data, in the engineering sciences at least:

$$\text{Sh} = 0{\cdot}66 \ \text{Re}^{0{\cdot}5} \left(\frac{\nu}{D}\right)^{0{\cdot}33} \tag{2.6}$$

$$\text{Nu} = 0{\cdot}66 \ \text{Re}^{0{\cdot}5} \left(\frac{\nu}{\kappa}\right)^{0{\cdot}33} \tag{2.7}$$

where κ is the thermal diffusivity. Derivation of 2.6 and 2.7 can be seen is suitable textbooks such as Welty, Wick & Wilson (1969). These equations imply that the ratio of resistances for different entities are related to each other as

$$\frac{r_a^{H_2O}}{r_a^{heat}} = \left(\frac{\kappa}{D_v}\right)^{0{\cdot}66} = 0{\cdot}93 \tag{2.8}$$

$$\frac{r_a^{CO_2}}{r_a^{heat}} = \left(\frac{\kappa}{D_c}\right)^{0{\cdot}66} = 1{\cdot}32 \tag{2.9}$$

$$\frac{r_a^{CO_2}}{r_a^{H_2O}} = \left(\frac{D_v}{D_c}\right)^{0{\cdot}66} = 1{\cdot}39 \tag{2.10}$$

and not in simple inverse proportion to the respective values of D as suggested from first principles (equations 2.1 and 2.2).

Formulae for turbulent boundary layers

As we have seen, boundary layers over leaves are frequently turbulent, or partly turbulent, and so these relationships (equations 2.6–2.10) cannot be expected to apply. Alternative expressions for *fully developed turbulent flow* are available in which the relationship between exchange rate and Reynolds number is steeper than for laminar flow (Bayley, Owen & Turner 1972):

$$\text{Sh} = 0{\cdot}03 \ \text{Re}^{0{\cdot}8} \left(\frac{\nu}{D}\right)^{0{\cdot}33} \tag{2.11}$$

$$\text{Nu} = 0{\cdot}03 \ \text{Re}^{0{\cdot}8} \left(\frac{\nu}{\kappa}\right)^{0{\cdot}33} \tag{2.12}$$

However, these expressions are valid only at high Reynolds numbers and do not provide reliable estimates when Re is less than 2×10^4, as is often the case for leaves in the natural wind. Moreover they may not include proper contribution for the laminar sublayer, nor for the buffer zone between this and the turbulent flow, especially for a hairy or rough leaf. Furthermore, we have seen that the boundary layer over a leaf may be laminar near the leading edge and on the flatter parts of the leaf, and turbulent where the air flow is tripped by uneven topography. In view of this complexity we cannot expect good agreement between experimental data and standard relationships, whether laminar or turbulent boundary layer theory is applied.

Experimental data

Experimental data are plotted in Fig. 2.2, along with lines drawn from standard relationships. Most results suggest that exchange rates for leaves are higher than predicted by the equations for laminar boundary layers, often by a factor of two. However, the slope of the relationship is not steep enough to indicate fully developed turbulent flow except at the three highest Reynolds

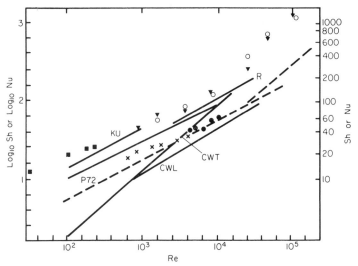

FIG. 2.2. Exchange rates for model leaves. The scatter about the standard relationship, shown by a dashed line, is considerable. Kuiper (KU), Rashke (R) and Slatyer & Bierhuizen (■), all reproduced from Monteith (1965); Thom (1968) (×), Parkhurst *et al.* (1968) (●), Pearman *et al.* (1972) (P72), Clark & Wigley (1975) with turbulent (CWT) and laminar (CWL) free streams and Grace & Wilson (1976) with turbulent (○) and laminar (▼) free streams.

numbers used by Grace & Wilson (1976). We may conclude that these data and the observations made with a hot-wire anemometer (Fig. 2.1), are consistent with the view that boundary layers over leaves are usually neither laminar nor strictly turbulent, but that a mixed regime prevails.

Estimates of r_a

The effect of wind speed on boundary layer conductance or resistance is shown in Fig. 2.3, on the assumption that the boundary layer is laminar. This graph may be used for estimation purposes, but in view of what has been said it must be realized that the conductance thus obtained may be an underestimate by a factor of two. Caution should also be applied when interconverting resistances for different entities using equations 2.8–2.10: in those parts of the leaf where the boundary layer is turbulent the resistances for different entities will not be as calculated.

A general point is evident from Fig. 2.3. It is that *leaf size* is important in determining the coupling of leaves to the atmosphere; small leaves have

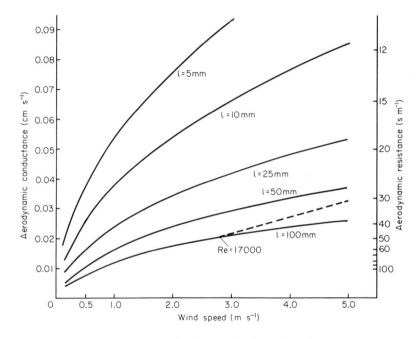

FIG. 2.3. Aerodynamic conductance for leaves of various dimension according to the standard relationships derived from engineering studies. For large leaves at high wind speed (3 m s⁻¹) the calculation for a turbulent boundary layer (broken line) gives a higher conductance than that for a laminar boundary layer (solid lines).

high boundary layer conductances and so are more closely coupled to the atmosphere than large ones.

There is another difficulty in applying Fig. 2.3 for specific estimation purposes. When one leaf is close to another, or close to another object such as a rock protruding from the ground, boundary layers will tend to coalesce.

There have been few studies of this phenomenon despite its ecological importance. In one case, Landsberg & Thom (1971) measured the boundary layer resistance of the bunches of spruce needles which form a shoot, and they found that the boundary layer resistance was twice that calculated from data for an isolated needle.

EFFECT OF WIND SPEED ON r_a OF VEGETATION

A more overall view of the exchange rates between vegetation and atmosphere is obtained if the vegetation is treated as a single rough surface which by virtue of its roughness, slows down the air flow. This results in a much-reduced wind speed immediately above the vegetation—in other words a boundary layer is formed. In this process momentum is transferred from the bulk of the atmosphere (the momentum concentration is ρu) to the leaves and branches in the vegetation. The air just above the vegetation has a reduced momentum and the air in contact with surfaces has no velocity, and hence no momentum at all. The whole process may be regarded as a downward flux of momentum from a source, the atmosphere, to an imaginary sink at height d in the vegetation. As the air over the vegetation is turbulent the basic mechanism involved here is turbulent transport, whereby momentum, heat and gases share the same vehicle of transport—the parcels of air whose motion constitutes the turbulence. It is worthwhile asking the question 'what attributes of the vegetation determine the rate of momentum exchange?' The answer should be found in a comparison of windspeed profiles of different sorts of vegetation.

Many workers have fitted the following equation to measurements of the wind speed made above the vegetation at several heights (z):

$$u(z) = \frac{u_*}{k} \ln\left(\frac{z - d}{z_o}\right) \tag{2.13}$$

where u_* is a parameter called the friction velocity (m s^{-1}), k is von Karman's constant (a pure number), d is the zero plane displacement (the apparent level inside the canopy where the wind speed is zero—the sink for momentum), and z_o is the roughness length.[†] For a good discussion of this equation, including a physical interpretation of the parameters, the reader is referred to Thom (1975).

Examples of wind profiles are given in Fig. 2.4a. The profiles start at a higher point above the ground for the tall vegetation because the roughness elements are held aloft. The dotted portions are extrapolated to a virtual

† To simplify this discussion, the meteorological conditions are assumed to be neutral (i.e. with a temperature gradient close to -0.01 °C m^{-1}). In unstable conditions, i.e. a gradient of 1 °C m^{-1}, turbulent transfer is enhanced by buoyancy. In 'stable' conditions, when temperature decreases with height by -1 °C m^{-1}, vertical motion is suppressed.

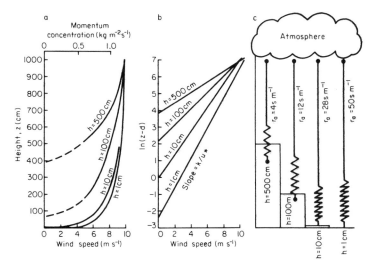

FIG. 2.4. Momentum exchange as affected by vegetation height: (a) vertical profile of momentum over vegetation of four heights, h, when the wind speed at 10 m corresponds to a strong breeze; (b) the logarithmic plot used to obtain the roughness parameter z_0; (c) calculated resistances to momentum transfer between 10 m above the ground and the hypothetical momentum sink in the vegetation. The data from which these curves were drawn were calculated using equation 2.13.

point of zero windspeed at d, which is normally between 0·6 to 0·8 of the vegetation height. Also the profiles are a different shape: immediately over short vegetation the wind speed changes very rapidly with height, implying that the mixing process is less effective than over tall vegetation.

The difference between the shapes of the profiles is better shown in Fig. 2.4b, where the ordinate is a logarithmic scale adjusted for differences in zero plane displacement. The slope of these lines, from equation 2.13, can be shown to be k/u_*. The intercept provides the important parameter z_0, the roughness length, which specifies the ability of the vegetation to capture momentum. Tall vegetation is always rougher than short vegetation, and for most crops $z_0 \simeq 0·1h$.

The more gradual change in wind speed with height over the rougher vegetation implies that the air is better mixed: the turbulent transfer coefficient must be high and the momentum resistance $r_a{}^m$ must be low. Monteith (1973) and Thom (1975) have explored the relationships between z_0, and $r_a{}^m$: an important result is

$$r_a{}^m = \{\ln (z - d)/z_0\}^2/(k^2 u(z)) \qquad (2.14)$$

This enables a resistance to be found provided that z_0 and d can be estimated. In Fig. 2.4c the resistance has been calculated for each vegetation height on the assumption that $z_0 = 0·1h$ and $d = 0·7h$—these values seem to be

approximately correct for most agricultural crops and for some wild vegetation. Resistance is very sensitive indeed to vegetation height, being least
for the tallest stands (Fig. 2.4c).

Wild vegetation may not conform to the relationship $z_o = 0.1h$, due to
the extreme range of plant spacement, lack of homogeneity (as in tropical
rain forest), or canopy smoothing (as in salt-sprayed coastal vegetation).
Recently Garratt (1977) has compiled data from a diversity of vegetation
types. From his graph it should be possible to select a suitable value of z_o,
and so estimate $r_a{}^m$, for any vegetation (Fig. 2.5).

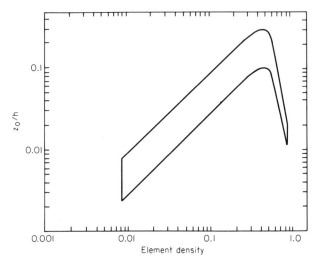

FIG. 2.5.   The influence of vegetation density on the fraction z_o/h, where z_o is the
roughness length and h is the plant height. Element density is defined as the plant
silhouette area normal to the wind per unit surface area occupied by each plant. Most
vegetation falls within the density range 0·05 to 0·15. The envelope contains the scatter
of observational data (Garrett 1977).

Equation 2.14 shows that $r_a{}^m$ is inversely proportional to windspeed,
assuming that z_o and d are insensitive to wind (some effect of wind on z_o
and d is likely as the canopy becomes smoothed or parted with changes in
the wind speed). Fig. 2.6 shows the effect of wind on the resistance $r_a{}^m$ and
conductance $g_a{}^m$ for vegetation of differing heights, again assuming $z_o = 0.1h$, $d = 0.7h$, and further assuming that z_o and d are not affected by
windspeed.

In the turbulent boundary layer over vegetation, the aerodynamic
resistance may be expected to be the same irrespective of the diffusing entity,
i.e.

$$r_a{}^m = r_a{}^{gas} = r_a{}^{heat} \tag{2.15}$$

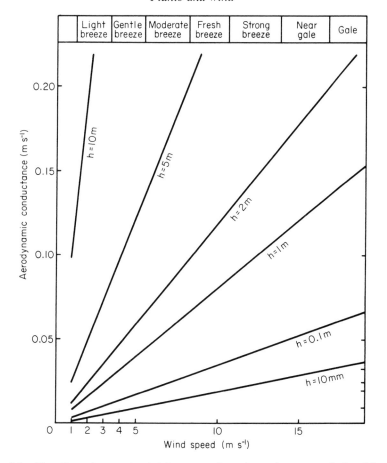

FIG. 2.6. The effect of vegetation height on aerodynamic conductance, calculated from equation 2.11 assuming $z_o = 0 \cdot 1h$. The figures on the lines are the heights of the vegetation (m).

In practice, however, the aerodynamic resistance for momentum transfer to the vegetation may differ substantially from the resistances for heat and gases. One reason for this is that the 'sinks' for momentum within the canopy are all the structural elements, living or dead, which slow down the air flow. The 'sources' and 'sinks' for gases and heat exchange may not be the same as this, and will be modified by physiological activity; for carbon dioxide transfer, for instance, only the leaves are active. There is another reason too. When air flows over a leaf surface momentum exchange across the boundary layer occurs by *skin friction*, a process analogous to the diffusion of heat or gas. If on the other hand the surface is held perpendicular to the flow the air is blocked and so very considerable momentum is exchanged, by *form drag*, without a corresponding increase in the exchange of heat or gas. The net effect is that the apparent sources and sinks for heat and water vapour occur at lower level in the canopy than the sink for momentum, and so— because of the increased path length—the aerodynamic resistances for them are greater. Monteith (1973, pp. 197–198) and Thom (1975, pp. 90–92) make suggestions about the quantitative relationships between these resistances.

The general point, that tall plants are more closely coupled to the atmosphere, seems important. That there are indeed substantial differences in the microclimate of tall and short vegetation is demonstrated by field observations: Salisbury and Spomer (1974) showed that the temperatures of cushion plants in the alpine environment were several degrees higher than those of nearby 'erect' plants. In further support, it is well known that species of wide geographical distribution are of shorter stature in their most 'exposed' situations, and that this short stature is maintained even when individuals are transplanted to a more mesic site (e.g. Clausen, Keck & Hiesey 1940). Natural selection evidently favours short stature at 'exposed' sites and this may result from the higher temperatures. Similar arguments may be made for coastal vegetation in relation to salt spray; for short plants, weakly coupled to the atmosphere, receipt of salt droplets will be less than for their taller neighbours.

WIND AND TRANSPIRATION

The influence of wind speed on transpiration rate and surface temperature cannot be estimated from a knowledge of r_a alone, because the extent to which its influence is felt depends on other factors, notably the available energy and stomatal conductance. The heat balance equation of a leaf may be stated

$$R + \lambda E + G + C + S + P = 0 \qquad (2.16)$$

where R is the net radiation, λE is the energy used in evaporating water (λ is the latent heat of vaporization of water), G is the rate of exchange with the other parts of the system by conduction, C is the convective heat exchange, S is the rate of change in stored heat, P is the rate of conversion of radiant energy to chemical energy by net photosynthesis.

In many cases G, S and P are small compared with R (e.g. see Cernusca & Seeber, this volume), and so the 'available' energy H is roughly equal to the net radiation. It is 'available' inasmuch as it may drive evaporation or be converted to sensible heat, warming the leaf.

$$H = C + \lambda E \qquad (2.17)$$

The convection of heat from the leaf proceeds at a rate which depends on the difference in the temperature of the surface and the ambient air $(T_s - T_a)$

$$C = \frac{\rho c_p (T_s - T_a)}{r_a^{heat}} \tag{2.18}$$

where r_a^{heat} is the boundary layer resistance for heat, and ρ is the density of air.

Similarly the evaporation rate depends on the difference between the saturation vapour pressure for water at the evaporating sites within the leaf $(e_s(T_s))$ and the vapour pressure of the ambient air, e

$$E = \frac{\rho c_p (e_s(T_s) - e)}{\lambda \gamma (r_s^{H_2O} + r_a^{H_2O})} \tag{2.19}$$

where γ is the psychrometric constant and $r_s^{H_2O}$ is the stomatal resistance.

In a well-watered plant with its stomata open the evaporation rate may be high, using up most of the available energy so that C is low. If stomata are shut then the evaporation rate is low, and most of the available energy will go towards increasing T_s.

The importance of these equations is that they may be solved to estimate the influence of wind on leaf surface temperatures or on evaporation rates from plant leaves (Linacre 1964; Monteith 1965; Gates & Papian 1971).

When this is done we obtain the somewhat unexpected result that in quite a lot of conditions an increase in the wind causes a *decrease* in the rate of transpiration, not an increase (Fig. 2.7). This is because the wind cools the leaf, and at this lower temperature the concentration of water vapour in water-saturated air is less: consequently the difference between the concentration of water in the sub-stomatal air cavities and that in the bulk of the atmosphere is less. Thus, with a smaller driving gradient, the diffusion rate of water molecules must be correspondingly slower. Yamaoka (1958) demonstrated this experimentally, but his paper has been overlooked until recently (compare Fig. 2.8 with Fig. 2.7).

WIND AND GAS EXCHANGE

The influence of wind speed on F the maximum rate of uptake of CO_2 may be estimated from a knowledge of r_a and r_s.

$$F = \frac{\chi_{air} - \chi_i}{r_a + r_s + r_m} \tag{2.20}$$

χ_{air} and χ_i are the concentrations of carbon dioxide in the ambient air and of the sites of photosynthetic carboxylation, respectively; and r_m is the resistance inside the leaf, discussed at length by Jarvis (1971). The value of

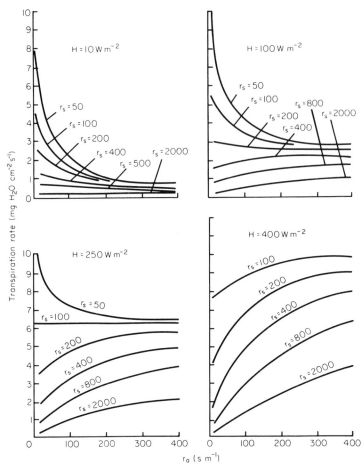

Fig. 2.7. Transpiration rate of leaves as influenced by aerodynamic resistance r_a, stomatal resistance r_s and available energy H. From Grace (1977), calculated from the Penman Monteith equation using an atmosphere saturation vapour deficit of 800 Pa and an air temperature of 15 °C. Units of r_s are sm[-1].

χ_i presumably depends on the affinity of the enzyme system for carbon dioxide and cannot be determined directly. Many workers have assumed a value of zero, as an expedient. Since the work of Gaastra (1959) many measurements of r_a, r_s, and r_m have been made; for tabulated examples see Whiteman & Koller (1967), Körner, Scheel & Bauer (1979).

Grace (1977) applied equation 2.20 and inserted various values of aerodynamic resistance to estimate the influence of wind on photosynthesis (using Fig. 2.3). The result suggested that wind had only a small effect on CO_2 exchange, except at extremely low values of wind speed where r_a may

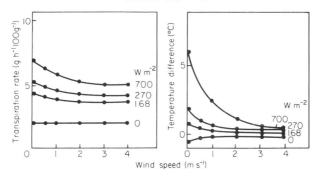

Fig. 2.8. Observed effect of wind speed on the rate of transpiration and leaf temperature in *Cryptomeria japonica*. Air temperature 20 °C, relative humidity 60%. After Yamaoka (1958).

become substantial in relation to $r_s + r_m$. In outdoor conditions it is doubtful whether r_a is often very high as still air rarely occurs—thermal gradients in plant canopies give rise to air movements by natural convection. In the laboratory, rather high values of r_a are apparent in unstirred photosynthesis chambers (Decker 1947; Warren-Wilson & Wadsworth 1958; Avery 1966; Parkinson 1968).

A similar approach to the estimation of the effect of wind speed on gas exchange might be applied to pollutant gases. Each gas is likely to have a somewhat different diffusion path inside the leaf however, so basic research in controlled environment chambers is required to elucidate the value of r_m and its sensitivity to environment and to the physiological conditions of the plant.

The influence of wind on transpiration rate has previously been discussed. The conclusion that wind does not always increase transpiration rate, holds for undamaged leaves. Wind damage to leaf surfaces is often apparent to the naked eye (Bauer 1966; Taylor & Sexton 1972), and Thompson (1974) has demonstrated significant wind damage on a micro-scopic scale to leaves of *Festuca arundinacea* exposed in a wind tunnel. This damage, resulting from collisions of leaves, involved alterations in the wax depositions on the cuticle and included the rupture of epidermal cells. Such leaves had a much higher conductance to water vapour, both cuticular and stomatal components being affected (Fig. 2.9). Similar abrasion has been reported by MacKerron (1976) in *Fragaria* and by Wilson (1979) in *Acer*, but seems to be negligible in *Picea* (Grace, Malcolm & Bradbury 1975) and *Pinus* (Rees 1979).

The integrity of the cuticle may be much more important than has been previously realized. In drought or in the winter when the ground is frozen, the cuticular resistance to water loss may be critical for the plant's survival. Tranquillini (1976) showed that trees which suffer from 'winter desiccation'

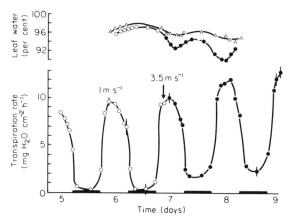

FIG. 2.9. Effect of an increase in the wind speed on the diurnal course of transpiration rate in *Festuca arundinacea* in a wind tunnel. Open symbols denote low wind speed (Grace 1974).

at the tree line do so because in the shortened growing season their cuticle does not fully develop—it remains thinner than in trees from lower altitude and the transpiration rate is consequently higher.

On the other hand, some species are said to shut their stomata in the wind (Martin & Clements 1935; Tranquillini 1969; Caldwell 1970; Davies, Kozlowski & Pereira 1974; Flückiger, Oertli & Flückiger-Keller 1978; Davies, Gill & Halliday 1978). This may be mediated by turgor changes caused by mechanical shock, as suggested to account for the occasional observation that stomata are sensitive to the shock incurred when a porometer cup is attached to the leaf (Knight 1916; Williams 1949; Fenton, Davies & Mansfield 1977).

SOME PHYSIOLOGICAL CONSIDERATIONS

Mechanical stimulus

Several authors have demonstrated that plant growth is sensitive to mechanical stimulus. Neel & Harris (1971) shook *Liquidambar* stems for 30 s daily and observed an 80 per cent reduction of height growth. Moreover these plants became dormant sooner. Rees (1979) describes a series of careful experiments like this on *Pinus contorta* in which the treatments included a continuous shake as well as a brief shake each morning. Although the results were less dramatic than those reported by Neel & Harris (1971), he found that brief shakes were as effective as continuous shaking: height growth and leaf extension were significantly reduced, and he concluded that 'shaking is a potent inhibitor of plant growth'.

How can shaking affect plant growth? The dwarfing effect of wind on plants resembles the effect of drought, and has led authors to think that water stress is the underlying cause of dwarfing, yet we have already seen that wind probably does not increase transpiration rate in most circumstances, and so is unlikely to have an effect on leaf water potential. Moreover, when water potential has been measured in shelter experiments, or in wind tunnel experiments, no cause-and-effect relationship between wind and water stress has usually been found (Russell & Grace 1978, 1979a).

On the other hand, it is well known that *pressure potential* of individual plant cells may change independently of the water potential of the tissue as a whole, as it does in guard cells, where the pressure potential is regulated by the flux of K^+ ions (Penny & Bowling 1974). In some plants, the same mechanism is developed to enable the plant to respond to tactile stimuli; in *Mimosa* and certain other genera the leaves fold up when touched. The cells of the leaf are able to transmit a message to pulvini at the base of each leaflet. When the message is received there is a flux of K^+ ions which in turn results in an osmotic flow of water within the tissue and a change in pressure potential, which causes the leaf to fold (Allen 1969; Findlay & Pallaghy 1978). It is not suggested that such a phenomenon is general to all plants, yet the example shows how pressure potential may be able to respond to a variety of external stimuli.

The importance of pressure potential is that it provides the driving forces for cell expansion (Hsiao *et al.* 1976). Much work on wind or shaking suggests that these treatments prevent the lamina from properly expanding (Martin & Clements 1935; Grace & Russell 1977; Russell & Grace 1978, 1979a). Good measurements of pressure potential on such leaves while they are expanding have not yet been made.

It should be pointed out that other plant responses to tactile stimuli have been reported, including a response in the rate of respiration (Audus 1935; Godwin 1935; Todd, Chadwick & Sing-Dao Tsai 1972), and an increase in the rate of production of the growth substance ethylene (Goeschl, Rappaport & Pratt 1966; Jaffe & Galston 1968).

Displacement from vertical

The action of wind on plants may be construed as a series of displacements of the shoot apex from the vertical. It has been known for a long time by horticulturists that pulling down branches of fruit trees results in the outgrowth of lateral buds which would otherwise have remained dormant. To explain such observations, Wareing & Nasr (1961) suggested that the tree may detect the shoot's orientation in relation to gravity. Longman (1968) measured the growth in length of shoots produced from cuttings of cassava (*Manihot esculenta*) which were inclined at various angles. Although the total

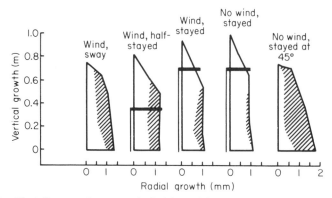

FIG. 2.10. The influence of sway on the height and distribution of wood increment in the stem of *Larix laricina*. Swaying was prevented in some trees by clamping with stays (shown in the diagram as solid bars). Shaded areas denote thick-walled (> 5 μm) tracheids. From Larson (1965).

shoot production was not much affected by angle, the *difference* between shoot growth at the apical end and that at the base was very sensitive to displacement from the vertical; as this displacement increased, the lower shoots outgrew the upper shoots. Such an effect in a mature tree would result in a bushy specimen with reduced leader growth.

The converse is apparently true: Jacobs (1954) supported trees over a period of years so that they could not be displaced from the vertical by the wind. At the end of the experiment they were tall and thin, and unable to support themselves when the stays were removed. A similar experiment on *Larix* by Larson (1965) included a treatment in which the plants were inclined at 45°. Inclined plants developed greater girth than the controls, but increased less in height (Fig. 2.10). He suggested that wind and displacement from the vertical exerted their effect by an influence on the downward gradient of auxin within the plant.

WIND AND STRUCTURAL FAILURE

From a practical point of view the structural failure of crops, caused by wind forces, is of great significance, to both agriculture and forestry (Holtam 1971; Pinthus 1973). An understanding of the principles involved in such wind damage is the essential prerequisite if crops are to be grown reliably in windy places.

The wind forces acting on individual plants that make up the canopy can be established from the within-crop wind profile if the vertical distribution of leaf area is also known (Grace 1977); for a given stratum of canopy the force acting laterally is $\frac{1}{2}\rho c_d u^2 a$ where ρ is the density of air, u and a are the wind speed and leaf area for that particular stratum and c_d is the drag coefficient for the leaves (a measure of momentum transfer across the boundary layer which must be determined in wind tunnel tests—for example see Thom 1968). This force results in a turning moment about the base of the stem. In addition, an extra turning moment arises when the stem is deflected due to the displacement of the centre of gravity from the vertical. The total turning moment for the plant can then be found by adding the calculated turning moments of all the strata.

Structural failure occurs when the total turning moment exceeds a critical value, when either the stem breaks or the plant is uprooted. This critical value may be found directly by pulling the plants over and measuring the turning moment required. This has been done by Fraser & Gardiner (1967) who found considerable between-site variation in the force required to uproot trees, depending on the soil type and consequent extent of root development.

From a knowledge of the critical turning moment of the plant, together with estimates of wind speeds within the vegetation, it should be possible to predict the threshold wind speed at which structural failure occurs; just as a civil engineer can estimate the critical wind speed for a building design. However, this approach has not been successful: Tani (1963) and Fraser (1964) both found that the wind speed calculated to cause the stem to break was larger than the actual destructive speed. It has been pointed out that plants are aeroelastic structures, in which the fluctuating property of the wind in relation to the natural period of oscillation should be taken into account (just as the suspension bridge of the civil engineer is prone to collapse at rather low windspeeds if the gust frequency happens to coincide with a natural period of swing).

Japanese workers have studied the phenomenon of waving plants (known as 'Honami'), although mainly from a theoretical point of view (e.g. Inoue 1955). At an observational level, Maitani (1979) reported the motions of wheat and *Juncus* plants, recorded with a cine camera, and tried to relate them to the statistics of wind turbulence obtained using a sonic anemometer near the top of the canopy. Displacements of the ear of wheat showed a natural period of oscillation of 0·8 or 0·9 s whilst those of *Juncus* were more variable according to the fluctuations in wind velocity. There was similarity between the power spectrum for wheat deflections and that of wind velocity, both showing a peak around 0·8 Hz—suggesting that eddies of this particular frequency were exciting the wheat ears. Similarity between power spectra in *Juncus* was not clear cut. Differences in the behaviour of the two species may be related to the aerodynamic differences between them: the inflorescence of each wheat plant may act as an efficient momentum-absorber, free to oscillate in simple harmonic motion at the end of an elastic member (the stem), whilst the tapering *Juncus* leaf may exhibit more complex modes of motion. The study underlines the difficulties in quantitatively understanding the aeroelastic behaviour of plants.

CONCLUSIONS

The application of physical principles to the study of exchange processes between plants and the atmosphere has been remarkably fruitful in advancing our understanding of the influence of wind upon vegetation. These conclusions may be drawn:

1. The boundary layer resistance r_a may be calculated from the wind speed and leaf dimension, although considerable divergence is evident when experimental observations are compared with the calculated values. This divergence may be due to irregularities in the topography of the leaf and turbulence in the air flow, both of which promote turbulence in the boundary layer over the leaf and so may reduce r_a.

2. From a knowledge of boundary layer resistance, the influence of wind on the exchange of gases and other entities between the leaf and the atmosphere may be investigated. The effect of wind on CO_2 exchange is normally rather small, because the boundary layer turns out to be only a small part of the total diffusion pathway. Application of the Penman-Monteith equation to investigate transpiration shows that the effect of wind on water loss is often small, and can be positive or negative. Exceptions to this occur in those leaves prone to surface damage, in which transpiration may be considerably increased by wind action; this seems to occur in grasses and broad-leaved trees but not in conifers. Such surface damage may also affect stomatal behaviour, so the influence of wind on transpiration may be very complex.

3. At a gross level, the micrometeorological approach provides a means of finding the boundary layer resistance over the vegetation. It can be shown that short vegetation is characterized by a high resistance over the vegetation. It is suggested how this may explain the apparent Darwinian fitness of prostrate life forms in cold and windy places; dwarf life forms by virtue of their high boundary layer resistance have the advantage of higher surface temperatures when net radiation is positive.

4. Recent studies suggest that shaking can be a potent inhibitor of plant growth. The mechanism of this effect is not yet known.

5. Catastrophic failure of plant structure as in windthrow of forest trees cannot be predicted from the force calculated from average wind speeds. It is suggested that a plant can be regarded as an aeroelastic structure which becomes excited when the gust frequency coincides with one of its natural modes of oscillation.

6. The term 'exposure' used by ecologists and foresters to describe the

stresses which plants experience in windy habitats cannot be concisely defined, although it may be understood as a close coupling of plant parts to the atmosphere, and may be analysed quantitatively in terms of diffusion resistances. A plant closely coupled to the atmosphere will experience (a) surface temperatures which are not much higher than the temperatures of the surrounding air; (b) fast deposition of any aerosols in the atmosphere such as salt spray; (c) considerable absorption of momentum which may cause structural damage or microscopic damage to the epidermis.

REFERENCES

Allen R.D. (1969) Mechanism of the seismonastic reaction in *Mimosa pudica*. *Plant Physiology*, **44**, 1101–1107.

Audus L.J. (1935) Mechanical stimulation and respiration rate in the cherry laurel. *New Phytologist*, **34**, 386–402.

Avery D.J. (1966) The supply of air to leaves in assimilation chambers. *Journal of Experimental Botany*, **17**, 655–677.

Bates C.G. (1911) *Windbreaks: Their Influence and Value*. U.S. Department of Agriculture, Forest Service Bulletin 86, 1–100.

Bauer von A.F. (1966) Windschadensymptome und landwirtschaflichen Kulturpflanzen. *Wissenschaftliche Zeitchrift der Universitate Rostock, Mathematisch—Naturwissenschaftliche Rieh*, **2**, 363–381.

Bayley F.J., Owen J.M. & Turner A.B. (1972) *Heat Transfer*. Nelson, London.

Caldwell M.N. (1970) Plant gas exchange at high wind speeds. *Plant Physiology*, **46**, 535–537.

Campbell G.S. (1977) *An Introduction to Environmental Biophysics*. Springer-Verlag, New York.

Chamberlain A.C. (1974) Mass transfer to bean leaves. *Boundary-Layer Meteorology*, **6**, 477–486.

Clausen J., Keck D.D. & Hiesey W.M. (1940) Experimental studies on the nature of species 1. The effect of varied environments on western North American plants. *Publications of the Carnegie Institution*, **520**, 1–452.

Clark J.A. & Wigley G. (1975) Heat and mass transfer in real and model leaves. *Heat and Mass Transfer in the Biosphere 1. Transfer Processes in the Plant Environment*. (Ed. by D.A. de Vries & N.H. Afgan), pp. 353–365. Wiley, New York.

Davies W.J., Kozlowski T.T. & Pereira J. (1974) Effect of wind on transpiration and stomatal aperture of woody plants. *Bulletin of the Royal Society of New Zealand*, **12**, 433–448.

Davies W.J., Gill K. & Halliday G. (1978) The influence of wind on the behaviour of stomata of photosynthetic stems of *Cytisus scoparius* (L.) Link. *Annals of Botany*, **42**, 1149–1154.

Decker J.P. (1947) The effect of air supply on apparent photosynthesis. *Plant Physiology*, **22**, 561–571.

Fenton R., Davies W.J. & Mansfield T.A. (1977) The role of farnesol as a regulator of stomatal opening in *Sorghum*. *Journal of Experimental Botany*, **28**, 1043–1053.

Findlay G.P. & Pallaghy C.K. (1978) Potassium chloride in the motor tissue of *Stylidium*. *Australian Journal of Plant Physiology*, **5**, 219–229.

Flückiger W., Oertli J.J. & Flückiger-Keller H. (1978) The effect of wind gusts on leaf growth and foliar water relations of aspen. *Oecologia*, **34**, 101–106.

Fraser A.I. (1964) Wind tunnel and other related studies of coniferous trees and tree crops. *Scottish Forestry*, **18**, 84–92.

Fraser A.I. & Gardiner J.B.H. (1967) *Rooting and Stability in Sitka Spruce*. Forestry Commission Bulletin 40. H.M.S.O., London.

Gaastra P. (1959) Photosynthesis of crop plants as influenced by light, carbon dioxide, temperature and stomatal diffusion resistance. *Mededelingen van de Landbouwhogeschool te Wageningen, Nederland*, **59**, 1–68.

Garratt J.R. (1977) *Aerodynamic Roughness and Mean Monthly Surface Stress over Australia*. CSIRO Australia, Division on Atmospheric Physics Technical Paper 29, 1–19.

Gates D.M. & Papian L.E. (1971) *Atlas of Energy Budgets of Plant Leaves*. Academic Press, New York.

Gimingham C.H. (1951) The use of life form and growth form in the analysis of community structure, as illustrated by a comparison of two dune communities. *Journal of Ecology*, **39**, 396–406.

Godwin H. (1935) The effect of handling on the respiration of cherry laurel leaves. *New Phytologist*, **34**, 403–406.

Goeschl J.D., Rappaport L. & Pratt H.K. (1966) Ethylene as a factor regulating the growth of pea epicotyls subjected to physical stress. *Plant Physiology*, **41**, 877–884.

Grace J. (1974) The effect of wind on grasses I. Cuticular and stomatal transpiration. *Journal of Experimental Botany*, **25**, 542–551.

Grace J. (1977) *Plant Response to Wind*. Academic Press, London.

Grace J. & Collins M.A. (1976) Spore liberation from leaves by wind. *Microbiology of Aerial Plant Surfaces*. (Ed. by C.H. Dickinson & T.F. Preece), pp. 185–198. Academic Press, London.

Grace J. & Russell G. (1977) The effect of wind on grasses III. Influence of continuous drought or wind on anatomy and water relations in *Festuca arundinacea* Schreb. *Journal of Experimental Botany*, **28**, 268–278.

Grace J., Malcolm D.C. & Bradbury I.K. (1975) The effect of wind and humidity on leaf diffusive resistance in Sitka spruce seedlings. *Journal of Applied Ecology*, **12**, 931–940.

Grace J. & Wilson J. (1976) The boundary layer over a *Populus* leaf. *Journal of Experimental Botany*, **27**, 231–241.

Holroyd E.W. (1970) Prevailing winds on Whiteface Mountain as indicated by flag trees. *Forest Science*, **16**, 222–229.

Holtam B.W. (1971) *Windblow of Scottish Forests in January 1968*. Forestry Commission Bulletin 45, H.M.S.O., London.

Hsiao T.C., Acevedo E., Fereres E. & Henderson D.W. (1976) Stress metabolism. Water stress, growth and osmotic adjustment. *Philosophical Transactions of the Royal Society of London*, **B273**, 479–500.

Inoue E. (1955) Studies of the phenomenon of waving plants ('Honami') caused by wind II. Studies of waving plant and plants vibration. *Journal of Agricultural Meteorology, Tokyo*, **11**, 87–90.

Jacobs M.R. (1954) The effect of wind sway on the form and development of *Pinus radiata*. *Australian Journal of Botany*, **2**, 35–51.

Jaffe M.J. & Galston A.W. (1968) The physiology of tendrils. *Annual Review of Plant Physiology*, **19**, 417–434.

Jarvis P.G. (1971) The estimation of resistances to carbon dioxide transfer. *Plant Photosynthetic Production, Manual of Methods*, pp. 566–631. (Ed. by Z. Šesták, J. Čatský, and P.G. Jarvis). Junk, Hague.

Knight R.C. (1916) On the use of the porometer in stomatal investigation. *Annals of Botany*, **30**, 57–76.

Körner Ch., Scheel J.A. & Bauer H. (1979) Maximum leaf diffusion conductance in vascular plants. *Photosynthetica*, **13**, 45–82.

Landsberg J.J. & Thom A.S. (1971) Aerodynamic properties of a plant of complex structure. *Quarterly Journal of the Royal Meteorological Society*, **97**, 565–570.

Larson P.R. (1965) Stem form of young *Larix* as influenced by wind and pruning. *Forest Science*, **11**, 412–424.

Leyton L. (1975) *Fluid Behaviour in Biological Systems*. Oxford University Press.

Linacre E.T. (1964) Calculations of the transpiration rate and temperature of a leaf. *Archiv für Meteorologie, Geophysik und Bioklimatologie*, **13**, 391–399.

Longman K.A. (1968) Effects of orientation and root position on apical dominance in a tropical woody plant. *Annals of Botany*, **32**, 553–566.

MacKerron D.K.L. (1976) Wind damage to the surface of strawberry leaves. *Annals of Botany*, **40**, 351–354.

Macleod N. & Todd R.B. (1973) The experimental determination of wall-fluid mass transfer coefficients using plasticized polymer surface coatings. *International Journal of Heat and Mass Transfer*, **16**, 485–504.

Maitani T. (1979) An observational study of wind-induced waving of plants. *Boundary-Layer Meteorology*, **16**, 49–65.

Martin E.V. & Clements F.E. (1935) Studies of the effect of artificial wind on growth and transpiration in *Helianthus annuus*. *Plant Physiology*, **10**, 613–636.

Monteith J.L. (1965) *Evaporation and Environment*. Symposium of the Society for Experimental Biology 19, pp. 205–234. Cambridge University Press.

Monteith J.L. (1973) *Principles of Environmental Physics*. Arnold, London.

Nägeli W. (1971) Der Wind als Standortsfaktor bei Anfforstungen in der subalpinen Stufe (Stillbergalp im Dischmatal Kanton Granbünden). *Schweizerische Anstalt für das Forstliche Versuchswesen*, **46**, 1–147.

Neel P.L. & Harris R.W. (1971) Motion-induced inhibition of elongation and induction of dormancy in *Liquidambar*. *Science*, **173**, 58–59.

Oke T.R. (1977) *Boundary Layer Climates*. Methuen, London.

Parkhurst D.F., Duncan P.R., Gates D.M. & Kreith F. (1968) Wind tunnel modelling of convection of heat between air and broad leaves of plants. *Agricultural Meteorology*, **5**, 33–47.

Parkinson K.J. (1968) Apparatus for the simultaneous measurement of water vapour and carbon dioxide exchanges of single leaves. *Journal of Experimental Botany*, **19**, 840–856.

Pearman G.I., Weaver W.L. & Tanner C.B. (1972) Boundary layer heat transfer coefficients under field conditions. *Agricultural Meteorology*, **10**, 83–92.

Penny M.G. & Bowling D.J.F. (1974) A study of potassium gradients in the epidermis of intact leaves of *Commelina communis* L. in relation to stomatal opening. *Planta*, **119**, 17–25.

Perrier E.R., Aston A. & Arkin G.F. (1973) Wind flow characteristics on a soybean leaf compared with a leaf model. *Physiologia Plantarum*, **28**, 106–112.

Pinthus M.J. (1973) Lodging in wheat, barley and oats: the phenomenon, its causes and prevention measures. *Advances in Agronomy*, **25**, 210–253.

Putnam P.C. (1948) *Power from the Wind*. Van Nostrand, New York.

Raunkiaer C. (1909) Formationsundersogelse og Formationsstatisk. *Botanisk Tidsskrift*, **30**, 20–132.

Raunkiaer C. (1934) *The Life Forms of Plants and Statistical Plant Geography*. University Press, Oxford.

Rees D. (1979) The effects of wind and shaking on the morphology, growth, gas exchange and water relations of *Pinus contorta* Douglas. Ph.D. thesis, University of Edinburgh.

Russell G. & Grace J. (1978) The effect of wind on grasses V. Leaf extension, diffusive conductance, and photosynthesis in the wind tunnel. *Journal of Experimental Botany*, **29**, 1249–1258.

Russell G. & Grace J. (1979a) The effect of shelter on the yield of grasses in southern Scotland. *Journal of Applied Ecology*, **16**, 319–330.

Russell G. & Grace J. (1979b) The effect of windspeed on the growth of grasses. *Journal of Applied Ecology*, **16**, 507–514.

Salisbury F.B. & Spomer G.G. (1964) Leaf temperatures of alpine plants in the field. *Planta*, **60**, 497–505.

Sekiguti T. (1951) Studies in local climatology—on the prevailing wind in early summer judged by bending shape of top twigs of persimmon trees at the Akaho Fan, Nagano Prefecture, Japan. *Papers in Meteorology & Geophysics*, **2**, 168–179.

Tansley A.G. (1939) *The British Islands and their Vegetation*. University Press, Cambridge.

Taylor S.E. & Sexton O.J. (1972) Some implications of leaf tearing in *Musaceae*. *Ecology*, **53**, 141–149.

Tani N. (1963) The wind over the cultivated field. *Bulletin of the National Institute of Agricultural Science*. **A10**, 1–99.

Thom A.S. (1968) The exchange of momentum, mass and heat between an artificial leaf and the airflow in a wind tunnel. *Quarterly Journal of the Royal Meteorological Society*, **94**, 44–55.

Thom A.S. (1975) Momentum, mass and heat exchange in plant communities. *Vegetation and the Atmosphere, Vol. 1. Principles*, (Ed. by J.L. Monteith) pp. 57–109. Academic Press, London.

Thompson J.R. (1974) The effect of wind on grasses II. Mechanical damage in *Festuca arundinacea* Schreb. *Journal of Experimental Botany*, **25**, 965–972.

Todd G.W., Chadwick D.L. & Sing-Dao Tsai (1972) Effect of wind on plant respiration. *Physiologia Plantarum*, **27**, 342–346.

Tranquillini W. (1969) Photosynthese und Transpiration einiger Holzarten bei verschieden starken Wind. *Zentralblatt für das gesamte Forstwesen*, **81**, 35–48.

Tranquillini W. (1976) Water relations and alpine timberline. *Water and Plant Life*. (Ed. by O.L. Lange, L. Kappen and E.D. Schulze), pp. 473–491. Springer-Verlag, New York.

Wardle P. (1968) Engleman spruce (*Picea engelmanii* Engel) at its upper limits on the front range, Colorado. *Ecology*, **49**, 483–495.

Wardle P. (1977) Japanese timberlines and some geographic comparisons. *Arctic and Alpine Research*, **9**, 249–258.

Wareing P.F. & Nasr T.A.A. (1961) Gravimorphism in trees I. Effects of gravity on growth and apical dominance in fruit trees. *Annals of Botany*, **25**, 321–340.

Warren-Wilson J. & Wadsworth R.M. (1958) The effect of wind speed on assimilation rate—a reassessment. *Annals of Botany*, **22**, 286–290.

Whitehead F.H. (1954) A study of the relation between growth form and exposure on Monte Maiella, Italy. *Journal of Ecology*, **42**, 180–186.

Whitehead F.H. (1959) Vegetational change in response to alterations of surface roughness on Monte Maiella, Italy. *Journal of Ecology*, **47**, 603–606.

Whiteman P.C. & Koller D. (1967) Species characteristics in whole plant resistances to water vapour and carbon dioxide diffusion. *Journal of Applied Ecology*, **4**, 363–377.

Williams W.T. (1949) Studies in stomatal behaviour III. The sensitivity of stomata to mechanical shock. *Annals of Botany*, **13**, 309–328.

Wilson J. (1979) Some physiological responses of *Acer pseudoplatanus* L. to wind at different levels of soil water, and the anatomical features of abrasive leaf damage. Ph.D. thesis, University of Edinburgh.

Yamaoka Y. (1958) Total transpiration from a forest. *Transactions of the American Geophysical Union*, **39**, 266–272.

Yoshino M.M. (1973) Some local characteristics of the winds as revealed by windshaped trees in the Rhone valley in Switzerland. *Erdkunde*, **18**, 28–39.

3. THE SIZE AND SIGNIFICANCE OF DIFFERENCES IN THE RADIATION BALANCE OF PLANTS AND PLANT COMMUNITIES*

G. STANHILL

*Institute of Soils and Water, Agricultural Research Organization
The Volcani Center, Bet Dagan, Israel*

SUMMARY

Measurements made during the last 25 years are presented which suggest that a marked similarity exists in the radiative properties of the leaves and stands of different plant species, especially agricultural ones. The ecological significance of the small differences that have been reported, and their exploitation, both by plants and by man, are discussed.

INTRODUCTION

Several excellent reviews dealing with the radiation balance of plant communities have been published during the last 25 years. A comparison of Sauberer and Hartel's book on the subject published in 1959 (Sauberer & Hartel 1959) with Ross's comprehensive review of radiative transfer in plant communities published in 1975 (Ross 1975) shows the considerable progress achieved during the intervening period in describing the interaction between radiation and plants in physical-mathematical terms.

Despite this progress, there remain many aspects of the subject which cannot yet be treated in this same rigorous fashion and yet are of considerable importance and interest to ecologists. One of these aspects—the size, significance and possible exploitation of differences that have been observed between the radiation balance of different plant species and plant communities—is the subject of this contribution.

METHODS OF MEASUREMENT

Awareness of the exaggeration of the differences in plants' radiative characteristics resulting from the bias inherent in human vision developed slowly

* Contribution from the Agricultural Research Organization, The Volcani Center, Bet Dagan, Israel. No. 129-E, 1979 series. This research was supported by a grant from the United States–Israel Binational Science Foundation (BSF), Jerusalem, Israel.

after the publication of Newton's 'Opticks' in 1704 and Herschel's discovery of infra-red radiation in 1800. The first measurements of the spectral characteristics of plants were made on extracted pigments; the optical properties of chlorophyll were described by Brewster in 1833 and by the end of the 19th century the absorption spectra of the different plant pigments were well established.

The spectral characteristics of intact plant tissue were first measured by Coblenz in the first decade of this century, using the improved spectroscopic techniques that he had developed. In the last two decades the results of a number of laboratory studies have been published in which the reflectivities and transmissivities of a wide variety of detached leaves have been compared over the solar spectrum (e.g. Kleshnin & Shulgin 1959; Gausman *et al.* 1973).

The first measurements of the spectral radiative characteristics of entire and undisturbed plant communities appear to be those of Knuchel who, in 1914, described the spectrum of light transmitted by various forest canopies. In 1937 Egle reported field measurements of the spectral reflectivity of a plant community. In the last decade a number of portable spectral radio-meters have become commercially available, enabling the spectral radiative characteristics of plant communities to be measured relatively easily in the field (e.g. Scott, Menalda & Brougham 1968; McCartney & Unsworth 1976).

Since 1972 continuous measurements of spectral irradiance from the earth's surface have been made from space by satellites of the U.S. Landsat series. Currently, data are obtained with a ground level resolution of 80 m in seven wavebands, three (0·48 to 0·57, 0·58 to 0·68 and 0·69 to 0·83 μm) by high resolution television type cameras and four (0·5 to 0·6, 0·6 to 0·7, 0·7 to 0·8 and 0·8 to 1·1 μm) by a multispectral scanning system. The next satellite to be orbited in this series, Landsat D, will have a considerably improved ground level resolution—30 m—and monitoring capacity using a multi-spectral scanner covering six narrower bands centred at 0·48, 0·56, 0·66, 0·83, 1·65 and 11·5 μm (Otterman, Lowman & Salomonson 1976).

A wide range of spectral radiometers have been used for *in situ* plant studies. All include light collection, spectrum dividing and detector systems.

Ground based instruments commonly have a wide (180°) angle, cosine corrected light collection system compared with the much narrower ($<5\%$) acceptance angle used for airborne systems. Spectral division may be by interference or diffraction-grating filters giving high resolution over narrow, 5–10 nm wide, bands and suitable for scanning systems, or by coloured glass filters which can be used to divide the spectrum differentially over much broader, approximately 100 nm wide, bands. Highly sensitive photocells are generally used to detect the low energy signals received in the narrow band, narrow aperture scanning systems. More robust, but less sensitive thermopile detectors can be used to detect the stronger signals obtained with wider spectral band and wider acceptance angle systems.

In addition to narrow-band spectral measurements, comparative studies with pyranometers, i.e. radiometer thermopiles responding equally to radiation throughout the entire solar spectrum, have been made of plant communities (e.g. Angström 1925; Monteith 1959), and in some cases of foliage only (Birkebak & Birkebak 1964). Similar comparative studies of plant communities have also been made with pyrradiometers, radiometers equally sensitive to radiation in both the solar and terrestrial wavebands (Monteith & Szeicz 1962; Stanhill, Hofstede & Kalma 1966).

DIFFERENCES IN THE RADIATION BALANCE OF PLANT FOLIAGE

Mean values of the spectral reflectivity and transmissivity of 20 agricultural species (Gausman *et al.* 1973) are presented in Fig. 3.1 together with values of the interspecific variability as represented by the standard deviations. The values for each crop species are based on the mean of ten replicate measurements of the adaxial surfaces of mature and healthy leaves taken from field grown plants.

Crop leaf reflectivity and transmissivity are clearly very similar and statistically indistinguishable throughout the solar spectrum. Moreover, because of the inverse relationship between these two radiative characteristics, the differences between the absorptivity of the individual crops were smaller than for either their reflectivities or transmissivities. Thus at 550 nm, the wavelength at which the difference between crops was greatest, the coefficients of variation for absorptivity, reflectivity and transmissivity were 17, 33 and 82%, respectively.

The ratio of red to near-infrared reflectivities obtained from multispectral satellite observations has been used to discriminate between different crop species in a number of crop inventory studies. Data from Gausman *et al.* (1973) on reflectivities of mature leaves from 20 agricultural species show no evidence of a larger range in values of the 650 to 850 nm ratio than in values at the maximum reflectivity wavelength, 650 nm; the coefficients of variation were 24 and 36% respectively. However, for measurements in the field the red to near-infrared ratio can be used to distinguish between reflection from plant and from soil and thus discriminate between different crop species on the basis of the size rather than the optical properties of their leaf cover.

Three sets of comparative data on leaf absorptivity in the visible and photosynthetically active wavelengths are presented in Table 3.1, in the form of means and standard deviations. The first and second data set, both from McCree (1971/1972), include 22 agricultural crop species—20 grown

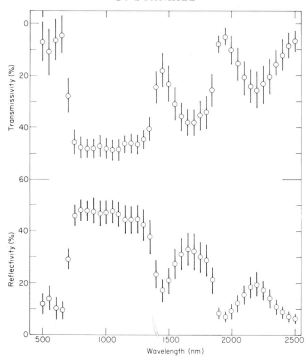

FIG. 3.1. Mean values of reflectivity and transmissivity of twenty agricultural crop species plus and minus one standard deviation of interspecific difference. Values from measurements of Gausman *et al.* (1973).

TABLE 3.1. Absorptivity of leaves of agricultural crop species in the visible and photo-synthetically active wavelengths; mean values and standard deviation between crops, both in per cent.

		Wavelength, nm								
		350	400	450	500	550	600	650	700	750
20 species from growth chamber	M.V.	94·6	92·6	92·7	91·2	74·4	84·3	90·4	78·9	6·5
2 replicates McCree, 1971/1972	S.D.	1·9	2·1	1·8	2·0	5·5	3·7	2·1	4·4	2·3
8 species from field	M.V.	95·8	95·3	93·6	92·6	76·4	85·7	91·5	80·3	7·3
2 replicates McCree, 1971/1972	S.D.	0·8	0·1	0·1	0·1	3·0	1·9	1·1	3·1	1·7
20 species from field	M.V.	—	—	—	80·7	74·3	83·0	85·4	43·1	7·5
10 replicates Gausman *et al.*, 1973	S.D.	—	—	—	11·7	12·8	12·1	10·7	7·8	2·1

in growth chambers and eight (six of them common to the first group) taken from field grown plants. The third data set, for 20 agricultural crops grown in the field, is that of Gausman *et al.* (1973) referred to previously.

The low values of intraspecific variation in Table 3.1 do not preclude the possibility that some small but consistent differences exist. Within each data set ranking of the different crop species indicated some small but non-random differences, but the order and pattern of absorptivity, as well as the absolute values for different crops, showed no agreement for the three data sets.

As mean curves of absorptivity, essentially similar to those presented in Table 3.1, have been reported by Kleshnin & Shulgin (1959) for 80, mainly non-agricultural, plant species, there seems considerable evidence for the conclusion that the scattering coefficient of healthy, mature leaves is the same, certainly for the major agricultural crops. The differences that do exist between species are smaller than those found in leaves from plants of the same species, but of different ages (e.g. Mooney, Ehleringer & Björkman

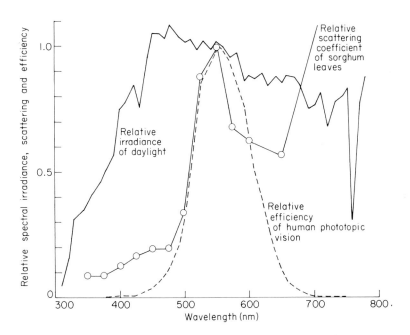

Fig. 3.2. Spectral irradiance of daylight, luminous efficiency of human phototopic vision and scattering coefficient of leaves, all values normalized to those at 550 nm.
——— Spectral energy distribution of daylight mean of 70 curves, from Henderson & Hodgkiss (1963).
- - - - Relative visual efficiency of human phototopic vision, from Altman & Dittmer (1966).
○———○ Scattering coefficient from mean of five upper surfaces of mature green sorghum leaves, unpublished data.

1977), or from plants of the same species but grown under different conditions (e.g. McCree 1971/1972).

The radiation balance of sunlit leaves could differ more than mono-chromatic laboratory measurements suggest, because the maximum flux density of solar radiation is carried in the same wavelengths in which the scattering coefficient of leaves is largest and most variable (Fig. 3.2). However, this enhancement of differences is likely to be of limited significance because of the small changes occurring in the spectral composition of global radiation during most of the day (McFarland & Munz 1976). Experimental confirma-tion that this is so is provided by Angström's (1925) comparison of the transmissivity of the leaves of eight north European trees and Birkebak & Birkebak's (1964) comparison of the reflectivity of leaves from 22 broad-leaved North American trees. Both sets of measurements were made on sunlit leaves with pyranometers. The transmissivity measurements had a standard deviation of 10% around the mean value, 0·24; the reflectivities had a stand-ard deviation of 11% around their mean, 0·28. These variabilities are com-parable to those obtained in monochrometer studies (Fig. 3.1, Table 3.1).

DIFFERENCES IN THE RADIATION BALANCE OF PLANT COMMUNITIES

Compared to measurements made on individual leaves, there is even less evidence for large and consistent differences between the radiative properties of stands of different crops and even more evidence for large differences within measurements for a given crop species. Values for the most widely measured radiative property, solar reflectivity, have been listed for both individual sunlit leaves and field stands of four very different crops in Table 3.2. In all four cases the seasonal and spatial variations in stand reflectivities are greater than those between the crops. There is generally a negative correlation between both crop and sun height and stand reflectivity (Ogun-toyinbo 1970; Kalma & Stanhill 1969). For the short and densely sown grain and pasture crops and the low sun angles typical of temperate zone agriculture, reflectivity values tend to approach those of individual sunlit leaves, i.e. between 0·25 and 0·30 (Monteith 1959), for much of the active growing season. For taller and more widely-spaced row crops and high sun angles typical of tropical and semi-arid zones, lower crop reflectivity values, between 0·15 and 0·20, are more common (Oguntoyinbo 1970).

A wider range of stand reflectivities, 0·04 to 0·35, has been observed for natural plant communities. An analysis of the sources of variation found in a series of reflectivity measurements made from a helicopter over eight different land surfaces, ranging from dense and high forest vegetation to almost bare sand dunes and mountain slopes, and replicated in both space and time, has been reported (Stanhill 1970). The major source was surface type, with a

TABLE 3.2. Solar reflectivity of the leaves and stands of four crop species.

Crop	Wheat	Cotton	Maize	Oranges
Upper surface of sunlit leaves	0·29	0·32	0·33	0·32
Stand mean	0·20	0·20	0·19	0·16
season and site range	±0·08	±0·08	±0·04	±0·05
No. of stands investigated	6	6	7	2
No. of measurements	58	77	44	15
Reference (see below)	1	1	2	3

1 Stanhill 1975.
2 Stanhill, Cox & Moreshet 1968.
3 Kalma & Stanhill 1969.

coefficient of variation of 61% for the mean of eight values. The spatial difference between replicate plots of the same surface type or between replicate measurements over one plot, was the next most important source with an average coefficient of variation of 18%. Too few seasonal replications were made to evaluate this source of variation statistically, but the seasonal range for five comparable sets of observations was only 0·03 for a mean reflectivity of 0·24, suggesting that the seasonal source of variation was small. A striking example of this was provided by one of the cover types, a mixed hardwood forest in eastern Tennessee with an average tree height of 24 m. Mean reflectivities and standard deviations for spatially replicated measurements at budbreak, full summer leaf cover and at the height of the visually brilliant autumn colouration, were $0·10 \pm 0·01, 0·11 \pm 0·01$ and $0·13 \pm 0·02$, respectively.

The solar reflectivity of the eight cover types compared also showed an inverse relation to stand height, with a reduction of 0·02 per metre height increase; a similar slope has been reported for crop surfaces (Kalma & Stanhill 1969).

Very high values of solar reflectivity for natural plant communities are generally caused by the effect of highly reflecting ground surfaces such as sand, snow or stones, below sparse vegetation cover. Very low values of reflectivity are usually found in tall, dense vegetation associations whose foliage forms a system of radiation scattering elements with a high cavity depth to scattering element radius ratio, enhancing internal trapping of scattered radiation—the so-called 'velvet-pile' effect.

Differences in the net, i.e. terrestrial plus solar, radiation balance of plant communities are determined largely by variations in their solar reflectivities. For example, among the nine natural and two agricultural cover types compared by Stanhill, Hofstede & Kalma (1966), the range in values of solar reflectivity was three times that for net long-wave emission.

Differences in the terrestrial long-wave radiation losses are mainly the result of variations in their radiative surface temperatures, the difference in long-wave emissivity being small at least for individual leaves. A comparison of the emissivity of leaves from 32 very diverse plant species (Idso *et al.* 1969) showed them all to be near black bodies with a mean emissivity of 0·972, a standard deviation *between* species of ±0·013 and an average standard deviation for *within* species replicates of ±0·005. Under clear sky conditions (ambient air temperature of 25 °C with a radiative sky temperature equivalent to −4 °C) an emissivity difference between plant stands equivalent to ±0·013, i.e. the one standard deviation inter-specific value found by Idso *et al.* (1969), would lead to a surface temperature difference of 1 °C.

Much larger variations in the radiative surface temperature of different plant stands have been reported under clear sky conditions. In part they are attributable to the differences in solar absorptivity previously discussed, but differences in rates of latent heat loss are a more important factor (Monteith & Szeicz 1962). These differences are the basis of current attempts to estimate evapotranspiration rates from remotely sensed observations of surface temperatures (Bunnik 1978; Otterman, Lowman & Salomonson 1976).

The general conclusion that can be drawn from this necessarily brief and selective review of the differences that have been found during the last 25 years in the radiation balance of plant leaves and communities was anticipated by Thoreau's observation, published in *Walden* more than one hundred years ago: 'We are wont to forget that the sun looks on our cultivated fields, and on the prairies and forests without distinction. They all reflect and absorb his rays alike, and the former makes but a small part of the glorious picture which he beholds in his daily course.'

RADIATIVE INTERACTIONS BETWEEN MAN AND PLANTS

Methods of observation

The reason that we are 'wont to forget' how small the differences are that exist between the radiative characteristics of plants is that the first and still most widely employed radiation detector for observing such differences, the human eye, is almost perfectly adapted to that purpose.

At least four characteristics of human vision and the radiative characteristics of plants enhance our discriminatory ability. Foremost is the coincidence of the action spectra for human phototopic vision with that for the radiative scattering properties of foliage elements. Our visual sensitivity to plant colour is further reinforced by the fact that for most of the day incident global solar radiation also has its peak flux densities within the same wavelength band (Fig. 3.2). Thus, radiation scattered by plants reaches the eye as a signal the maximum intensity of which is in wavelengths to which the eye is most sensitive. As the flux densities of global radiation fall to the illuminance levels occurring at sunset and during twilight, the peak sensitivity of the human eye falls to around 507 nm, matching the blue-bluegreen peak intensities of twilight (McFarland & Munz 1976). The apparent shift in colour—the Purkinje effect—is well illustrated by the blue vegetation depicted in Samuel Palmer's twilight landscapes.

A second highly relevant physiological feature of human vision is that the ability of the eye to discriminate differences in colour is at its maximum, $\Delta\lambda$ 1–5 nm, in the 480–630 nm wavelength band that is most strongly scattered by vegetation (Altman & Dittmer 1966).

A third characteristic of the system is that the radiative properties of plants are at their most diverse at the same wavelength at which the human eye is most sensitive and discriminating. Thus, the data for agricultural crop species listed in Table 3.1 show a maximum diversity (coefficient of variation) between species at 550 nm.

Finally, the abundant detail of form and contrast typical of plant communities greatly enhances the human eye's sensitivity to colour, with patterns in time as well as space being important for this enhancement (Clayton 1971). Similarly, colour contrast produces an enhancement of perceived differences between neighbouring colours (Anon 1953).

The significance of the radiative interaction between plants and man extends beyond man's dependence on present and past photosynthesis for food, fibre and fossil fuels. For most of the world's population, radiation scattered by plant communities forms a major element in their visual and thermal environments. Urbanized man, deprived of this background, seeks substitutes for it in ornamental horticulture, landscape architecture and tourism, both actual and vicariously through coloured communication media.

Radiative interaction between man and agricultural plant communities

In pre-agricultural food-gathering societies the ability to distinguish quickly and accurately between edible and non-edible or even poisonous plants was vital and almost certainly based largely on colour differences. The number of plant species exploited in food gathering societies was far greater than that cultivated today in modern agricultural systems. Thus, even today, in Ethiopia only 11 of the 33 root and tuber crops exploited for food are cultivated (Getahun 1974). The same reference lists more than 190 wild edible Ethiopian plants currently providing leaf vegetables, edible fruits and seeds and starchy roots and tubers. Similarly, Gliessman, Garcia & Amador (in press) list more than 70 plant species cultivated in Indian food gardens in Mexico which are relicts of Mayan agriculture, indicating that this was not the virtual maize monoculture that has been suggested.

It is not known whether the wider range of food plants once exploited had a greater diversity in their radiative characteristics than the previously demonstrated near-uniformity of the remarkably few crop species cultivated today. It is clear, however, that there are a number of other factors leading to the near uniformity of modern agricultural landscapes.

One such factor, operating on a macro-scale, is the increase in the area and regularity of the shape of the fields and the elimination of natural boundaries associated with mechanized agriculture. Although the overall effect of eliminating hedges on the regional radiation balance appears to be rather small (Guyot & Seguin 1978), obviously the immediate local effect is very large, as is the overall visual effect.

The diversity of the radiation balance of modern agricultural crop stands has also been reduced by the increase in the amount of solar radiation absorbed by their canopies—the main physiological basis for the improvement in crop yields (Watson 1947). The greater radiation absorption has been achieved by increasing the size of the crop canopies and extending their duration through the introduction of varieties whose development is less sensitive to temperature and photoperiod, and whose growth is more responsive to fertilization, irrigation and pest control. This intensification of cultural practices still further reduces diversity by the near elimination of spectral changes resulting from nutrient and water deficiencies and pest damage. As a consequence of the larger leaf area duration, the diverse radiative properties of soil surfaces (Bowers & Hanks 1965) play a much smaller role in the radiation balance of agricultural plant communities.

As might be expected in intensive agriculture which aims at light-limited rates of photosynthesis, variations between crops in their rates of total dry matter production are rather small. Thus Sibma (1968) has shown that in the Netherlands, after a closed green crop surface has been attained, the differences in the total dry matter production of the major crops are very small and the rate of production is similar to that achieved in algal culture— 250 kg ha^{-1} day^{-1}. Data presented for four very different crops in the United Kingdom by Monteith (1977) show a maximum variation of less than 15% in their mean dry matter production rate—1·8 g per MJ intercepted solar radiation.

This near uniformity also appears to be the case, and for similar reasons, with regard to the rate of water loss from different crops, at least in temperate zones, when growing under potential evapotranspiration conditions— defined as referring to 'an extended surface of short green crop, actively growing, completely shading the ground, of uniform height and not short of water' (Anon. 1956).

Thus, the high and nearly uniform absorptivity displayed by the major crops cultivated in modern intensive agriculture is a major factor in their high and similar rates of both dry matter production and water loss.

Modification of the radiative properties of agricultural surfaces

Fig. 3.3, based on McCree's (1971/1972) measurements for 20 agricultural crop species, shows that the mean spectral curves for absorptivity and photosynthetic action (CO_2 absorption per unit light energy absorbed) are dissimilar. The considerable difference suggests that, at least theoretically, the radiant heat load on crop canopies, and hence their water loss, could be reduced by decreasing absorptivity in short wave bands of the visible spectrum without a corresponding reduction in photosynthesis. The morphological and biochemical changes needed to achieve a closer coincidence between the two spectral curves cannot be specified at present.

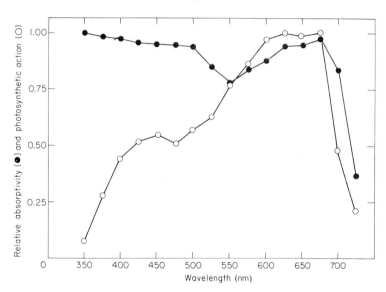

FIG. 3.3. Mean values of absorptivity and photosynthetic action for 20 agricultural crop species grown in growth chambers: values from McCree (1971/1972), normalized to maximum of 1·00. ●——● absorptivity, ○——○ relative action (μmol CO_2 absorbed per J light absorbed).

The few cases in which foliage absorptivity has been artificially reduced with some agricultural success by applying reflectant sprays refer to crops growing under widely spaced, light-saturated and water-limiting conditions (e.g. Stanhill, Moreshet & Fuchs 1976; Moreshet, Stanhill & Fuchs 1977).

Redistribution of light within a canopy which incompletely covers the ground can be achieved by increasing soil reflectivity and has been used experimentally to advantage with some high value crops (Moreshet, Stanhill & Fuchs 1975; Stanhill *et al.* 1975). Changing soil reflectivity before crop establishment is also a potentially useful technique. In arid zones, whitening the soil can reduce supra-optimum summer soil temperatures (Stanhill 1965); in temperate zones, darkening the soil can advance spring germination and growth (Watts 1976). In arid zones soil darkening treatments have been used very successfully for thermal sterilization of soils prior to the establishment of autumn crops.

Agricultural applications of remotely sensed differences in surface radiation balance

Despite the reduced diversity in the radiative properties of modern intensive agricultural landscapes, unkindly referred to as the 'green emptiness' by one Dutch ecologist, it has been possible to 'teach' satellite-borne, multispectral radiometer systems to substitute for the farmer's eye in interpreting changes in the amounts of radiation scattered from crop surfaces.

Under certain conditions, e.g. large fields and crops of different structures, individual crops can be identified with considerable accuracy using known areas as training sets with a rather elaborate interactive system of data reduction (Otterman, Lowman & Salomonson 1976). Phenological calendars have been drawn up and biomass assessed by remote sensing, exploiting the considerable contrast that exists between the scattering properties of soil and vegetation to obtain data on crop development. Recent work suggests that it may be possible to exploit these differences to obtain information on the structure (Bunnik 1978) as well as relative size (Pearson, Miller & Tucker 1976) of vegetation canopies. The large variation existing in the spectral reflectivity of soils, depending on their water content, mineral composition and particle size (Bowers & Hanks 1965), indicates that any analytical method to be used in interpreting radiation reflected from crops with incomplete soil cover will require specific information on the scattering properties of each soil type, and that any empirical relationships used for such interpretation will need to be established separately for each soil type.

Other agricultural applications of remotely sensed radiation scattered by agricultural surfaces include the prediction of yield, recognition of plant nutrient and water deficiencies and of damage resulting from salinity, impeded drainage and pest infestation. In many cases the change in spectral 'signature' of the affected crop results from a reduction in the radiation reflected from the vegetation and an increase in that from the larger area of exposed soil surface. In other cases the change is directly caused by changes in the spectral properties of the affected vegetation. When the latter is the case, it should be possible to develop an operational control system in which appropriate agricultural operations, such as the application of specific fertilizers, pesticides or herbicides, would be implemented according to remotely sensed information on specific changes in the crop radiation balance.

A domestic example of such a system is provided by grass lawns. Their visible reflectivity has been shown to be negatively but highly correlated with subjective quality assessments by lawn experts and with the amounts of irrigation water and nitrogen fertilizer applied (Mantell & Stanhill 1966). Thus it would be relatively simple to install an automatic system in which lawn irrigation and liquid fertilization would be controlled by a light sensor.

Many other plants, apart from lawn grass, are grown, to a considerable extent, for their visual reflective properties. These include many food crops, including most fruits and vegetables, the colour of which is an important marketing feature. With the recent banning, for health reasons, of a number of synthetic dyes previously widely used in food processing, a variety of plants, including sweet pepper and *Tagetes*, are now grown exclusively to provide caratenoid pigments to replace the synthetic food colourants.

The energy cost of 'colour coding' plant tissues is very inadequately known, as are the ecological consequences of pigmentation. Although the chemical energy content of plant pigments is high, the amounts are small; moreover, their biosynthetic pathways are uncertain, as are their effects on the rate of photosynthesis (Goodwin 1976). Their only ecological function that appears certain is the attraction that coloured fruits and flowers have for man, bird and insect.

Radiative interactions between natural plant communities and man

In environments allowing the development of plant communities with a large, evergreen biomass, the influence of their radiative properties on man's visual and thermal environment can be all-pervasive, e.g. pigmies within dense African forests or summer housewives in the 'suburban forests', the low density housing within heavily deciduous wooded area which have replaced the agricultural communities over much of New England.

In environments unfavourable for the development of high or closed communities, especially in arid hot deserts, significant modification of man's radiative environment by plants is rare, but especially appreciated or even venerated where it does occur. Thus, the original Middle Eastern meaning of paradise is a watered grove of fruit trees.

Most natural plant communities, including the increasing area managed by man for recreation and/or water catchment, are intermediate between the extreme examples given. In such varied landscapes Waggoner (1963) has shown that the radiative properties of the plant and land surfaces are of major importance in controlling man's energy balance and hence comfort, primarily through screening both the short wave gain from the sun and the long wave loss to the sky. Even within a single, non-homogeneous plant community such as the open oak scrub-forest typical of Mediterranean regions, man is able to maximize his comfort by choice of microenvironment, level of metabolic activity and dress according to the time of day and year. Based on a study of microclimate and heat stress variations in two Mediterranean forests, Schiller and Karschon (1973) have recommended management practices to be followed to enhance the extent and range of microclimate modifications within recreational forests.

The radiative properties of different natural plant communities comprising the cover of water catchments have been investigated for their possible influence on the energy balance and hence water loss by evapotranspiration. A comparison of a 4-m high oak scrub forest with pasture grasses on the Carmel Mountain, known from soil and rock water content measurements to differ markedly in their evapotranspiration rates, showed considerable differences in their radiation balance (Stanhill, Israeli & Rosenzweig 1973). The annual radiation balance of the scrub forest was 11% larger than that of the pasture, the difference increasing to a maximum of 27% in late spring. The main cause of this considerable difference was the larger long wave emission from the pasture, which had a radiative surface temperature averaging 4 °C more than that of the scrub vegetation in late spring. As the measured differences in the water loss during the six rainless summer months were much larger than the differences in both the radiant and sensible heat terms of the energy balance, it was concluded that differences in the rate of water loss were the cause and not the consequence of the differences in radiation balance. The deeper rooted forest vegetation was able, for two months after the last spring rains, to maintain its canopy in a more active, darker, cooler state with a lower surface diffusive resistance to water flux than the shallow rooted grass cover.

A comparison between the radiation balance of a 14 m high pine plantation and a nearby grassland surface in S.E. Australia gave essentially similar results (Moore 1976). The slightly larger differences in the radiation balance also reached their peak during the period of maximum water stress.

Without attempting to generalize on the basis of these two studies, there is much evidence to show that, relative to incident solar radiation, both the long wave emission and short wave reflection from vegetation increase with aridity, leading to a reduction in the radiation balance. The wider implications of such changes in the radiation balance occurring on a large scale for the rainfall regimes of arid zones have attracted considerable attention recently and Otterman (1974) and Charney (1975) have suggested desertification mechanisms based on vegetation-induced changes in the radiation balance. In brief, the hypothesis is that the reduction of vegetation cover in arid zones by overgrazing, etc., will increase surface reflectivity through the exposure of soils of greater reflectivity. The subsequent reduction in solar absorptivity will increase the heat loss from the bared surface, leading to subsidence of air masses over the cooled surface. The consequent reduction in rainfall will further accelerate the desertification process by inhibiting the recovery of the depleted vegetation cover. Computer tests of this hypothesis with a zonal atmospheric model for the latitudes encompassing the Sahara (Ellsaesser *et al.* 1976) confirm a negative feedback effect of changes in surface reflectivity on precipitation. Thus at $20\,^{\circ}$N, an increase in surface reflectivity from 14 to 35%, e.g. from that of a tall irrigated row crop to that of a low sparse desert association, indicated a precipitation reduction of 22%.

CONCLUSIONS

The last aspect of this subject reviewed provides an excellent example of the progress that has been made during the last 25 years in measuring and modelling the radiative interactions between plant communities and their atmospheric environment. Our ability to appreciate the significance of the general uniformity and occasional differences found in the radiative properties of plant parts and communities, both to the plants themselves and in their interaction with man, have progressed far less. The wide range and growing appreciation of such radiation interactions with man, both for agricultural and natural vegetation, suggests that further research in this rather neglected field of applied ecology might be most rewarding.

REFERENCES

Altman P.L. & Dittmer D.S. (Eds.) (1966) *Environmental Biology*, Federation of American Societies for Experimental Biology, Bethesda, Maryland. p. 14.

Ångström A. (1925) The albedo of various surfaces of ground. *Geografiska Annaler*, **7**, 323–342.

Anon. (1953) *The Science of Colour*, Committee on Colorimetry, Optical Society of America. Thomas Y. Crowell Company, New York.

Anon. (1956) Conclusions reached after discussions concerning evaporation. Proceedings of informal meeting on physics in agriculture. Wageningen. The Netherlands 7–13 September 1955. *Netherlands Journal of Agricultural Science*, **4**, 95–97.

Birkebak R. & Birkebak R. (1964) Solar radiation characteristics of tree leaves. *Ecology*, **45**, 646–649.

Bowers S.A. & Hanks R.J. (1965) Reflection of radiant energy from soils. *Soil Science*, **100**, 130–138.

Bunnik N.J.J. (1978) The multispectral reflectance of short wave radiation by agricultural crops in relation to their morphological and optical properties. *Communication of Agricultural University, Wageningen*, **78-1**, 175 pp.

Charney J.G. (1975) Dynamics of deserts and drought in the Sahel. *Quarterly Journal of the Royal Meteorological Society*, **101**, 193–202.

Clayton R.K. (1971) *Light and Living Matter: A Guide to the Study of Photobiology, Volume 2: The Biological Part.* McGraw-Hill, New York.

Ellsaesser H.W., McCracken M.C., Potter G.L. & Luther F.M. (1976) An additional model test of positive feedback from high desert albedo. *Quarterly Journal of the Royal Meteorological Society*, **102**, 655–666.

Gausman H.W., Allen W.A., Wiegand C.L., Escobar D.E., Rodriquez R.E. & Richardson A.J. (1973) The leaf mesophylls of twenty crops, their light spectra, and optical and geometrical parameters. *Technical Bulletin* 1465, U.S. Department of Agriculture, 59 pp.

Getahun A. (1974). The role of wild plants in the native diet of Ethiopia. *Agro-Ecosystems*, **1**, 45–56.

Gliessman S.R., Garcia E. & Amador M. The ecological basis for the application of traditional agricultural technology in the management of tropical agroecosystems. *Agro-Ecosystems* (in press).

Goodwin T.W. (Ed.) (1976) *Chemistry and Biochemistry of Plant Pigments.* 2nd ed. Vol. 1. 870 pp. Academic Press, London.

Guyot G. & Seguin B. (1978) Influence du bocage sur le climate d'une petite region: resultats des mesures effectuees en Bretagne. *Agricultural Meteorology*, **19**, 411–430.

Henderson S.T. & Hodgkiss D. (1963) The spectral energy distribution of daylight. *British Journal of Applied Physics*, **14**, 125–131.

Idso S.B., Jackson R.D., Ehrler W.L. & Mitchell S.T. (1969) A method for determination of infrared emittance of leaves. *Ecology*, **50**, 899–902.

Kalma J.D. & Stanhill G. (1969) The radiation climate of an irrigated orange plantation. *Solar Energy*, **12**, 491–508.

Kleshnin A.F. & Shulgin I.A. (1959) The optical properties of plant leaves. *Doklady Akademie S.S.R.*, **125**, 1158–1160. (Bot. Sci. Soc. translation 108–110).

Mantell A. & Stanhill G. (1966) Comparison of methods for evaluating the response of lawngrass to irrigation and nitrogen treatments. *Agronomy Journal*, **58**, 465–468.

McCartney H.A. & Unsworth M.H. (1976) A spectroradiometer for measuring the spectral distribution of radiation in plant canopies. *Light as an Ecological Factor: II.* Symposia of the British Ecological Society, 16. (Ed. by G.C. Evans, R. Bainbridge & O. Rackham). pp. 565–568. Blackwell, Oxford.

McCree K. (1971/1972) The action spectrum, absorptance and quantum yield of photo-synthesis in crop plants. *Agricultural Meteorology*, **9**, 191–216.

McFarland W.N. & Munz F.W. (1976) The visible spectrum during twilight and its implications to vision. *Light as an Ecological Factor: II.* Symposia of the British Ecological Society, 16. (Ed. by G.C. Evans, R. Bainbridge & O. Rackham), pp. 249–270. Blackwell, Oxford.

Monteith J.L. (1959) The reflection of short wave radiation by plants. *Quarterly Journal of the Royal Meteorological Society*, **85**, 386–392.

Monteith J.L. (1977) Climate and the efficiency of crop production in Britain. *Philosophical Transactions of the Royal Society London B.*, **281**, 277–294.

Monteith J.L. & Szeicz G. (1962) Radiative temperature in the heat balance of natural surfaces. *Quarterly Journal of the Royal Meteorological Society*, **88**, 496–507.

Mooney H.A., Ehleringer J. & Björkman O. (1977) The energy balance of leaves of the evergreen desert shrub *Atriplex hymenelytra*. *Oecologia (Berl.).*, **29**, 301–310.

Moore C.J. (1976) A comparative study of radiation balance above forest and grassland. *Quarterly Journal of the Royal Meteorological Society*, **102**, 889–899.

Moreshet S., Stanhill G. & Fuchs M. (1975) Aluminum mulch increases quality and yield of 'Orleans' apples. *Horticultural Science*, **10**, 390–391.

Moreshet S., Stanhill G. & Fuchs M. (1977) Effect of increasing foliage reflection in the CO_2 uptake and transpiration resistance of a grain sorghum crop. *Agronomy Journal*, **69**, 246–250.

Oguntoyinbo J.S. (1970) Reflection coefficient of natural vegetation, crops and urban surfaces in Nigeria. *Quarterly Journal of the Royal Meteorological Society*, **94**, 430–441.

Otterman J. (1974) Baring high-albedo soils by overgrazing: a hypothesized desertification mechanism. *Science*, **86**, 531–533.

Otterman J., Lowman P.D. & Salomonson V.V. (1976) Surveying earth resources by remote sensing from satellites. *Geophysical Surveys*, **2**, 431–467.

Pearson R.L., Miller L.D. & Tucker C.J. (1976) Hand-held spectral radiometer to estimate gramineous biomass. *Applied Optics*, **15**, 416–418.

Ross J. (1975) Radiation transfer in plant communities. *Vegetation and the Atmosphere*, *Vol.* 1 *Principles*. (Ed. by J.L. Monteith), pp. 13–55. Academic Press, London.

Sauberer F. & Hartel O. (1959) *Pflanze und Strahlung*. Probleme der Bioklimatologie, Band 5, Geest & Portig, K.G. Leipzig.

Schiller G. & Karschon R. (1973) Microclimate and thermal stress of man in an Aleppo pine plantation and an oak scrub. *Israel Journal of Agricultural Research*, **23**, 79–90.

Scott D., Mendalda P.H. & Brougham R.W. (1968) Spectral analysis of radiation transmitted and reflected by different vegetation. *New Zealand Journal of Botany*, **6**, 427–429.

Sibma L. (1968) Growth of closed green crop surfaces in the Netherlands. *Netherlands Journal of Agricultural Science*, **16**, 211–214.

Stanhill G. (1965) Observations on the reduction of soil temperature. *Agricultural Meteorology*, **2**, 197–203.

Stanhill G. (1970) Some results of helicopter measurements of albedo of different land surfaces. *Solar Energy*, **11**, 59–66.

Stanhill G. (1975) Solar radiation effects and crop production. *Progress in Biometeorology*. Division C. Volume 1. Period 1963–1974. (Ed. by L.P. Smith) pp. 58–72. Swets & Zeitlinger, Amsterdam.

Stanhill G., Cox J.T.H. & Moreshet S. (1968) The effect of crop and climatic factors on the radiation balance of an irrigated maize crop. *Journal of Applied Ecology*, **5**, 707–720.

Stanhill G., Hofstede G.J. & Kalma J.D. (1966) Radiation balance of natural and agricultural vegetation. *Quarterly Journal of the Royal Meteorological Society*, **92**, 128–140.

Stanhill G., Israeli M. & Rosenzweig, D. (1973) The solar radiation balance of scrub forest and pasture on the Carmel mountain, Israel: A comparative study. *Ecology*, **54**, 819–828.

Stanhill G., Moreshet S. & Fuchs M. (1976) Effect of increasing foliage and soil reflectivity on the yield and water use efficiency of grain sorghum. *Agronomy Journal*, **68**, 329–332.

Stanhill G., Moreshet S., Jurgrau M. & Fuchs M. (1975) The effect of reflecting surfaces on the solar radiation and carbon dioxide fixation of a glasshouse rose crop. *Journal of the American Society of Horticultural Science*, **100**, 112–115.

Waggoner P.E. (1963) Plants, shade and shelter. *Connecticut Agricultural Experiment Station Bulletin*, **656**, 16 pp.

Watson D.J. (1947) Comparative studies on the growth of field crops. 1. Variation in the N.A.R. and L.A.R. between species and varieties, and within and between years *Annals of Botany, London, N.S.*, **11**, 41–76.

Watts W.R. (1976) Soil reflection coefficient and its consequences for soil temperature and plant growth. *Light as an Ecological Factor*. II. Symposia of the British Ecological Society, 16. (Ed. by G.C. Evans, R. Bainbridge & O. Rackham). pp. 409–421. Blackwell, Oxford.

4. CANOPY STRUCTURE, MICROCLIMATE AND THE ENERGY BUDGET IN DIFFERENT ALPINE PLANT COMMUNITIES

ALEXANDER CERNUSCA AND MARION CLAUDIA SEEBER

*Institut für Botanik der Universität, Sternwartestraße 15,
A-6020 Innsbruck, Austria*

SUMMARY

Canopy structure along an altitudinal transect, microclimate and energy fluxes were analysed at three sites on a south-facing slope in the Central Alps (Hohe Tauern, Austria): a meadow at 1635 m, an alpine pasture near the timberline at 1915 m and a natural alpine sedge mat (*Caricetum curvulae*, 'Curvuletum') at 2300 m.

1. With increasing altitude, canopy height decreased from 90 cm (meadow) to 30 cm (alpine pasture) and 12 cm (sedge mat). The green leaf area index of the phanerogams decreased from 7 to 1·9, and the standing amount of dead plant parts increased from 12% of the total phytomass in the meadow to 48% in the sedge mat.

2. The extinction coefficients for photosynthetically active radiation ranged between 0·34 and 0·54. In the meadow 82%, in the alpine pasture 78% and in the sedge mat only 32% of the incident photosynthetically active radiation was absorbed by the green plant parts.

3. The driving forces for energy transfer increased with altitude. Global radiation increased by 20%, net radiation by 27%, and the temperature difference (between canopy and screen level) by 100%. In contrast to these trends the average water vapour pressure difference decreased from the meadow to the sedge mat by 7%. Evapotranspiration at the three sites was almost the same (about 260 W m^{-2} = 3·7 mm) on clear days in August. The sensible heat flux, however, increased from 100 W m^{-2} in the meadow to 196 W m^{-2} in the sedge mat, resulting in an increase in the Bowen ratio from 0·38 to 0·75. These differences in the energy balance are the result of decreasing amounts of green leaf area and larger amounts of attached dead plant parts at the higher altitude.

4. Along this altitudinal transect the *total* canopy resistance for water vapour transfer increased whereas the average canopy aerodynamic resistance and the leaf resistance decreased.

INTRODUCTION

Altitudinal variation of macroclimate in mountain regions induces significant changes in aspects of the vegetation structure such as growth form, plant height, leaf area development and stratification of phytomass. Previous investigations on the microclimate and energy budget of dwarf shrub communities and alpine pastures at and above the timberline in the Central Alps have shown that significant differences exist between the microclimate above the canopy and within it (Cernusca 1976a, b; Cernusca et al. 1978). The microclimate above the canopy is affected primarily by the different exposure of the leaves to solar radiation and wind, whereas the microclimate within the canopy depends largely on canopy structure. In order to obtain more detailed information on these differences we have investigated the microclimate and heat exchange processes above and within three grassland canopies along an altitudinal transect on a south-facing slope of the Central Alps (Hohe Tauern, Austria, 47°2′ N, 12°51′ E). The sites were a meadow at 1635 m above sea level dominated by *Dactylis glomerata, Arrhenatherum elatius* and *Poa pratensis*, an alpine pasture at 1915 m, dominated by *Nardus stricta*, and a sedge mat at 2300 m dominated by *Carex curvula*. These measurements are part of an integrated ecosystem study within the Austrian MaB-program (Cernusca 1975). Results presented here are preliminary, because some of the micrometeorological investigations are not yet completed.

METHODS

Water vapour and heat flux were calculated by the Bowen ratio method (Thom 1975) using profiles of temperature and water vapour pressure measured between the canopy surface and a height 60 cm above it (Fig. 4.1). The total canopy resistance, r_c (aerodynamic and leaf resistance) and the aerodynamic resistance, r_a, for water vapour transfer were calculated from the equations

$$r_c = \frac{\rho \cdot c_p (e - e_s(T_\phi))}{\gamma \lambda \mathbf{E}_T}$$

$$r_a = \frac{\rho \cdot c_p (T - T_\phi)}{\mathbf{C}}$$

$\mathbf{E}_T$ is the transpiration rate of the canopy (measured by weighing lysimeters with the soil surface sealed), $\mathbf{C}$ is the sensible heat flux, e and T the water vapour pressure and the air temperature above the canopy, $e_s(T_\phi)$ the saturation vapour pressure at the leaf temperature T_ϕ, inside the canopy, and γ is the psychrometric constant ($= 0.66$ mb °C^{-1}). The factor 0.93 converts the aerodynamic resistance for heat to the appropriate resistance for water vapour.

Fig. 4.1.   Canopy structure, microclimate profiles and energy budget for the investigated sites. Microclimate profiles are mean values for the time interval 12.00 to 13.00 on a bright day in August. Energy fluxes are mean values for the time interval 8.00 to 18.00 for all bright days in August 1978. For symbols see text. Windspeed denoted as *u*, heat flux to ground is **G**.

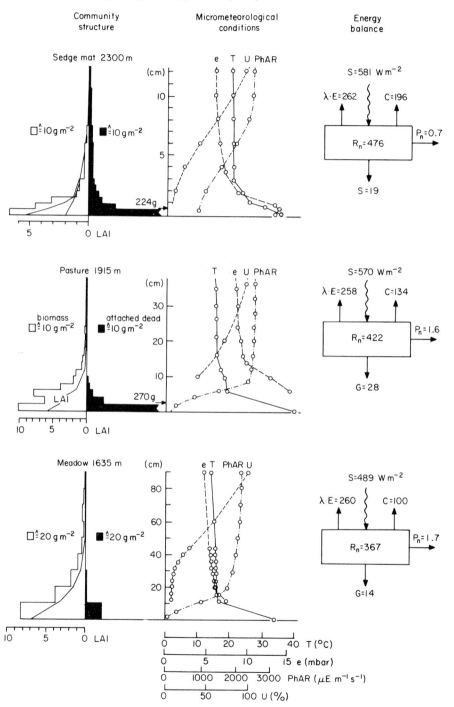

The micrometeorological measurements were made at intervals ranging from 1 min for rapidly changing variables like radiation, temperature and water vapour pressure to 12 min for soil temperature and soil heat flux. The measurements were recorded by a portable data acquisition system powered by batteries. This system contains a microprocessor to convert the raw data to appropriate units, calculate energy balance fluxes, and print out summarized values for each hour (Cernusca, unpublished).

The following sensors were used: Star pyranometer and net radiometer (Schenk, Vienna), small thermocouples for leaf and air temperature (home made: copper/constantan, 0·08 mm diameter), thermocouple psychrometers (home made: ventilated by a micro vane), heat flux plates (Keithley Instruments, S.A.), cup anemometers (Contact Type R/AMC, Rauchfuss Instruments, Burwood, Australia) and hotwire anemometers (home made, Cernusca 1967), photon sensor (Lambda, Nebraska, U.S.A.) for measurement of quantum flux density.

RESULTS AND DISCUSSION

1. Phytomass stratification and community structure (Fig. 4.1, left)

The height of the canopy in the meadow at 1635 m was 90 cm, the leaf area index was 7·0 and there was 463 g of dry matter per m². Of the total above-ground phytomass 46% was composed of grasses, 42% of herbs, and 12% of dead plant parts. Eighty-one per cent of the photosynthetically active biomass was concentrated in the lowermost 20 cm of the plant cover.

The alpine pasture community at 1915 m reached a height of 38 cm. The leaf area index was 6·0 and there was 642 g of dry matter per m². Of the above-ground phytomass, 34% were grasses, 18% herbs and 48% dead plant parts. The alpine pasture was well adapted to grazing by cattle which has been going on for over a century. Fifty-two per cent of the phytomass and 35% of the green plant parts was concentrated in the lowermost 2 cm of the canopy and could not, therefore, be reached by the grazing cattle, as the maximal grazing depth of cattle is only 2 cm. The maximum loss of biomass by grazing of the alpine pasture comes close to the maximum value of 50% recommended by Odum (1967) for grazed grassland. The phytomass of an alpine pasture at 1850 m in the Gastein valley (Province Salzburg, Austria) was also concentrated similarly to the lowermost canopy layer (Cernusca et al. 1978).

The sedge mat at 2300 m canopy reached a height of only 12 cm. The leaf area index (phanerogams only) was 1·9 and there was 775 g of dry matter per m². The leaf area index of cryptogams was 3·5! The proportion of photosynthetically active plant parts, 48%, was remarkably large. The sedge mat was also characterized by the presence of two distinct layers: a dense lower layer with relatively weak convective exchange (0 to 1·5 cm with rosette plants, mainly *Primula minima*) and an upper layer, consisting of sedges and grasses, which was subjected to atmospheric turbulence.

2. *Extinction of photosynthetically active radiation* (Table 4.1)

Extinction coefficients of quantum flux density have been calculated in relation to the projected green leaf area. Average extinction coefficients of 0·34 for the sedge mat, 0·36 for the alpine pasture, and 0·54 for the meadow were obtained. In dwarf shrub communities at the alpine timberline (Mt. Patscherkofel near Innsbruck, Austria) *Loiseleuria procumbens* yielded values of 0·66, *Calluna vulgaris* 0·65, *Vaccinium myrtillus* 0·87 and *Rhododendron ferrugineum* 1·13, respectively (Cernusca 1976b).

TABLE 4.1. Distribution of photosynthetically active radiation to different plant components. Values in % of incoming quantum flux density at canopy surface (p.p. is plant parts).

| Vegetation | Altitude | Reflected | Green p.p. | Absorbed by | | |
				Non-green p.p.	Other p.p.	Soil surface
Sedge mat	2300 m	7·0	31·6	28·3	4·3	29·0
Pasture	1915 m	12·8	77·6	6·6	—	3·0
Meadow	1635 m	14·0	82·3	2·8	—	0·9

To evaluate the efficiency of energy utilization for dry matter production it is necessary to know the proportion of the total incident radiation absorbed by green plant parts. The absorption of photosynthetically active radiation by green plant parts amounted to 82% in the meadow, 78% in the alpine pasture and only 32% in the alpine sedge mat. In the above mentioned dwarf shrub heath, values between 65% and 85% were measured by Cernusca (1976b).

3. *Microclimate above and within the plant canopy*

Figure 4.1 shows characteristic microclimatic profiles for sunny days in August. The following points are apparent.
Wind: Wind velocity decreased to 50% at the canopy surface and to 15–30% in the middle layer of the canopy as compared with that measured at a reference height of 2 m. Over 50% of the green plant material was exposed to wind velocities which reach only 5–10% of the values measured at 2 m.
Temperature: Temperature profiles show that most heat exchange occurred in the lowermost canopy layer and at the soil surface. Temperatures measured in the lowest canopy layer sometimes exceeded screen temperatures by 10 °C (meadow) and 25 °C (alpine pasture and sedge mat). During clear nights in

summer the temperature minimum, however, occurred in the uppermost third of the canopy.

Vapour pressure: Vapour pressure profiles were similar to the temperature profiles with a steep vapour pressure gradient in the lowermost third of the canopy. Remarkably, the relative humidity in this layer rarely decreased below 50%.

Thus, it can be concluded that above the vegetation the microclimate mainly reflects the exposure to wind and radiation. The sedge mat is windswept and cool, the pasture is calm and warm. The climate within the plant cover clearly shows the buffering effect of the canopy structure. In the lowermost third of each canopy, where more than 50% of the total biomass occurs, there are only slight differences in microclimate amongst these three grassland communities.

4. Energy balance

The input-output diagrams presented in Fig. 4.1 show the average energy fluxes during clear days in August. There was 20% more global radiation at 2300 m than at 1635 m and the net radiation was increased by 27%. Surprisingly, the additional energy supply at the higher elevation sites is used in increased sensible heat exchange (C), whereas evapotranspiration remains almost unchanged along this altitudinal transect. The rates of evapotranspiration determined by the Bowen ratio method (about 3·9 mm) from 0800 to 1800 h, and values obtained from weighing lysimeters by Körner & Mayr (1979) (4·0 to 4·5 mm) from 0700 to 1900 h, corresponded remarkably well. The altitudinal increase in sensible heat flux evidently finds expression in the Bowen ratio which increased from 0·38 at the meadow to 0·75 at the sedge mat. A similar altitudinal increase in the Bowen ratio was found in dwarf shrub communities at Mt. Patscherkofel near Innsbruck (Cernusca 1976b) (Table 4.2).

The plant communities investigated were always well supplied with water. Plant water potentials hardly fell below −1·5 MPa and leaf diffusion resistances generally were low (Körner & Mayr 1979). Therefore the increase in

Table 4.2. Bowen ratio $C/\lambda \cdot E$ and R_n/S ratio for the plant communities on bright days in August with values for some other ecosystems.

		$C/\lambda \cdot E$	R_n/S	
Sedge mat	2300 m	0·75	0·81	
Pasture	1915 m	0·52	0·78	
Meadow	1635 m	0·38	0·75	
Dwarf shrub zone on Mt. Patscherkofel:				
Loiseleuria heath	1950 m	0·45	0·50	Cernusca 1976b
Loiseleurietum	2175 m	0·70	0·40	Cernusca 1976b

the Bowen ratio is not a result of impaired water supply. Rather, the drastic decrease in the green leaf area of the phanerogams (from LAI 7 to 1·9) and the large increase in the amount of standing, dead plant parts explain this phenomenon.

Total canopy resistance, r_c, was found to increase, whereas the canopy aerodynamic resistance, r_a, decreased with elevation (Table 4.3). The difference $r_c - r_a$ gives the canopy leaf resistance, r_l, which was lowest at the highest site.

TABLE 4.3. Total water vapour transfer resistance for the canopy, r_c, aerodynamic resistance, r_a, canopy leaf resistance, r_l, and leaf conductance g_l, for the investigated sites. Leaf area index in brackets.

			r_c s cm^{-1}	r_a s cm^{-1}	r_l s cm^{-1}	g_l s cm^{-1}
Sedge mat	2300	(1·9)	0·86	0·12	0·63	0·42
Pasture	1915 m	(6·1)	0·79	0.15	0.56	0.15
Meadow	1635 m	(7·0)	0·63	0·16	0·42	0·14

The increase in r_l, however, is not caused by increased leaf resistance of the individual plant species. Leaf conductance values calculated from r_l and leaf area index as $g_l = 1/(2 \cdot L \cdot r_l)$ show an opposite trend, corresponding well with the altitudinal increase of leaf conductance determined for a number of plant species by diffusion porometry (Körner & Mayr 1979). Thus the increase in r_l along the altitudinal transect is the result of the decrease in amount of green plant parts with altitude.

REFERENCES

Cernusca A. (1975) MaB-6-Projekt 'Pilotprojekt Alpine Ökosysteme'. Sitzungsberichte der Österr. Akademie der Wissenschaften, Mathematisch-naturwissenschaftliche Klasse, *Abteilung I*, **184**, 133–141.

Cernusca A. (1976a) Energie- und Wasserhaushalt eines alpinen Zwergstrauchbestandes während einer Föhnperiode. *Archiv für Meteorologie, Geophysik und Bioklimatologie*, **B24**, 219–241.

Cernusca A. (1976b) Bestandesstruktur, Bioklima und Energhiehaushalt von alpinen Zwergsträuchern. *Oecologia Plantarum*, **11**, 71–102.

Cernusca A., Seeber M.C., Mayr R. & Horvath A. (1978) Bestandesstruktur, Mikroklima und Energiehaushalt von bewirtschafteten und aufgelassenen Almflächen in Badgastein. *Ökologische Analysen von Almflächen im Gasteiner Tal* (Ed. by A. Cernusca), pp. 47–66, Veröfflichungen des Österreichischen MaB-Hochgebirgsprogramms Hohe Tauern, Bd. 2, Universitätsverlag Wagner, Innsbruck.

Odum E.P. (1967) *Ökologie*. Aus der Reihe *Moderne Biologie*, Bayerischer Landwirtschaftsverlag, München-Basel-Wien.

Thom A.S. (1975) Momentum, mass and heat exchange of plant communities. *Vegetation and the Atmosphere*, Vol 1. (Ed. by J.L. Monteith), pp. 57–109. Academic Press, New York.

5. SHOOT EXTENSION AND WATER RELATIONS OF *CIRCAEA LUTETIANA* IN SUNFLECKS

F.I. WOODWARD*

Department of Plant Science, University College, P.O. Box 78, Cardiff, CF1 1XL

SUMMARY

1. The physiological effects of the passage of natural sunflecks was studied on *Circaea lutetiana* growing on a woodland floor.

2. Temperature, transpiration and convection increased rapidly in the first minute of a 7-minute sunfleck. The diffusion resistance increased throughout the remainder of the sunfleck and transpiration was reduced below the level of shade leaves.

3. Leaf water potential declined rapidly in the first 2 minutes of the sunfleck, and led to a reduction in pressure potential.

4. Shoot extension was reduced during and after the sunfleck. The rate before the sunfleck was not regained, by 30 minutes after the termination of the sunfleck, in 75% of the observations.

INTRODUCTION

Flecks of direct solar radiation are characteristic features of the woodland floor and they may contribute a large proportion of integrated daily irradiance even though sunflecks are most frequently short lived (Evans 1956). Long lived sunflecks can exert a substantial influence on the water and thermal balance of a plant. Ellenberg (1963) demonstrated a rapid increase of the transpiration of *Asperula odorata* L. during sunflecks. Rackham (1975) showed that temperature excesses of *Mercurialis perennis* L. leaves could cause heat damage.

Qualitatively, radiation in sunflecks differs from that transmitted by leaves, particularly in the ratio of energy in the photomorphogenically active wavelengths of 660 and 730 nm. A ratio of 1 is typical of natural radiation while a ratio of 0·2 is characteristic of radiation transmitted by leaves.

*Present address: Department of Botany, University of Cambridge, Downing Street, Cambridge, CB2 3EA.

A change in the ratio from 0·8 to 0·3 can induce etiolation responses, e.g. in the shoot extension of the arable weed *Chenopodium album* L. (Morgan & Smith 1978; Smith, this volume). The long term extension of the true shade plant *Circaea lutetiana* L. is unaffected by differences in the 660 : 730 nm ratio (C.D. Pigott, personal communication) indicating that species can differ in their photomorphogenic responses.

The aim of the work presented here is to examine the effects and inter-actions of the sunfleck microclimate on the physiological responses of *C. lutetiana*.

MATERIALS AND METHODS

Site

Measurements were made in a pure stand of *C. lutetiana* (4 m diameter) at an early stage of flowering, in a *Fagus sylvatica* L. woodland on a brown earth soil, at Tongwynlais, Cardiff, between the 18th and 21st July, 1978. The centre of the stand of *C. lutetiana* was at a mean distance of 3 m from the surrounding trees which had a uniform girth at breast height of 0·3 m.

Apparatus

Measurements of microclimate were made in two areas of *C. lutetiana* (2 metres apart). In each, the measuring apparatus was mounted within a small circular area (0·01 m²), to ensure that all the sensors were simultaneously exposed to the direct radiation of at least the larger sunflecks. A solarimeter and a net radiometer were positioned at the height of the youngest fully expanded leaf (200 mm). Two chromel-constantan thermocouples, with a junction diameter of 100 μm, were each embedded into separate leaves from the abaxial surface with the reference junctions positioned under a radiation screen at air temperature. Shoot elongation was measured on adjacent plants with linear variable differential transformers (LVDT, type ND2, 5·00 mm, Sangamo Weston Controls) (Morgan & Smith 1978).

Aspirated wet and dry bulb temperatures were measured with solid state temperature sensors (LM 3911, National Semiconductor Corp.) above the *C. lutetiana* at 300 mm and under radiation shields. Wind speed was measured with a heated thermistor anemometer (type ETA 3000, Airflow Developments Ltd). Leaf diffusion resistance (r_s) was measured with an automatic diffusion porometer (Crump Scientific Instruments) on plants adjacent to the areas selected. Similarly leaf water potential (Ψ_L), pressure potential (Ψ_P) and osmotic potential (Ψ_π) were determined for detached leaves, with a pressure bomb (Crump Scientific Instruments) using a modification of the pressure/volume technique of Tyree and Hammel (1972).

Sampling

LVDTs were attached to single plants for the course of an experimental period. Previous measurements have indicated that a simulated 7 minute sunfleck with a mean irradiance of 620 W m^{-2} causes errors in the LVDT and clamp apparatus (due to expansion) equivalent to a plant shrinkage of 0·05 mm h^{-1}. This represents the likely errors in the measurement of stem extension.

The initial settling period after connecting the LVDT to the plant was between 20 and 90 minutes. The LVDTs were sufficiently sensitive to unequivocally detect extension after a period of 2 to 3 minutes. Measurements were then taken when the wind speed at plant height was 0·1 m s^{-1} or less; this avoided errors of measurement due to the wind. Furthermore the wind speed was generally at 0·1 m s^{-1} or less so that sampling was not a problem.

Measurements of r_s take a minimum of 15 seconds, that of Ψ_L, Ψ_P, Ψ_π take 120 seconds. This created a logistic problem if all these variables were to be measured during a single sunfleck. Plants were therefore measured from the time sunflecks were observed on the plants attached to the LVDTs and then after fixed intervals, often in different sunflecks. This method provides a composite picture of the time course of these variables in a single sunfleck. Three replicates were taken for each period (1 minute for r_s and 2 minutes for Ψ). Irradiance and leaf temperature also vary between sunflecks and so these were monitored at the time of measurement.

Leaf net radiation balance

The net radiation balance of the leaf was determined for the youngest fully expanded leaf, L_1, adjacent to the radiometers. The radiation balance has been determined from the fluxes shown on Fig. 5.1 and presented in the following equation:

$$\mathbf{R}_n = \mathbf{S} \cdot (a) + \mathbf{S} \cdot (r \cdot a) + \mathbf{S} \cdot (t \cdot r \cdot a) + \mathbf{L}(\varepsilon) +$$
$$\varepsilon \sigma T_{L2}{}^4 \cdot (\varepsilon) + \varepsilon \sigma T_{L1}{}^4 \cdot 2\varepsilon(1 - \varepsilon) - \varepsilon \sigma T_{L1}{}^4 \quad (5.1)$$

Absorbed fluxes are positive in sign, emitted fluxes are negative in sign. a, r, t, ε are the coefficients of radiation absorption, reflection, transmission and emission. $\mathbf{S}$ and $\mathbf{L}$ are the short- and long-wave radiation fluxes and T is the leaf temperature (K). The equation accounts for reflected fluxes of both short- and long-wave radiation. Leaf L_2 was 30 mm beneath leaf L_1.

The coefficients r (0·24) and t (0·21) were determined with appropriately oriented miniature solarimeters. ε (0·95) was determined in the laboratory by the method of Gates and Tantraporn (1952).

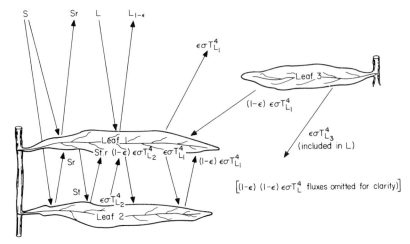

FIG. 5.1. Radiation fluxes and the net radiation balance of the youngest expanded leaf.

The energy budget of the leaf was determined according to the equation:

$$\mathbf{R}_n = \lambda\mathbf{E} + \mathbf{C} + \mathbf{J}$$

where $\mathbf{E}$ and $\mathbf{C}$ are respectively the transpiration and convection fluxes and $\mathbf{J}$, the heat storage. $\mathbf{J}$ was determined from the rate of change of leaf temperature and from a knowledge of the heat capacity of the leaf ($501 \cdot 6$ J m^{-2} °C^{-1}) and assuming the leaf ($0 \cdot 12 \pm 0 \cdot 05$ mm thick) has the specific heat of water ($\mathbf{J}$ is positive when leaf storage is positive). $\lambda\mathbf{E}$ and $\mathbf{C}$ were determined using the equation presented by Monteith (1973). The boundary layer resistance (r_a) was estimated from these equations and was the only unknown. The boundary layer resistance to heat (r_a^H) and to water vapour (r_a) were inter-converted by the method of Monteith (1973).

RESULTS AND DISCUSSION

Figure 5.2 demonstrates the change in the leaf temperature (Fig. 5.2b), energy budget (Fig. 5.2d, e), r_a, r_s and r_a^H (Fig. 5.2e) during the passage of a 7 minute sunfleck. $\mathbf{R}_n$ (Fig. 5.2a) increased by an order of magnitude between the shade and sunfleck. Leaf temperature increased most rapidly in the first minute, and since r_a and r_s did not increase $\lambda\mathbf{E}$ increased. Throughout the remainder of the sunfleck $\lambda\mathbf{E}$ declined because of the increase in r_s, to significantly less than that of shade plants. $\mathbf{J}$ was small in magnitude and generally positive in the sunfleck. $\mathbf{C}$ accounted for the greater part of $\mathbf{R}_n$, and was an

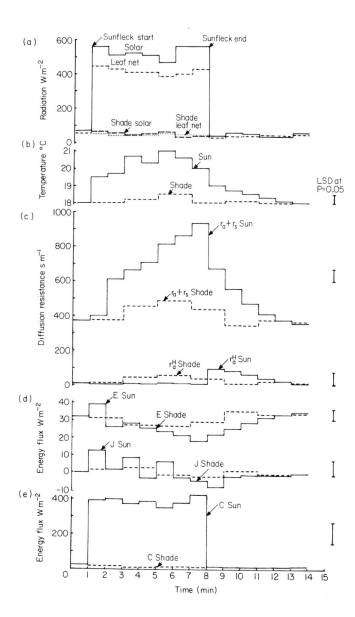

Fig. 5.2. Plant responses to a sunfleck: (a) radiation balance; (b) leaf temperature; (c) leaf diffusion resistances to water vapour ($r_a + r_s$) and to heat (r_a^H); (d) transpiration (λE) and heat storage (**J**); (e) convection (**C**). Least significant differences shown at $P = 0.05$.

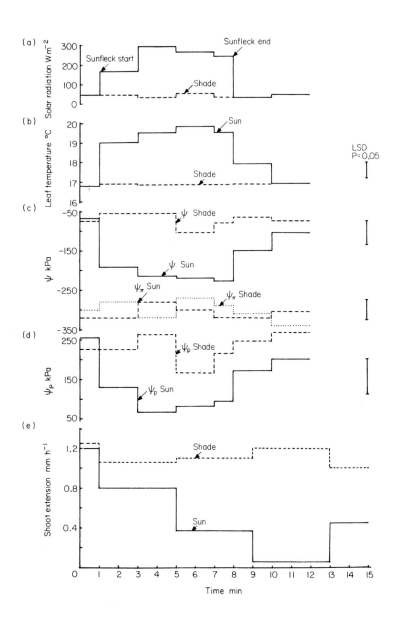

FIG. 5.3.   Plant responses to a sunfleck: (a) solar radiation input; (b) leaf temperature; (c) leaf water potential (ψ_L) and osmotic potential (ψ_π); (d) leaf pressure potential (ψ_p); (e) shoot extension. Least significant differences shown at $P = 0.05$.

order of magnitude greater than λE, which was limited by r_s and by the high ambient humidity (ambient vapour pressure deficit, 6 mb).

The post-sunfleck period was characterized by a rapid decline in R_n and C, and a slower decline in leaf temperature, r_a and r_s, until 4 min after the sunfleck when the shade and post-sunfleck plants were not significantly different.

In *C. lutetiana* net radiation is dissipated by transfer of sensible, C, rather than latent, λE, heat. λE is strictly controlled by r_s. This response accounts for the high leaf temperatures observed by Rackham (1975) and indicates a problem for the survival of shade species in sunflecks.

Photosynthetic rates (**P**) were not measured in the field. However the influence of changes in r_s on **P** can be estimated using the following equation:

$$\mathbf{P} = \frac{\chi_a - \chi_c}{r_a{}^{CO_2} + r_s{}^{CO_2} + r_m} \tag{5.2}$$

where χ_a and χ_c are the CO_2 concentrations of the ambient air and at the site of carboxylation respectively. χ_c is taken to be zero and χ_a as 0.594 g m^{-3}.

The resistances terms $r_a{}^{CO_2}$ and $r_s{}^{CO_2}$ are the boundary and stomatal resistances to carbon dioxide transfer, while r_m is the mesophyll or residual resistance to transfer (Gaastra 1959).

Data of Holmgren, Jarvis & Jarvis (1965) have been used to estimate a maximum rate of photosynthesis, $\mathbf{P}_{max}$ of 4.4 mg dm^{-2} h^{-1}, at a saturating irradiance of 73 W m^{-2}. This indicates that **P** is not saturated in the observed shade conditions but will be in the sunfleck. r_a and r_s are converted to $r_a{}^{CO_2}$ and $r_s{}^{CO_2}$ respectively by multiplying by the ratio of the diffusion coefficient of water vapour and CO_2 at ambient temperature (1.65). Substituting these values into equation 5.2 provides an estimate of 3884 s m^{-1} for r_m.

If it is assumed that r_m remained constant through the sunfleck then r_m and $r_a{}^{CO_2} + r_s{}^{CO_2}$ at the end of the sunfleck can be inserted into equation 5.2, to determine the influence of the changes of $r_s{}^{CO_2}$ on **P**. This technique predicted that $\mathbf{P}_{max}$ would decrease by 10% during the sunfleck, a rate which is still greater than that in the shade. λE decreased by 54% during the sunfleck, indicating that the plant benefits photosynthetically from the increased irradiance but not at the expense of greater water loss. For the first 2 min of the sunfleck, Ψ'_L declined rapidly, remained constant for the remainder of the sunfleck and then increased steadily from the termination of the sunfleck (Fig. 5.3c). The changes in Ψ_L (Fig. 5.3c) were entirely the result of changes in Ψ_P (Fig. 5.3d) with Ψ_π remaining constant throughout.

Shoot extension is very sensitive to Ψ_P (Bunce 1977) and this was reflected in the rapid reduction of extension during the sunfleck. Twenty-four sunflecks were observed over 2 days of study, and in 75% of the cases the

TABLE 5.1. Steady post-sunfleck rates of shoot extension as a percentage of the pre-sunfleck rate.

% of pre-sunfleck rate	% of samples
100	25
50–99	25
< 50	50

TABLE 5.2. Multiple regression analysis of the reduction in extension rate (y) resulting from a sunfleck, and pre-sunfleck growth rate (x_1) and sunfleck duration (x_2).

$$y = 0.702.x_1 + 0.075.x_2 - 0.2083$$
$$\quad\text{xxx} \qquad\quad \text{xxx}$$

(a) xxx = significant at $P = 0.001$
(b) $x_1, y(\text{mm h}^{-1})$, x_2 (min)
(c) $r^2 = 0.951$

post-sunfleck rate was less than the pre-sunfleck rate (Table 5.1), after 30 minutes from the termination of the sunfleck. The observed sunflecks were variable in duration and the maximum depression of extension rate occurred over the longest period (Table 5.2). However, this effect was influenced by the pre-sunfleck rate, with the highest initial rates being the most severely depressed.

ACKNOWLEDGEMENTS

I am grateful to Mr. N. Jones for assistance with the field measurements.

REFERENCES

Bunce J.A. (1977) Leaf elongation in relation to leaf water potential in soybean. *Journal of Experimental Botany*, **28**, 156–161.

Ellenberg H. (1963) *Vegetation Mitteleuropas mit den Alpen.* Stuttgart.

Evans G.C. (1956) An area survey method of investigating the distribution of light intensity in woodlands with particular reference to sunflecks. *Journal of Ecology*, **44**, 391–428.

Gaastra P. (1959) Photosynthesis of crop plants as influenced by light, CO_2, temperature and stomatal diffusion resistance. *Mededelingen van de Landbouwhogeschool te Wageningen*, **59**, 1–68.

Gates D.M. & Tantraporn W. (1952) The reflectivity of deciduous trees and herbaceous plants in the infra-red to 25 microns. *Science*, **115**, 613–616.

Holmgren P., Jarvis P.G. & Jarvis M.S. (1965) Resistances to carbon dioxide and water vapour transfer in leaves of different plant species. *Physiologia Plantarum*, **18**, 557–573.

Monteith J.L. (1973) *Principles of Environmental Physics.* Edward Arnold, London.

Morgan D.C. & Smith H. (1978) Simulated sunflecks have large, rapid effects on plant stem extension. *Nature*, **273**, 534–536.

Rackham O. (1975) Temperatures of plant communities measured by pyrometric and other methods. *Light as an Ecological Factor. II.* Symposia of the British Ecological Society 16. (Ed. by G.C. Evans, R. Bainbridge & O. Rackham), pp. 423–449. Blackwell, Oxford.

Tyree M.T. & Hammel H.T. (1972) The measurement of the turgor pressure and the water relations of plants by the pressure-bomb technique. *Journal of Experimental Botany*, **23**, 267–282.

6. LIGHT QUALITY AS AN ECOLOGICAL FACTOR*

HARRY SMITH

Department of Botany
The University
Leicester

SUMMARY

1. The importance of light quality as a potential ecological factor is demonstrated.

2. It is argued that phytochrome, because it exists in two isomeric forms, is the most likely pigment to perceive changes in light quality.

3. The two isomeric forms of phytochrome exist in tissue in proportions which depend upon the spectral quality of incident radiation. The amounts at photoequilibrium are usually expressed as the ratio of the amount of the far-red form to the total (i.e. Pfr/Ptotal, or ϕ).

4. Because of the difficulty of measuring ϕ, a useful parameter is the ratio of the photon fluence rates in the red (655–665 nm) and far-red absorbing (725–735 nm) spectral regions (i.e. ζ).

5. A relationship between ϕ and ζ is described. This shows that ϕ varies most steeply over the natural range of ζ.

6. Experiments in specially constructed growth cabinets showed that elongation growth decreased exponentially with increasing ζ and linearly with increasing ϕ in ruderal herbs. Shade-tolerant species showed much less decrease in extension growth rate with increasing ϕ.

7. The evidence for the hypothesis that phytochrome perceives light quality through the establishment of photoequilibria is critically discussed.

INTRODUCTION

The natural radiation environment is highly complex and variable. Although the solar radiation reaching the earth's atmosphere is, at least within a biologist's terms of reference, constant as far as irradiance and spectral distribution are concerned, these parameters as experienced at the surface of

---

* In accordance with recent practice (Mohr & Schäfer 1979) the term 'fluence rate' is used in this article in preference to the synonymous term 'flux density'. *Abbreviations:* PAR, photosynthetically active radiation (400–700 nm); Pr, red absorbing form of phytochrome; Pfr, far-red absorbing form of phytochrome; Ptotal, total quantity of phytochrome present in a given tissue sample.

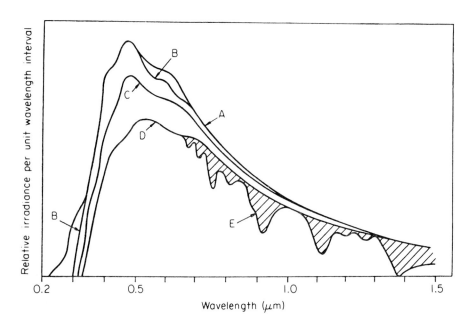

FIG. 6.1. Spectral energy distribution of solar radiation as affected by various atmospheric factors (from Henderson 1970): A, extra terrestrial sunlight; B, after ozone absorption; C, after molecular scattering; D, after aerosol scattering; E, at the earth's surface, after water and oxygen absorptions.

the earth are subject to wide-ranging modifications and fluctuations. Radiation is selectively absorbed by atmospheric constituents, scattered by atmospheric molecules and aerosols, and attenuated and reflected by clouds (Fig. 6.1). Consequently, weather and pollution may significantly affect the radiation reaching a growing plant. In addition, the extent to which the above factors modify the global radiation varies regularly with the daily march of the sun, and with latitude and time of year, as a result of the variation in path length of the direct beam through the atmosphere, and the consequent changes in the relative proportions of direct and diffuse radiation incident upon a particular surface. Of greater ecological significance, within plant communities themselves, absorption, reflection and transmission of radiation by vegetation result in profound variations in the radiation microclimate.

For an individual plant to adjust physiologically to the changing aspects of its environment, it needs to be able to obtain reliable information from that environment; in other words, perception mechanisms are required which allow the plant to react, metabolically and developmentally, in a manner appropriate to the changed, or changing, conditions. In this article I wish to present a particular view of the ecological and physiological significance of the perception of light quality by higher plants. This view is based largely on my own work with a number of colleagues, and no attempt is made here to offer a comprehensive coverage of the previous literature, of current work going on in other laboratories, or of related topics, such as the perception of light quantity. The latter subject is covered in a further article (Smith 1980), and a comprehensive review of the perception of light quality is in preparation (Morgan D.C. and Smith H., Encyclopedia of Plant Physiology). In the present article, therefore, I shall restrict myself to an attempt (a) to demonstrate how systematically informative light quality is as a parameter of the radiation environment; (b) to present an outline of the theoretical basis for the perception of light quality; (c) to comment on the likely significance of the perception cf light quality in plant ecology; and (d) to speculate on the possible perception mechanisms.

LIGHT QUALITY IN THE NATURAL ENVIRONMENT

In temperate latitudes, the spectral photon distribution of the global radiation during the day (i.e. solar angle $> 10°$) is relatively constant (Fig. 6.2a); although clouds have large effects on the quantity of light reaching the earth's surface, their effects on light quality are small. When the solar angle is less than $c.$ $10°$ (i.e. at sunrise and sunset or at high latitudes) substantial changes occur, with a relative increase in the long-wavelength contribution from the direct beam, as a result of refraction by the atmosphere, and a relative enrichment of the blue, because of the larger contribution of scattered radiation to the global radiation (Fig. 6.2b). However, the most striking changes in light quality occur upon attenuation of the radiation by vegetation; the red and blue wavelength regions are strongly absorbed by the photosynthetic pigments whereas the far-red is largely transmitted through vegetation (Fig. 6.2c). A more remarkable situation occurs on the shaded side of a hedgerow or stand of vegetation, where the direct beam is filtered through the vegetation, giving a substantial drop in the red : far-red ratio, but where relative enrichment in the blue occurs as a result of the large amount of unfiltered skylight (Fig. 6.2d). Under water, the ratio of red : far-red radiation is increased because of selective attenuation of the far-red wavelengths (Fig. 6.2e). Finally, moonlight, which is principally reflected sunlight, may conceivably be of physiological and ecological importance (Fig. 6.2f).

Figure 6.2, therefore, illustrates how differences in light quality may convey important information to the plant. In principle, the possession of

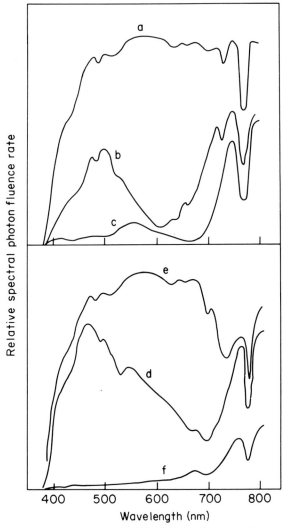

FIG. 6.2. Spectral photon fluence rate distributions of natural global radiation under various terrestrial conditions; (a) mid-day, clear skies; (b) sunset, clear skies; (c) under a vegetation canopy; (d) on the shaded side of a hedgerow; (e) as (a) but filtered through 30 cm of pure water; (f) moonlight ($\times 10$). (All spectra obtained with a Gamma Scientific Spectroradiometer.)

sensitive detection mechanisms capable of monitoring the relative proportions of blue, red and far-red radiation, would enable plants to obtain reliable information on daylength, position in the canopy, depth of immersion, etc., and this information could be of significant value in acclimation to the environmental conditions.

PERCEPTION OF LIGHT QUALITY

Theoretical aspects of the detection of light quality

It is a *sine qua non* of photobiology that the perception of a light stimulus requires the involvement of a specific photoreceptor. In the case of perception of light quality—which may be thought of as a primitive form of colour perception—a comparative element must be introduced. Obviously it would not be possible to distinguish changes in light quality and light quantity using a single photoreceptor operating with a single absorption band. Consequently, the perception of light quality must involve the biological comparison of photon fluence rates in at least two separate wavelength bands. Although there may be many complex ways in which such a comparative mechanism could be organized, the two most obvious possibilities are (a) an interaction between two separate and independent photoreceptors, and (b) the operation of a single photoreceptor with independent absorption bands at two rather widely-separated wavelengths. A crucial requisite in both cases is that absorption of light at the two wavelengths should result in *different* photochemical reactions which are somehow capable of interacting. As we shall see, there is strong evidence for the involvement of the second type of mechanism, although interactions between separate photoreceptors may yet be shown to be important.

Plants possess four functional photoreceptors, whose absorption spectra are shown in Fig. 6.3. Chlorophyll, present in various forms (principally chlorophyll *a* and chlorophyll *b* in higher plants) is concerned with energy capture and on *a priori* grounds, seems unlikely to be involved in the perception of environmental signals (but see below). Protochlorophyll, and its various forms, undergoes photochemical reduction to chlorophyll and, although it may be implicated in plastid development, it also seems a rather unlikely candidate for a light quality perceptor.

The remaining photoreceptors, i.e. phytochrome, and the blue-absorbing photoreceptor sometimes known as cryptochrome (Presti & Delbrück 1978), have been demonstrated to exert extensive control over a wide range of developmental processes and thus may be considered as prime contenders for perception of light quality. Indeed the absorption maxima (in the blue for cryptochrome, and in the red or far-red for phytochrome) are exactly as one might predict for perceptors of light quality, bearing in mind the known variations in the spectral distribution of natural radiation. Cryptochrome, however, having only one major absorption band in the visible, could not differentiate changes in overall irradiance, as would be brought about, for example, by cloud cover, from relative changes in the blue and other wavelength regions resulting from sunset or partial vegetational shading. Crypto-

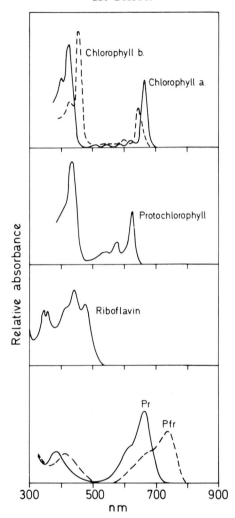

FIG. 6.3. Absorption spectra of the four principal plant photoreceptors. Riboflavin is included as being the most likely candidate for the blue-light absorbing photoreceptor (cryptochrome).

chrome could only act as a light quality perceptor, therefore, if it were able to interact directly, or possibly indirectly through secondary processes, with another photoreceptor having an absorption band at a higher visible wavelength. Phytochrome, on the other hand, exists in two isomeric forms with quite widely separated absorption bands, and thus could very effectively act to perceive changes in light quality in the red and far-red regions.

The remaining part of this chapter is concerned with summarizing the evidence for the view that the principal function of phytochrome is the detection of vegetational shade and the initiation of appropriate modulations of development so that acclimation to the shaded conditions may take place.

Phytochrome and perception of light quality

The property which allows phytochrome to act as a light quality perceptor is its photochromicity. By this is meant the capacity of phytochrome to exist in two photoconvertible, isomeric forms, as follows:

$$\text{Pr} \xrightleftharpoons[hv; \lambda_{max} 730 \text{ nm}]{hv; \lambda_{max} 660 \text{ nm}} \text{Pfr.}$$

Pr and Pfr are relatively stable and the photoconversions pass through different, highly unstable, intermediate states in the two directions. Since the absorption spectra of Pr and Pfr overlap below *c.* 730 nm, simultaneous photoconversion in both directions—sometimes known as 'cycling'—occurs with broad band visible irradiation. At 'natural' fluence rates, photo-equilibrium is rapidly achieved, and the proportions of Pr and Pfr present at equilibrium are determined by the spectral photon distribution of the radiation. Since the extinction coefficients of Pr and Pfr are highest in the red ($\lambda_{max} \approx 660$ nm) and far-red ($\lambda_{max} \approx 730$ nm) wavebands, respectively, the proportions of Pr and Pfr present at equilibrium may be seen to reflect principally the relative amounts of red and far-red wavelengths in the incident radiation. Thus, if phytochrome were to modulate metabolism and development as a function of the proportions of the two forms present, usually described as the 'photoequilibrium' and expressed as Pfr/Ptotal, or simply as ϕ, it could act as a direct perceptor of light quality in the 600–800 nm wavelength region. Fig. 6.2 shows that this is a very important waveband in ecological terms.

Phytochrome photoequilibrium in the natural environment

As yet, the only reliable method for measuring phytochrome photoequilibrium (i.e. Pfr/Ptotal or ϕ) is via *in vivo* dual wavelength spectrophotometry. This spectrophotometric assay is currently not possible in plant materials which contain more than a trace of chlorophyll, because of the overriding absorptance of chlorophyll in the red, and its fluorescence at higher wavelengths. Consequently, evidence on the actual relationship between natural light quality and phytochrome photoequilibrium in light-grown plants is necessarily indirect.

In order to construct such an indirect relationship, two practical possibilities exist. In the first, Pfr/Ptotal may be measured in etiolated tissue, or in samples of extracted phytochrome, exposed to the incident natural radiation. Alternatively, Pfr/Ptotal may be calculated from measured distributions of spectral photon fluence rate and the extinction coefficients and quantum efficiencies of photoconversion for the two forms of phytochrome. Neither of these approaches attempts to account for the selective attenuation of radiation which occurs within the perceiving organ, and at present this problem may only be approached from a theoretical standpoint (Holmes & Fukshansky 1979).

Using the first method outlined above, it is necessary to introduce a simple but arbitrary parameter to describe the distribution of the spectral photon fluence rate. Since the absorption peaks of phytochrome are at 660 and 730 nm, and since the major natural variations in light quality are found in the red and far-red, the following parameter, now known as ζ (zeta), was chosen:

$$\zeta = \frac{\text{photon fluence rate (655–665 nm)}}{\text{photon fluence rate (725–735 nm)}}$$

The relationship between ζ and ϕ in etiolated tissues is shown in Fig. 6.4 (Smith & Holmes 1977). This Figure was constructed from many determinations of ϕ made under natural and artificial light sources, and provides a valuable calibration curve for the conversion of ζ values to ϕ_e, where $\phi_e =$ Pfr/Ptotal in etiolated tissues.

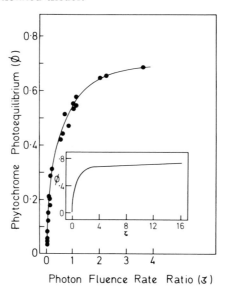

FIG. 6.4.   Relationship between ζ and ϕ (from Smith & Holmes 1977).

It must be stressed at this stage, that ϕ_e is not intended as an estimate of the Pfr/Ptotal actually present within *green* tissues. Nevertheless, the photo-equilibrium at any point within green tissues, being determined primarily by the distribution of the spectral photon fluence rate of the incident radiation and secondarily by the optical properties of the tissues, must be systematically related to ϕ_e. Consequently, ϕ_e should be regarded as a physiologically meaningful descriptor of the radiation quality, rather than an estimate of internal photoequilibrium (see Smith 1978 for further discussion).

An important aspect of Fig. 6.4 is that ϕ_e varies most steeply over the natural range of ζ. Except underwater, the ζ values of natural radiation in temperate regions are never higher than 1·2, which is the value found in mid-day sunlight. Lower values are found at sunset and sunrise and in the shade of vegetation. A small drop in ζ corresponds to a large change in ϕ_e, indicating that phytochrome has the capacity to be a very sensitive perceptor of changes in natural light quality.

The derivation of ϕ_e takes into account only the spectral quality in the red and far-red regions. Using the second method referred to above, a parameter—ϕ_c—may be calculated which takes into account the distribution of spectral photon fluence rate throughout the whole of the visible wavelength region. The equation for this calculation is:

$$\phi c_{400-800} = \cfrac{1}{1 + \cfrac{\sum_{\lambda=400}^{\lambda=800} E_\lambda \, \varepsilon_{fr\lambda} \, \Phi_{fr\lambda}}{\sum_{\lambda=400}^{\lambda=800} E_\lambda \, \varepsilon_{r\lambda} \, \Phi_{r\lambda}}}$$

where:

E_λ = photon fluence rate in each ten nm bandwidth segment from 400 nm–800 nm,

$\varepsilon_{fr\lambda}, \varepsilon_{r\lambda}$ = molar extinction coefficient of Pfr and Pr respectively at wavelength λ,

$\Phi_{fr\lambda}, \Phi_{r\lambda}$ = quantum yield at wavelength λ of the photoconversion of Pfr → Pr, and Pr → Pfr, respectively.

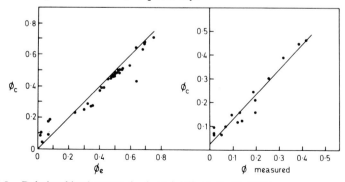

FIG. 6.5. Relationships between ϕ_c, ϕ_e and ϕ (from Smith 1978).

The parameter, ϕ_c, is the theoretical phytochrome photoequilibrium established by the incident radiation in etiolated test material, or in phytochrome extracts. The relationships between ϕ_c and ϕ (i.e. between calculated and observed phytochrome photoequilibria) and between ϕ_c and ϕ_e (derived from ζ) are given in Fig. 6.5, showing that, at least for ζ values above 0.1, ϕ_e is a good estimate of ϕ_c and of ϕ. Indeed, we have found empirically that ϕ_e is only a poor estimate of ϕ when there is a substantial contribution of blue light to the incident radiation.

LIGHT QUALITY AND GROWTH

Any serious attempt to relate light quality to plant growth and development must ensure that the experimental results are not confounded with effects of light quantity. Consequently, it is necessary to construct artificial light sources in which the total radiation fluence over the 400–700 nm photosynthetically active region (PAR) remains constant, but in which various amounts of added far-red may be supplied. The construction of such cabinets is a technically demanding exercise, since achieving sufficient modulation of

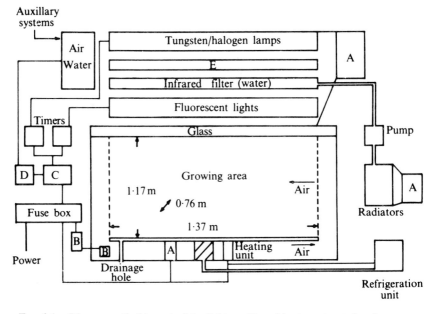

FIG. 6.6.   Diagrammatical layout of the light quality cabinet constructed at Sutton Bonington. A = blower units for cooling; B = temperature sensor and control unit; C = main isolating relay; D = safety devices control unit; E = 3·2 mm thick red 400, plus 3·2 mm thick green 600 'Perspex' (ICI Ltd., Welwyn Garden City, England) (from Heathcote, Bambridge & McLaren 1979).

the red : far-red ratio to provide ζ levels which simulate those found in natural plant communities, whilst maintaining PAR constant and relatively high, requires the input of very large amounts of added far-red radiation. Over a period of several years, one such cabinet, based on the requirements of the author and his colleagues, was designed and constructed by the technical staff (Heathcote, Bambridge & McLaren 1979) at the University of Nottingham, Sutton Bonington (Fig. 6.6). The cabinet provides four separate chambers with uniform PAR from fluorescent tubes supplemented with quartz-iodide lamps, filtered through a Perspex far-red filter, and gives ζ values from 3·0 (without added far-red), down to approximately 0·2, thus covering most of the range of ζ found in nature.

When aggressive ruderal herbs are grown in the cabinets, extensive modulation of growth and development may be observed. Although a number of different developmental responses have been obtained (Morgan & Smith 1978; McLaren & Smith 1978; Whitelam, Johnson & Smith 1979), the most striking is an increase in extension growth with decrease in ζ (Holmes & Smith 1975, 1977; Morgan & Smith 1976, 1978). The relationship between ζ and growth rate (expressed as logarithmic stem extension rate) is a curve, as can be seen for four ruderal herbs in Fig. 6.7.

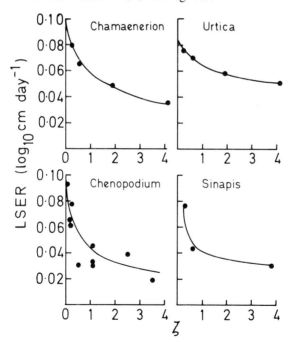

FIG. 6.7. Relationship between ζ and stem growth in four ruderal herbs (*Chamaenerion angustifolium* L., *Urtica dioica* L.; *Chenopodium album* L., *Sinapis alba* L.). LSER = logarithmic stem extension rate.

Although Fig. 6.7 shows that growth rate is related to ζ, the shape of the curve provides little information on the mechanism. Transformation of ζ to ϕ_e, using the calibration curve in Fig. 6.4, on the other hand, reveals a perfect linear relationship as shown in Fig. 6.8 for *Chenopodium album* L. Similar linear relationships are also generated if ϕ_c is calculated from distribution of spectral photon fluence rate, or indeed if etiolated test materials are used to measure ϕ directly.

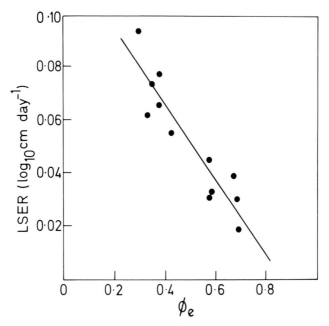

FIG. 6.8. Relationship between stem growth rate and ϕ_e for *Chenopodium album* L. (from Morgan & Smith 1976).

The linear relationship of Fig. 6.8 is strong, if circumstantial, evidence that phytochrome operates through the establishment of photoequilibrium, to measure the spectral proportions of red and far-red wavelengths and to modulate growth and development appropriately.

ECOLOGICAL SIGNIFICANCE OF LIGHT QUALITY

A range of different weed species (Table 6.1) have been grown in the light quality cabinets (Morgan & Smith 1979) and the results show that the response of extension growth rate (and a number of other responses) is systematically related to the normal ecological niche of the particular species (Fig. 6.9). Typical shade-avoiders, such as *Chenopodium album* L., *Sinapis*

TABLE 6.1. The habitat and collection site for each of the species used in the experiments. Nomenclature and typical habitat are taken from Clapham *et al.* (1962). (From Morgan & Smith 1979.)

Species	Habitat	Collection site
Open habitat		
Chamaenerion angustifolium (L.) Scop.	Rocky places, scree slopes, wood margins, wood clearings, disturbed ground	Arable headland, Sutton Bonington, Notts. (SK 514 261)
Chenopodium album L.	Waste places, cultivated land	Arable land,Sutton Bonington Notts. (SK 506 262)
Medicago arabica (L.) Huds.	Grassy places, waste ground, light soils near sea	Arable land, Sutton Bonington, Notts. (SK 506 262)
Sinapis alba L.	Arable and waste land	Commercial stock (arable land)
Senecio vulgaris L.	Cultivated land, waste places	Arable land, Sutton Bonington, Notts. (SK 506 262)
Tripleurospermum maritimum ssp *inodorum* (L.) *Hylex Vaarama*	Arable and waste land	Arable land, Sutton Bonington, Notts. (SK 506 262)
Intermediate canopy habitat		
Urtica dioica L.	Hedge-banks, woods, grassy places, fens	Hedgerow, Sutton Bonington, Notts. (SK 507 258)
Closed canopy habitat		
Circaea lutetiana L.	Woods and shady places	Closed oak woodland, Swithland Wood, Leics. (SK 537 117)
Geum urbanum L.	Woods, scrub, hedge-banks, shady places	Closed oak woodland, Swithland Wood, Leics. (SK 537 117)
Mercurialis perennis L.	Woods, and shady mountain rocks	Closed ash woodlands, nr. Bowers Hall, Derbys. (SK 233 652)
Oxalis acetosella L.	Woods, hedge-banks, shady rocks	Closed oak woodland, Swithland Wood, Leics. (SK 537 129)
Silene dioica (L.) Clairv.	Woods, rocky-slopes, hedgerows, cliff ledges	Closed mixed woodland, Domleo Spinney, Notts. (SK 514 263)
Teucrium scorodonia L.	Woods, grassland, heaths, dunes	(i) for survey: Closed oak woodland, Swithland Wood, Leics. (SK 539 122) (ii) Ecotypes: open: short calcareous grassland, Lathkilldale, Derbyshire (SK 176 654) shade: closed oak woodland, Swithland Wood, Leics. (SK 543 118)

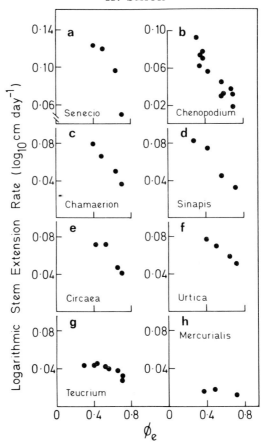

Fig. 6.9. The relationship between stem growth rate and ϕ_e for a range of weed species characteristic of open and shaded habitats (from Morgan & Smith 1979).

alba L. and *Chamaenerion angustifolium* (L.) Scop., show very marked increases in extension growth rate with decreasing ϕ_e. Normal woodland plants, which are shade-tolerators (e.g. *Mercurialis perennis* L., *Circaea lutetiana* L. and *Teucrium scorodonia* L.), showed much less increase in extension rate with decreasing ϕ_e. These responses would, in each case, be of particular value for growth and survival in the particular ecological niches of the different species.

PHYSIOLOGICAL IMPLICATIONS

The data summarized here are consistent with the hypothesis (Smith 1973) that phytochrome acts as a perceptor of light quality in the red and far-red spectral regions. Essentially, we have demonstrated a far-red mediated increase in extension rate which happens to exhibit a linear relationship to a highly indirect parameter of photoequilibrium. Indeed, such relationships would be expected of any far-red absorbing photoreceptor which showed a log-linear relationship between fluence rate and response, as this response does (Smith 1978). Thus, the alternative hypothesis of an as yet uncharacterized non-phytochrome, far-red absorbing photoreceptor cannot be ruled out.

Certain other evidence seems to indicate that such an alternative photoreceptor might exist but definitive identification has not yet been achieved. For example, Schneider & Stimson (1972) proposed that ATP production via cyclic-photophosphorylation in photosystem I of photosynthesis was involved in the effects of prolonged irradiation on anthocyanin synthesis in etiolated turnip seedlings. This view was based on results of the application of photosynthetic inhibitors to etiolated seedlings given prolonged far-red irradiation and could be misleading because of unknown side effects of the inhibitors. On the other hand, cyclic photophosphorylation has been reported to have an action maximum at 715 nm (Avron & Hayyim 1969) and Fisher & Shropshire (1979) have recently observed a non-reversible effect of far-red light, with a single peak of action at 711 nm, on fern spore germination.

In the case of the *S. alba* seedlings studied here, however, the involvement of ATP production via photosynthesis seems unlikely. Firstly, the plants were growing in white light at an irradiance well above compensation point and should not be in any way energy-restricted in growth. Secondly, the response to far-red light (although I have stressed here the extension growth only) is an overall change in developmental pattern, not a uniform increase in growth rate; e.g. although stem extension was markedly increased, leaf growth was, if anything, reduced. Thirdly, recent experiments in which the growth of the seedlings was recorded continuously with a position-sensitive transducer, have shown that added red light has no sustained effect on growth rate, whilst added far-red raises extension rate by a factor of up to 4-fold within 10–15 minutes (Morgan, O'Brien & Smith, in press); if photosynthesis were involved, an effect of red light would be expected. Even so, the extreme view that photosystem I absorption is responsible for the growth effects has not yet been completely eliminated; the necessary experiments are possible, but are rather complex and will take time to set up.

Consequently, the most likely current explanation of the data summarized here is that phytochrome acts to perceive changes in light quality in the red/far-red spectral regions and to modulate development in an appropriate manner. I suggest that this is the major function of phytochrome in light-grown plants and that further intensive investigation of the phenomena outlined here should be of great importance in elucidating the mechanism of action of phytochrome.

It is only in the last year or two that interest in the role of phytochrome in light-grown plants has become generally fashionable and the currently accepted views of phytochrome action are based on data derived solely from dark-grown plants. It must be admitted that it is at present difficult to reconcile the data summarized here, and the hypothesis based on those data, with the current phytochrome dogmas. The main problem is that in the hypothesis advanced here, phytochrome is proposed to perceive light quality through the establishment of photoequilibria, i.e. directly through its property of photochromicity. The generally accepted view from dark-grown plants is that Pfr/Ptotal is not in itself a crucial factor in phytochrome action, but that that action is dependent directly on the concentration of Pfr (or [Pfr]) (Steinitz *et al.* 1979). As long as Ptotal is constant with time and wavelength, then [Pfr] is equivalent to Pfr/Ptotal, but in dark-grown plants, Ptotal is subject to time-dependent, and wavelength-dependent fluctuations, as a result of the highly unstable nature of Pfr (Pr being much more stable) thus:

$$\xrightarrow{\text{synthesis}} \text{Pr} \underset{h\upsilon}{\overset{h\upsilon}{\rightleftharpoons}} \text{Pfr} \xrightarrow{\text{degradation}}$$

If similar differences in stability between Pr and Pfr exist in light-grown plants, and this is by no means certain as yet, then the relationship in Fig. 6.8 cannot be seen as a relationship between [Pfr] and growth.

Several different models exist for phytochrome action, but the only one which appears to account for the direct linear relationship between Pfr/Ptotal and extension growth shown in Fig. 6.8, is that proposed by the present author in 1970. This model gave a central role to the photoconversion processes at specific photoequilibria. The importance of a model, however, is not that it should be proved correct or incorrect, but that it should direct ideas and experiments along logical lines. What is needed at the present time is much more information on the biophysical and biochemical properties of phytochrome in light-grown plants. I suggest that we now know that the major function of phytochrome is to enable plants to perceive light quality changes, in particular those due to mutual shading; we are still, however, a very long way from knowing *how* phytochrome performs this function.

ACKNOWLEDGEMENTS

The author is grateful for the collaboration, over several years, with Dr M.G. Holmes, Dr D.C. Morgan, Dr R. Tasker, Dr C.B. Johnson, Dr J.C. McLaren, Mr T. O'Brien, Mr P. Foulkes and Mr R. Child, all of whom have contributed significantly to this work.

REFERENCES

Avron M. & Ben-Hayyim G. (1969) Interaction between two photochemical systems in photoreactions of isolated chloroplasts. *Progress in Photosynthesis Research*, **3**, 1185–1196.

Clapham A.R., Tutin T.G. & Warburg E.F. (1962) *Flora of the British Isles.* Cambridge University Press.

Fisher R.W. & Shropshire W.Jr. (1979) Reversal by light of ethylene-induced inhibition of spore germination in the sensitive fern *Onoclea sensibilis.* An action spectrum. *Plant Physiology*, **63**, 984–988.

Heathcote L., Bambridge K.R. & McLaren J.S. (1979) Specially constructed growth cabinets for simulation of the spectral photon distributions found under natural vegetation canopies. *Journal of Experimental Botany*, **30**, 347–353.

Henderson S.T. (1970) *Daylight and its Spectrum.* Adam Hilger Ltd., Bristol.

Holmes M.G. & Fukshansky L. (1979) Phytochrome photoequilibrium in green leaves under polychromatic irradiation: a theoretical approach. *Plant, Cell & Environment*, **2**, 59–65.

Holmes M.G. & Smith H. (1975) The function of phytochrome in plants growing in the natural environment. *Nature*, **254**, 512–514.

Holmes M.G. & Smith H. (1977) The function of phytochrome in the natural environment· IV Light quality and plant development. *Photochemistry and Photobiology*, **25,** 551–557.

McLaren J.S. & Smith H. (1978) The function of phytochrome in the natural environment. VI Phytochrome control of the growth and development of *Rumex obtusifolius* under simulated canopy light environments. *Plant, Cell and Environment*, **1**, 61–67.

Mohr H. & Schäfer E. (1979) Uniform terminology for radiation: a critical comment. *Photochemistry and Photobiology*, **29**, 1061–1062.

Morgan D.C. & Smith H. (1976) Linear relationship between phytochrome photoequilibrium and growth in plants under simulated natural radiation. *Nature*, **262**, 210–212.

Morgan D.C. & Smith H. (1978) The function of phytochrome in the natural environment· VII The relationship between phytochrome photoequilibrium and development in light-grown *Chenopodium album. Planta*, **142**, 187–194.

Morgan D.C. & Smith H. (1979) The function of phytochrome in the natural environment. VIII A systematic relationship between phytochrome-controlled development and species habitat, for plants grown in simulated natural radiation. *Planta*, **145**, 253–258.

Presti D. & Delbrück M. (1978) Photoreceptors for biosynthesis, energy storage and vision. *Plant, Cell and Environment*, **1**, 81–100.

Schneider M.J. & Stimson W. (1972) Phytochrome and Photosystem I interaction in a high-energy photoresponse. *Proceedings of the National Academy of Science, USA*, **69**, 2150–2154.

Smith H. (1970) Phytochrome and photomorphogenesis in plants. *Nature*, **227**, 665–668.

Smith H. (1973) Light quality and germination: ecological implications. *Seed Ecology* (Ed. W. Heydecker), pp. 219–231. Butterworths, London.

Smith H. (1978) Light quality and phytochrome action in the natural environment. *Plant Growth and Light Perception* (Eds. B. Deutch, B.I. Deutch, & A.O. Gyldenholm). Institute of Ecology and Genetics, University of Aarhus, Denmark.

Smith H. (1980) Adaptation to shade. *Physiological Factors Limiting Plant Productivity* (Ed. C.B. Johnson). Butterworths, London.

Smith H. & Holmes M.G. (1977) The function of phytochrome in the natural environment. III Measurement and calculation of phytochrome photoequilibrium. *Photochemistry and Photobiology*, **25**, 547–550.

Steinitz B., Schäfer E., Drumm H. & Mohr H. (1979) Correlation between far-red absorbing phytochrome and response in phytochrome-mediated anthocyanin synthesis. *Plant, Cell & Environment*, **2**, 159–163.

Whitelam G.C., Johnson C.B. & Smith H. The control by phytochrome of nitrate reductase in the curd of light-grown cauliflower. *Photochemistry and Photobiology* (in press).

7. THE EXCHANGE OF CARBON DIOXIDE AND AIR POLLUTANTS BETWEEN VEGETATION AND THE ATMOSPHERE

M.H. UNSWORTH

University of Nottingham School of Agriculture,
Sutton Bonington, Loughborough, Leics. LE12 5RD

SUMMARY

Methods for studying the gas exchange between vegetation and the atmosphere are reviewed. Techniques include the use of cuvettes and chambers for enclosing single leaves, whole plants or communities, and micrometeorological methods whereby fluxes are deduced from measurements in the atmosphere. Resistance models for describing gas exchange are discussed, and resistances to pollutant deposition on single leaves and plants are estimated. The gas exchange of canopies of vegetation and the soil below may also be described by resistance analogues, and advantages and limitations of this approach are discussed. Mathematical models of canopy exchange are often limited by lack of knowledge of fluxes to and from the soil. New developments are necessary before much further progress can be made in studying gas exchange within canopies by micrometeorological methods; in the meantime, cuvettes operated in the field are likely to yield most information.

INTRODUCTION

Laboratory studies of CO_2 exchange have been a well-established tool in plant physiology for many years; the development of sensitive infra-red gas analysers for CO_2 in the 1950s allowed field measurements of gas exchange to be made much more easily than before, both by micrometeorological methods (reviewed by Denmead & McIlroy 1971) and by using enclosures or cuvettes (Jarvis & Čatský 1971). More recently there has been interest in applying techniques developed for studying CO_2 exchange to measure fluxes of various pollutant gases to vegetation, with two objectives: to evaluate the role of vegetation as a sink for pollutants, and to relate physiological and biochemical responses to the absorbed pollutant. Progress in the former has been rapid, but in general not particularly thorough. The second objective has received little attention from physiologists, who still commonly relate plant response to the pollutant exposure defined as the time integral of pollutant concentration.

111

In this paper I will review some of the recent trends and developments in the measurement and analysis of the gas exchange of vegetation, in particular illustrating how ideas first developed for studying CO_2 may be applied to study pollutant uptake.

METHODS FOR MEASURING GAS EXCHANGE

Single leaves

Methods for measuring CO_2 exchange in laboratory and field have been thoroughly reviewed by Šesták, Čatský & Jarvis (1971). For studying the net exchange of CO_2 of single leaves, cuvettes with differential analysis of CO_2 on entry and exit are often used. Recently, increasing numbers of cuvettes of this type have been built for field use (Mooney et al. 1971; Biscoe et al. 1975; Marshall & Biscoe 1977), and the development of new and more robust CO_2 IRGAs (e.g. 'Binos' gas analyser) will undoubtedly lead to further progress in field measurements. Cuvettes have also been built for measuring pollutant uptake of ozone (Rich, Waggoner & Tomlinson 1970; King & Smith, personal communication) and SO_2 (Garland & Branson 1977; Black & Unsworth 1979c). Because there are no differential gas analysers for pollutants, such measurements of concentration differences between entry and exit are less precise than for CO_2 flux measurements, but, as will be seen later, they yield important information about sink strengths for pollutants and physiological control of uptake.

Turner, Rich & Waggoner (1973) deduced fluxes of ozone to soils and other materials in chambers by measuring the rate of decay of ozone when the chambers were sealed. With most pollutants, allowances must be made for the uptake of pollutant on the chamber walls; for SO_2 the uptake by surfaces depends on relative humidity (Spedding 1969).

Measurements of CO_2 fixation using the radioactive isotope $^{14}CO_2$ are convenient in the field, although when small areas of leaves are exposed many replicates are necessary because of the variability of uptake. Incoll (1977) recently reviewed the technique. There is still debate over whether ^{14}C uptake is a true measure of gross photosynthesis or whether there are some respiratory losses during the period of exposure. The answer may well depend on the design and operation of the exposure system.

Garland & Branson (1977) modified a $^{14}CO_2$ assimilation chamber designed for use on shoots of *Pinus sylvestris* (Neilson 1977) to allow exposure of shoots to $^{35}SO_2$. Raybould, Unsworth & Gregory (1977) made use of the same apparatus to study $^{35}SO_2$ uptake by wheat leaves. In such measurements it is necessary to use gas concentrations typical of normal ambient conditions to avoid stomatal responses to the pollutant (Black & Unsworth

1979a). Garsed & Read (1977) exposed plants in laboratory experiments to $^{35}SO_2$, but their gas concentrations were more than 50 times those observed in urban areas. The same criticism can be levelled at the $^{34}SO_2$ measurements of Belot *et al.* (1974).

Canopies

In a canopy of vegetation there may be several sources and sinks for CO_2 exchange. For pollutant gases the situation is usually simpler because the canopy and soil are only sinks and not sources of the gas. Fig. 7.1 shows the directions of CO_2 fluxes in a barley canopy during the day and night. Because the vertical concentration profile of CO_2 had a minimum near the top of the canopy by day, Biscoe *et al.* argued that the respiratory fluxes $\mathbf{R}_c$ and $\mathbf{R}_r$ from the canopy and roots, respectively, represented recycled CO_2, and consequently the net rate of uptake of CO_2 in the light was $\mathbf{P}_a + \mathbf{R}_s$. At night, when CO_2 concentration decreased monotonically from the soil upwards,

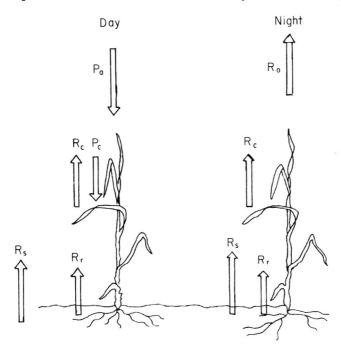

FIG. 7.1. Directions of CO_2 fluxes in a barley crop during the day and night (after Biscoe, Scott & Monteith 1975). By day, $\mathbf{P}_a$, the net CO_2 flux measured above the canopy is the difference between the flux, $\mathbf{P}_c$, into leaves and the canopy respiration, $\mathbf{R}_c$. Respiratory fluxes from roots, $\mathbf{R}_r$, and soil, $\mathbf{R}_s$, are assimilated within the canopy. At night, the sum of canopy, root and soil respiration is the flux, $\mathbf{R}_a$, measured above the canopy.

the net rate of loss of CO_2 was $R_a - R_s$. The measurement of R_s, and in particular the separation of soil respiration from root respiration is very uncertain in the field; Biscoe *et al.* concluded that for most of the season R_s was less than 10% of net photosynthesis when weekly totals were considered. The relative magnitude of R_s may be greater however in soils rich in organic matter, in natural ecosystems where decaying vegetation is present, or at times when net photosynthesis is limited by low light or senescence.

There are three methods by which the exchange of gases between vegetation and the atmosphere may be estimated: (i) enclosure of an area of the canopy in a transparent chamber; (ii) measurement of the gas exchange of individual organs combined with a model of the crop environment and (iii) micrometeorological methods.

The enclosure method is attractive for short dense canopies, e.g. grassland (Ripley & Redmann 1976; Louwerse & Eikhoudt 1975) or for isolated bushes or small trees of complex shape (Eckardt 1968). With such enclosures there is substantial modification of the aerial environment, and, in particular, control of temperature and humidity is essential. Further, the soil respiratory flux may be influenced by pressure gradients between the enclosure and outside air. More than forty years ago, Thomas and his co-workers established enclosures over alfalfa growing in the field, and, by fumigating with SO_2 and using chemical methods of gas analysis for CO_2 and SO_2, they investigated the dependence of CO_2 exchange on SO_2 uptake (Thomas & Hill 1937); Katz (National Research Council of Canada 1939) made similar studies in Canada at about the same time. Both of these pioneering investigations used much higher SO_2 concentrations than are commonly found today, and it is unfortunate that their experiments have not been repeated for lower gas concentrations. Hill (1971) described a chamber in which he could assess pollutant uptake of plants in the laboratory by measuring the rate at which the gas needed to be added to maintain a constant concentration; his work with a number of gases indicated that sites of uptake and sink strengths differed between gases.

The integration of measurements of gas exchange by individual organs to estimate the exchange of the whole canopy is a useful technique for canopies where direct measurement of fluxes is difficult, e.g. forests. Biscoe and co-workers tested the validity of such integration for the carbon assimilation of a barley crop. Photosynthesis light-response curves for organs from various levels within the crop were combined with measurements of the leaf area distribution and of the radiation intercepted at each level to estimate canopy photosynthesis. These estimates agreed well with micrometeorological measurements of flux on an hourly basis and, when suitably summed, with weekly estimates (Biscoe *et al.* 1975). In a later paper (Biscoe *et al.* 1977), measurements with the $^{14}CO_2$ technique compared satisfactorily with those from a leaf chamber.

Jarvis & Čatský (1971) emphasized the importance of ensuring that cuvette environments were close to the environment of the rest of the plant. Recent work by Rawson & Begg (1977) illustrates the false conclusions which may arise when this warning is not heeded: they found substantial differences in the responses of photosynthesis to vapour pressure deficit (v.p.d.) between whole plants exposed at various values of v.p.d. and single leaves exposed in a similar manner whilst the rest of the plant was at a different v.p.d.

Micrometeorological methods for measuring fluxes above extensive canopies are well established and have been reviewed frequently (Denmead & McIlroy 1971; Thom 1975). Provided that the rather rigorous requirements for crop uniformity and fetch are met, fluxes between the atmosphere and the canopy can be measured without interference with the vegetation. Although such measurements do not reveal the gas exchange of individual organs in a canopy in the way that cuvette studies do, they have the potential advantages that measurements can be made continuously and automatically for long periods and that the methods give good spatial averages. However, in practice micrometeorological projects have too often been a field exercise in engineering and physics rather than a tool for physiologists, and there have been few research programmes in which the gas exchange of vegetation has been followed for long enough and in sufficient depth to establish relationships between, for example, CO_2 exchange and dry matter production, or to link flux measurements with environmental variables and physiological responses. Some of the exceptions have been the studies of the carbon dioxide exchange of cereals at Sutton Bonington (Biscoe *et al.* 1975), of forests at Fetteresso (Jarvis, James & Landsberg 1976) and of natural grassland in Saskatchewan (Ripley & Redmann 1976).

Measurements of fluxes of pollutants by micrometeorological methods have become possible only relatively recently as the sensitivity of methods of gas analysis has improved, and in general there has been little investigation of the dependence of the flux on properties of the vegetation. Exceptions to this statement are the work of Fowler on fluxes of SO_2 to wheat (Fowler & Unsworth 1979; Fowler, this volume), Garland & Branson (1977) on SO_2 to pine forests, Wesely *et al.* (1978) on ozone to maize, and Leuning *et al.* (1979a, b) on ozone to maize and tobacco. In each of these studies there were attempts at identifying the various sinks for pollutants in the canopy.

Several recent developments in micrometeorological methods have important implications for studies of gas exchange of vegetation. To put them into context a brief description of some principles of atmospheric transfer is necessary. Transfer of gases between the atmosphere and vegetation takes place in turbulent eddies generated in two ways: (i) by friction at surfaces, and (ii) by buoyancy resulting from gradients in air density near the ground, most commonly generated when there is strong surface heating ('instability'). There are two standard micrometeorological methods of measuring vertical fluxes. In the more direct method, eddy correlation, instruments with fast response times ($<0\cdot1$ s) are used to measure the turbulent fluctuations in vertical wind speed, w', and gas concentration, c', associated with individual eddies. The mass flux density, $\mathbf{F}$, is given by the time-average of the product of the two measurements, i.e.

$$\mathbf{F} = \overline{w'c'} \tag{7.1}$$

In the second method, flux-gradient analysis, it is assumed, by analogy with molecular diffusion, that the flux density is proportional to the vertical gradient of mean gas concentration, $\partial C/\partial z$ (averaged over 10–30 minutes). The constant of proportionality is the eddy diffusivity for mass, i.e.

$$\mathbf{F} = -K\,\partial C/\partial z \tag{7.2}$$

where the minus sign indicates that the flux is towards the surface when the concentration increases with increasing height. The determination of K is a central problem in micrometeorology. Generally, empirical relationships are assumed between eddy diffusivities for heat, mass and momentum, so that flux-gradient methods require measurements of vertical gradients of gas concentration and estimation of the appropriate eddy diffusivity, which is usually found by one of two methods. The energy balance method assumes that K is the same for heat and mass and partitions the available energy at the surface into sensible and latent heat fluxes to find K; the assumptions appear to be valid over a wide range of stability, but the method is inaccurate when available energy is small, e.g. at night. The aerodynamic method derives K for momentum from wind profiles and assumes empirical relationships derived over short grass or bare soil between K for momentum and mass (Businger 1975). It is the validity of these empirical relationships over vegetated surfaces that has recently been questioned.

For short and homogeneous agricultural crops in Britain, mass fluxes calculated by the aerodynamic method generally agree well with fluxes determined otherwise, e.g. Biscoe *et al.* (1975) over barley, Penman & Long (1960) over wheat. In areas where there is stronger surface heating and consequently greater atmospheric instability, the aerodynamic method is less accurate for estimating fluxes to canopies; over natural grassland in Canada, Saugier & Ripley (1978) found that the aerodynamic method generally had errors of up to 20%, but errors were larger in extremes of stability; in Australia, when the atmosphere over a wheat crop was strongly unstable, Denmead (1976) showed that the method underestimated fluxes by about 50%. The most serious discrepancies arise over tall, rough vegetation. Over forests, it is becoming increasingly clear that the aerodynamic method severely underestimates fluxes (by a factor of 2 or 3) in neutral and unstable conditions when the equations established over smooth surfaces are used (Thom *et al.* 1975; Jarvis, James & Landsberg 1976). Empirical corrections for forests have been derived, and there appears to be some consistency between sites (Thom *et al.* 1975; James 1977; T. Sinclair, unpublished). It seems generally agreed that the problem arises when measurements are taken close to canopies where the distance between plants is quite large relative to their height.

Until a theoretical explanation of these discrepancies is developed, the use of the aerodynamic method seems likely to introduce uncertainties which may be acceptable over short crops in a restricted range of stability but not over forests. However, in using the aerodynamic method to measure pollutant fluxes, one of the main sources of uncertainty is in determining gradients of very low concentrations of gases that may be unevenly distributed in space and time; in such circumstances, uncertainties introduced in the aerodynamic method are probably no larger than inherent uncertainties in the measurements. The aerodynamic method is the only viable flux-gradient approach at night, when pollutant fluxes can be significant.

The second important development is of new sensors for eddy correlation. Because instrumentation with sufficiently rapid response time for eddy correlation is usually delicate and expensive, this technique has seldom been used for studying exchanges over crops for longer than a few hours. A further problem is that gas analysers with fast responses have rarely been developed. Several years ago Desjardins (1974) modified a CO_2 analyser for fast response and used it in conjunction with a propeller anemometer in a modified eddy correlation approach. Inoue *et al.* (1969) measured momentum flux using eddy correlation and by simultaneously measuring gradients of mean wind speed they deduced K for momentum. Then, assuming relationships between K for momentum and K for mass they derived K for mass and used it to estimate CO_2 fluxes from gradients of CO_2 (equation 7.2). Recently two types of open-path infra-red CO_2 analysers have been incorporated in eddy correlation systems (Leuning 1976; Jones, Ward & Zwick 1978) and fast-response gas analysers have been developed for eddy correlation measurements of ozone (Wesely *et al.* 1979) and SO_2 (Galbally, Garland & Wilson 1979). Simpler anemometers for measuring wind fluctuations in vertical wind speed have also been developed (Campbell & Unsworth 1979), using sonic transducers and integrated electronic circuits. There is little doubt that in the next ten years the availability of reliable sensors coupled to cheap computers for on-line analysis will make the eddy correlation technique a useful tool for environmental physiologists.

The final developments concern transfer within canopies. Ten years ago the concept of treating fluxes within crop canopies as one-dimensional and evaluating them by flux-gradient methods was enthusiastically pursued (Lemon & Wright 1969). However when Legg & Monteith (1975) reviewed the subject they pointed out substantial objections to this approach, the main one being the variability in, for example, gas concentration or temperature in a canopy on a scale comparable with the size of the turbulent eddies, implying that values of K are not a unique function of the rate of turbulent mixing but differ for each entity. In the long run, the future of micro-meteorology as a tool for investigating the distribution of sources and sinks within canopies may depend on finding new theories for understanding transfer in canopies, and new techniques (possibly eddy correlation) (den Hartog & Shaw 1975) for measurements. The situation is not entirely gloomy; in many cases the major control of gas exchange is located at vegetated surfaces, and this means that measurements with cuvettes and porometers can be used to develop models of gas and water vapour exchange in crops, as will be discussed later.

ANALYSIS OF GAS EXCHANGE

Resistance analogues for describing mass exchange between vegetation and the atmosphere are well established now as a convenient approach of formal-izing the way in which fluxes are driven by potential differences (e.g. gas con-centrations) and are controlled by resistances. Jarvis (1971) comprehensively reviewed the subject. Analogues are especially useful for distinguishing between external resistances which describe characteristics of transfer through boundary layers (Monteith, this volume), and internal resistances which describe gaseous diffusion through stomata and cuticle. Less rigorously, resistances are sometimes attributed to mechanisms involving diffusion in the liquid phase and/or chemical reactions (see Jarvis 1971).

Single leaves

Figure 7.2 shows two resistance analogues; the first, Fig. 7.2a, describes the exchange of CO_2 between one surface of a single leaf and the atmosphere, and the second, Fig. 7.2b, is a general model of pollutant uptake by a leaf. Unsworth, Biscoe & Black (1976) discussed such models in detail and re-viewed some methods of determining the resistances. For carbon dioxide, the concentration difference $(\phi - \phi_0)$ between the atmosphere and the leaf surface leads to a flux $\mathbf{F} = (\phi - \phi_0)/r_b$ to the leaf, where r_b is the resistance of the leaf boundary layer. From the leaf surface there are two parallel diffusion pathways in the gas phase, through the stomata (resistance r_s) or through the cuticle (r_c). Little is known about the magnitude of r_c, although Jarvis (1971) reviewed evidence that cuticular uptake of CO_2 may be signi-ficant in some species. Stomatal resistances for CO_2 transfer are often found from measurements of water vapour diffusion, applying a factor for differences

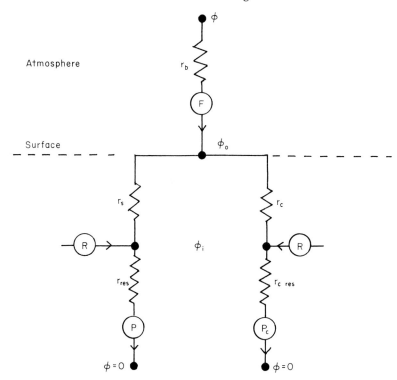

Fᴵɢ. 7.2a.   A resistance analogue describing exchanges of CO_2 between one surface of a leaf and the atmosphere. In this and other resistance analogues in this paper the following symbols apply, ○ flux; ● concentration; ⌇ resistance. A flux of CO_2, F, is transferred through the boundary layer resistance, r_b, between concentrations ϕ in the atmosphere and ϕ_0 at the leaf surface. Parallel paths account for gaseous diffusion through stomata, resistance r_s, and cuticle, resistance r_c, to the mesophyll where CO_2 concentration is ϕ_i. Residual resistances r_{res} and $r_{c\ res}$ represent all further restrictions to CO_2 fixation, i.e. liquid transfer in cell walls and cells, excitation and carboxylation.

in diffusivity. In principle r_c can be found in the same way, provided that it is regarded strictly as a resistance to gas-phase diffusion.

Within cell walls and cells, transport is in liquid-phase, and resistance relationships as defined here are inappropriate forms of description. For practical purposes, to quantify the control of stomata and boundary layer on fluxes, it is useful to define an internal resistance to transfer as a residual. Jarvis (1971) pointed out the hazards of attempting to attribute to such residual resistances any forms of response to the environment; it is safer to interpret residual resistances as mathematical expedients rather than physiological entities. In principle there are two residual resistances to CO_2 transfer (Fig. 7.2a), corresponding to paths from the substomatal cavity and through

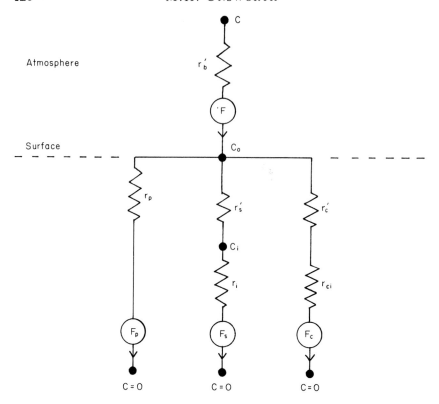

Fig. 7.2b. A resistance analogue for air pollutant uptake by one side of a leaf, showing three possible paths: stomatal diffusion (resistance r_s') and reaction in substomatal cavities and mesophyll (r_i); cuticular diffusion (r_c') and reaction within the epidermis or mesophyll (r_{ci}); and surface reaction (r_p). The total flux, $\mathbf{F}$, through the boundary layer resistance, r_b', establishes a concentration difference, $C - C_0$, between the atmosphere and the surface. The sum of the fluxes through each path, $\mathbf{F}_p + \mathbf{F}_s + \mathbf{F}_c$, is $\mathbf{F}$.

the epidermis. In practice, cuticular transfer of CO_2 has usually been ignored in resistance analysis, and so at light saturation

$$r_{res} = r_{CO_2} - r_s - r_b$$

(Gaastra 1959; Jarvis 1971) where r_{CO_2} is defined from the gross photosynthesis $\mathbf{P}_g$ as

$$r_{CO_2} = \phi/\mathbf{P}_g$$

Examples of resistance analogues applied to CO_2 transfer from single leaves were discussed fully by Jarvis (1971).

Fig. 7.2b shows a resistance analogue for uptake of an air pollutant by a leaf. The boundary layer and stomatal resistances r_b' and r_s' are related to those for water vapour and CO_2 by the diffusivities of the gases, as discussed by Unsworth, Biscoe & Black (1976). Within the substomatal cavity there will be chemical reactions with cell walls and with fluid in the cell walls and there may be gas or liquid phase diffusion to eventual sinks. Liss & Slater (1974) discussed the resistances imposed by air-pollutant liquid interfaces. In Fig. 7.2b a bulk internal resistance, r_i, to pollutant uptake is shown between the potential, C_i, in the substomatal cavity and the sink ($C = 0$). There are two further paths by which air pollutants may be taken up by leaves. The first involves diffusion through the cuticle (r_c') and internal reaction (r_{ci}). The second allows for the ability of many reactive pollutants to be taken up on natural materials presumably by absorption and adsorption on surface layers or by similar mechanisms on dust, micro-organisms etc lying on the surface. Spedding (1969) investigated SO_2 uptake by natural and man-made materials and showed this type of surface uptake; Bennett, Hill & Gates (1973) discuss similar mechanisms for ozone. In Fig. 7.2b the resistance offered to surface uptake is termed r_p.

From Fig. 7.2b, the total flux of pollutant may be written

$$\mathbf{F} = (C - C_0)/r_b' \tag{7.3}$$

or, denoting the resultant of the three parallel resistance paths as r', where

$$r' = [(r_s' + r_i)^{-1} + (r_c' + r_{ci})^{-1} + (r_p)^{-1}]^{-1} \tag{7.4}$$

as

$$\mathbf{F} = (C_0 - O)/r' \tag{7.5}$$

Eliminating the surface concentration, C_0, gives

$$\mathbf{F} = C/(r_b' + r') = C/r_t \tag{7.6}$$

where the total resistance $r_t = r_b' + r'$.

In studies of pollutant uptake, the deposition velocity, v_g, defined as $v_g = \mathbf{F}/C$, is commonly used to estimate fluxes when concentrations are known (Chamberlain 1975). Equation 7.6 shows that $v_g = (r_b' + r')^{-1}$, and demonstrates that in general it is wrong to assume that there is a constant value for v_g. Clearly v_g depends on wind speed, physiological responses and chemical reactivity. The paper by Fowler at this meeting demonstrates that v_g to vegetation may vary by an order of magnitude depending on the canopy environment. Bennett *et al.* (1973) proposed a model of pollutant uptake similar to the one in Fig. 7.2b and showed, with hypothetical resistances, how it could be used to estimate ozone uptake and internal ozone concentrations. The determination of resistances from real measurements of pollutant fluxes does not seem to have been fully exploited.

To illustrate the principles of pollutant uptake shown in Fig. 7.2b,

Table 7.1. Resistances to SO_2 uptake by leaves, used to illustrate the application of equation 7.7.

		s m⁻¹	Reference
Cuticle	$r_c' + r_{ci}$	5×10^4	Jarvis 1971
Surface	r_p	4×10^3	Spedding 1969
Boundary layer	r_b	100	Spedding 1969
Internal	r_i	100	Hypothetical
Stomatal	r_s	50–500	—

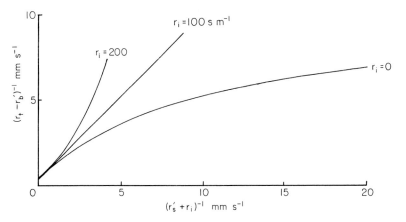

Fig. 7.3. An example of a method of determining internal and cuticular resistances (equation 7.7) using values of resistances from Table 7.1. The value $r_i = 100$ s m⁻¹ yields a straight line graph and the intercept is an estimate of $r^{-1}{}_p$.

previously unpublished measurements will now be analysed. Cuvette measurements with pollutants allow estimates of fluxes at specific concentrations and hence $r_t (= C/F)$ may be found. Standard methods for determining boundary layer and stomatal resistances, r_b' and r_s', may also be used (Unsworth et al. 1976). From equations 7.4 and 7.6,

$$(r_t - r_b')^{-1} = (r_s' + r_i)^{-1} + (r_c' + r_{ci})^{-1} + r_p^{-1} \qquad (7.7)$$

Consequently a graph of $(r_t - r_b')^{-1}$ against $(r_s' + r_i)^{-1}$ should be a straight line with slope unity and intercept $(r_c' + r_{ci})^{-1} + r_p^{-1}$. Values of r_i are not known and so estimates must be made until a straight line graph is obtained. To illustrate the method, Table 7.1 shows a hypothetical set of resistances derived from the literature and used to estimate SO_2 uptake. Fig. 7.3 indicates that the choice of $r_i = 100$ s m⁻¹ yields a straight line, and the intercept, 0.27×10^{-3} m s⁻¹, corresponds to a resistance of 3.7×10^3 s m⁻¹. Because

$(r_c' + r_{ci})^{-1}$ was assumed much smaller than $(r_p)^{-1}$ (Table 7.1) this intercept is a good approximation of r_p^{-1}. It is implicit in this form of analysis that r_i and the resistances determining the intercept are independent of r_s and that environmental factors influencing the resistances are held constant.

King & Smith (personal communication) modified a leaf cuvette designed by Beadle *et al.* (1973) to allow simultaneous measurement of ozone, carbon dioxide and water vapour exchange. Fig. 7.4 shows a set of their measurements obtained on a leaf of *Zea mays* cv. Coop 3990 exposed to 400 parts 10^{-9} by volume of ozone. Boundary layer resistance to ozone uptake was 10 s m^{-1} and stomatal resistances to ozone diffusion, r_s', were altered by changing the irradiance, giving values ranging from about 300 to 1500 s m^{-1} (the ratio of the diffusivities of water vapour and ozone is about 1·6). The

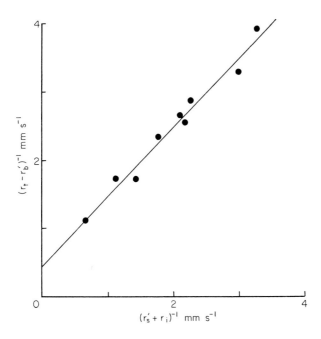

FIG. 7.4. The relationship between resistances to ozone uptake by a leaf of *Zea mays* in a cuvette (from unpublished measurements of King & Smith). A straight line fits the observations with r_i taken as zero.

relatively large minimum values of r_s' are because ozone induced partial stomatal closure. The points closely fit a straight line with $r_i = 0$, suggesting that the walls of the substomatal cavities were a perfect sink for ozone. It is likely therefore that r_{ci} was also zero. The intercept of 0·46 mm s^{-1} corresponds to a resistance of 2180 s m^{-1}. The slope is not significantly different

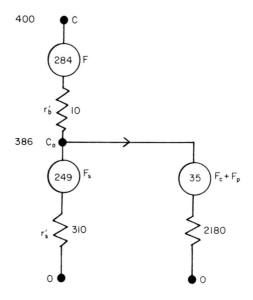

Fig. 7.5. Potentials, resistances and fluxes corresponding to the point on Fig. 7.4 where $r_s' = 310$ s m^{-1}. Units: potentials, parts 10^{-9} by volume (1 part 10^{-9} O$_3 \simeq 2$ μg m^{-3}); resistances, s m^{-1}; fluxes μg m^{-2} s^{-1}. Symbols correspond to Fig. 7.2b.

from unity, showing that the measurements are self-consistent. Fig. 7.5 summarizes this analysis, showing the fluxes, potentials and resistances when r_s' was at its minimum value, 310 s m^{-1}, and indicating that nearly 90% of the ozone flux entered the leaf.

The appropriate measurements for analysis of SO$_2$ fluxes are more difficult to obtain because the gas reacts readily with many natural surfaces, and surface resistances depend strongly on relative humidity (Spedding 1969). Garland and Branson (1977) exposed shoots of *Pinus sylvestris* L. to ^{35}SO$_2$ in a portable exposure chamber in the field and deduced values of r_t^{-1} (equation 7.6). They also measured the stomatal resistance r_s for water vapour diffusion from the shoots. Their measurements are shown in the paper by Jarvis (p. 182). The mean ratio r_t/r_s was 1·98, close to the ratio of the diffusivities of water vapour and SO$_2$, suggesting that stomata were the main limitation to uptake of SO$_2$. Because there is much scatter in their results it is not possible to determine r_i and r_p from these field measurements. Recent analysis of laboratory experiments (Black & Unsworth 1979c) on SO$_2$ uptake by *Vicia faba* has shown that $r_i \simeq 0$ and $r_p \simeq 3600$ s m^{-1}. It will be interesting to learn whether the same magnitudes apply to plants grown and exposed in different environments.

Canopies

An analysis of gas exchange between the atmosphere and crop canopies in terms of resistance analogues was proposed by Monteith (1963) who extended the concepts already applied to single leaves by defining a fictitious 'effective crop surface' where exchange took place and where the gas concentration could be determined by extrapolation from measurements above the canopy. This elegant approach aroused the wrath of micrometeorologists (for example Swinbank & Businger in discussions in Monteith 1963, and Philip 1966), who justifiably pointed out that the method is strictly valid only when source and sink distributions in the canopy are identical for each entity used in evaluating the resistances. This, and a second objection concerning the differences between boundary layer resistances for momentum and mass transfer were discussed by Monteith (1973) and Thom (1975). Fig. 7.6 illustrates the basis of Monteith's method and shows the meanings of the resistances in the crop analogue. In neutral conditions, average wind speed u increases logarithmically with height above the zero plane, the displacement of which, d, is usually about 0·6–0·8 of crop height above the ground. By

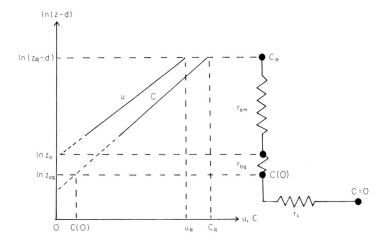

FIG. 7.6. Typical profiles of windspeed u and gas concentration C over a canopy of vegetation in neutral conditions, showing the linear relationship between u or C and the logarithm of height above the zero plane displacement d (height above the ground is z). When windspeed is extrapolated, $u = 0$ when $z - d = z_0$, the roughness length for momentum transfer. The roughness length for gas transfer, z_{0g} is smaller than z_0, and at the height $z - d = z_{0g}$, the gas concentration is $C(0)$. In the equivalent resistance analogue for gas transfer, the flux between concentrations C_R and $C(0)$ is restricted by the aerodynamic resistance for momentum r_{am} and the additional aerodynamic resistance for mass transfer r_{bg}. Between concentration $C(0)$ and sinks where $C = 0$, flux is restricted by the canopy resistance r_c.

extrapolation, wind speed is zero at a height z given by $z - d = z_0$ where z_0 is the roughness length. Because the extrapolated wind speed is zero when $z = d + z_0$, this level may be regarded as the height above the ground of a fictitious 'effective surface' containing the sink for momentum. The value of z_0 depends on structure of the crop and is usually about 0·1 of crop height. The rate of transfer of momentum from height z_R to the effective surface where $z = d + z_0$, is restricted by the aerodynamic resistance r_{am}.

Because the turbulent eddies that transfer momentum also transfer mass, the average gas concentration for a gas absorbed by the crop also increases logarithmically with height above the zero plane. However, because momentum is absorbed more easily than mass, the effective surface where sinks for the gas are located is lower than that for momentum, at a height $z = d + z_{0g}$ (Fig. 7.6). Differences between the heights of the effective surfaces for momentum and mass may be allowed for by the additional aerodynamic resistance for mass transfer r_{bg} (Monteith 1973; Thom 1975; Shuttleworth 1976).

At the effective surface for mass transfer, the gas concentration is zero only if there are perfect sinks for mass associated with the surface. In general, by analogy with single leaves (Fig. 7.2) there are additional resistances with physiological and chemical components restricting transfer from the surface to sinks where $C = 0$. Consequently the gas concentration at the effective surface usually has a finite value $C(0)$, as shown by extrapolating the profile of gas concentration to $z = d + z_{0g}$ in Fig. 7.3. The final stage of transfer is through a bulk canopy resistance r_c (Fig. 7.3), the resultant of the various parallel paths between $C(0)$ and sinks where $C = 0$.

If the appropriate values for aerodynamic resistances are known or can be estimated (Thom 1975), values of canopy resistance for water vapour transfer may be derived from measurements of heat and water vapour fluxes and available energy (Jarvis *et al.* 1976) or directly from water vapour flux as a residual. When the leaf area index L is large, the main determinant of r_c is stomatal resistance. It may be shown from a theoretical model of canopy transfer that $r_c \simeq r_l/L$ (Shuttleworth 1979) where r_l is the mean stomatal resistance of leaves in the canopy, a result expected when resistances of L leaves per unit ground area are combined in parallel; several experimental studies have found that $r_c \simeq r_l/L$ (Monteith, Szeicz & Waggoner 1965; Szeicz, Van Bavel & Takami 1973; Tan & Black 1976). For evaporation measurements, this interpretation of r_c is successful for a number of reasons: the distributions of sources or sinks of heat, water vapour and momentum in dense canopies are similar; cuticular resistances are much larger than stomatal resistances by day; evaporation from the soil is usually restricted by the small amount of available energy for evaporation and/or the dry soil surface.

In principle a similar approach is applicable to the analysis of carbon dioxide fluxes, but in practice the method proposed by Monteith (1963) has seldom been used. A resistance model of CO_2 transfer in a canopy must allow

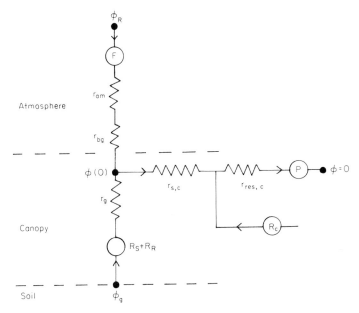

FIG. 7.7. A resistance analogue for CO_2 exchange in a crop canopy by day (after Monteith, Szeicz & Yabuki 1964). Aerodynamic resistances to CO_2 flux, **F**, from above the canopy are as in Fig. 7.6. Additional CO_2 fluxes, $\mathbf{R}_s$ from the soil and $\mathbf{R}_R$ from roots, establish the concentration difference, $\phi_g - \phi(0)$, between the soil surface and the apparent sink for CO_2. The canopy resistance between $\phi(0)$ and $\phi = 0$ has two components, a bulk stomatal resistance $r_{s,c}$ and a residual resistance $r_{res,c}$. The respired flux of CO_2 from the canopy, $\mathbf{R}_c$ is released in the 'bulk mesophyll'.

for respiratory fluxes of CO_2 from the soil, and must include bulk internal resistances to CO_2 transfer as well as bulk stomatal resistances. Fig. 7.7 adapted from Monteith, Szeicz & Yabuki (1964) illustrates these points. Important features of this analogue are: the CO_2 concentration, $\phi(0)$, at the effective crop surface depends on (i) **F** and $(r_{am} + r_{bg})$ which determine the potential difference, $\phi_R - \phi(0)$, and (ii) the respiratory flux from the soil, $\mathbf{R}_s + \mathbf{R}_r$, essentially a constant current and hence independent of the transfer resistance, r_g, from the soil to the effective crop surface. Because soil respiration is usually a small fraction of **F** by day, and $(r_{am} + r_{bg})$ is also small, it can be shown that the influence of $\mathbf{R}_s$ and $\mathbf{R}_r$ on $\phi(0)$, and hence on gross photosynthesis, **P**, is usually negligible (Monteith, Szeicz & Yabuki 1964).

Although analysis like that in Fig. 7.7 can be used to derive a value for the bulk internal resistance, $r_{res,c}$, it would be inadvisable to attempt to relate variations in $r_{res,c}$ to environmental factors. For example, the assumptions inherent in Fig. 7.7 (and Fig. 7.2a), that respired CO_2 is recycled, do not hold when photosynthesis is not light-saturated. Consequently $r_{res,c}$ derived

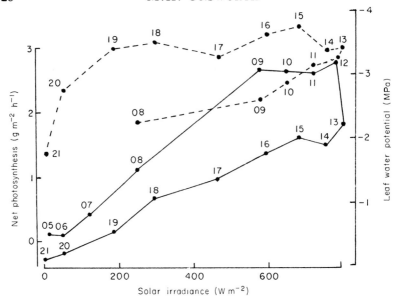

FIG. 7.8.   The relation between net CO_2 fixation of a barley crop (○), water potential of the flag leaf (●) and irradiance. The figure beside each point is the time of measurement and the lines connect consecutive measurements. (From Biscoe, Scott & Monteith 1975.)

when a canopy is not light saturated will inevitably show a strong dependence on irradiance irrespective of true changes in resistances associated with photochemical reactions. Nevertheless, physiologists could use resistance analogues of canopy transfer to investigate the role of stomata in controlling CO_2 uptake in the field. For example, Fig. 7.8 shows diurnal variations in canopy photosynthesis and leaf water potential in barley (Biscoe *et al.* 1975). During the afternoon, net photosynthesis was lower than in the morning at the same irradiance, and the authors speculated that the difference was a result of the increased negative water potential in the afternoon. They concluded that 'Further analysis is needed to discover whether photosynthesis was restricted by stomatal closure, by reduced enzyme activity or by a combination of physiological mechanisms'. If they had calculated the bulk stomatal resistance of their crop from analysis of water vapour flux, one of these hypotheses could have been tested.

Canopy resistances to pollutant uptake are more amenable to analysis than those for CO_2 transfer because sources and sinks are generally not present together. Fig. 7.9 shows an analogue first proposed by Fowler for SO_2 (1975, 1978). The canopy resistance, r_c, may be interpreted as the resultant of up to four bulk resistances acting in parallel, stomatal resistance r_{c1}, surface and cuticular resistances r_{c2}, resistances to soil uptake r_{c3}, and

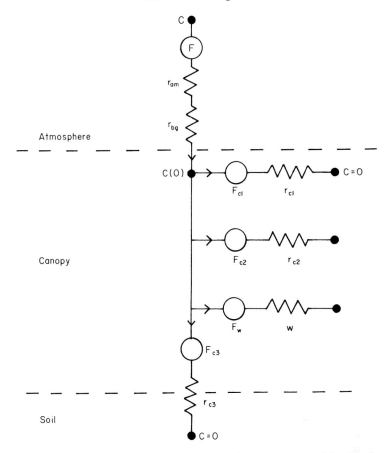

FIG. 7.9. A resistance analogue for pollutant uptake by a crop canopy (after Fowler 1978). The canopy resistance is the resultant of the resistances in four parallel paths: stomata (r_{c1}), cuticle and dry surfaces (r_{c2}), soil (r_{c3}), and moisture trapped on leaves (w). Fluxes $\mathbf{F}_{c1}$, $\mathbf{F}_{c2}$, $\mathbf{F}_{c3}$ and $\mathbf{F}_w$ travel along each path. Other symbols correspond to Fig. 7.6.

resistances of surface moisture w. Interpretation of SO_2 fluxes on a diurnal and seasonal basis allowed estimation of each component (Fowler 1978; Fowler & Unsworth 1979; Fowler, this volume). It is implicit in the analysis of components of canopy resistance that the bulk resistances thus derived depend partly on the efficiency of turbulent transfer within the canopy and near the soil, but, as will be seen later, aerodynamic resistances associated with transfer in the canopy are usually much smaller than those attributed to diffusion and reactions at surfaces so that the influence of such resistances on the gas exchange of vegetation is usually small.

Wesely *et al.* (1978) used the concept of canopy resistance to analyse their measurements of ozone flux to a maize crop. They found that r_c for ozone uptake varied from 400 to 50 s m^{-1} during the daytime. Stomatal diffusion resistance was the most important determinant of r_c, but they concluded that about 20–50% of the total ozone flux was non-stomatal uptake. On the basis of Figs. 7.4 and 7.5, this large non-stomatal uptake cannot be accounted for by deposition of ozone on the cuticle, and so the flux of ozone to the soil appears to have been substantial, in contrast with Fowler's results for SO_2 uptake. Soil surfaces are known to be good sinks for ozone (Galbally 1971; Macdowall 1974); from field and laboratory experiments Turner *et al.* (1973) concluded that soil resistances to ozone uptake were typically 20–50 s m^{-1}, increasing with increasing water content, and that in the field the major restriction on ozone uptake by soil was the resistance of the boundary layer close to the soil surface.

Leuning *et al.* (1979b) measured fluxes of ozone to a tobacco crop, and estimated stomatal uptake by combining porometer measurements of stomatal resistance with a model of canopy transfer described in the next section. From an analysis of gradients of vapour pressure and temperature between the soil and the air near the bottom of the leaf layer they concluded that the resistance of the soil boundary layer to ozone uptake was about 100 s m^{-1}. Using this resistance to calculate daily uptake of ozone by the soil they concluded that 40–50% of the daily total ozone flux to the canopy was to the soil. It is possible that this proportion is overestimated because no allowance was made for a soil resistance in series with the soil boundary layer resistance.

The role of the soil as a sink for air pollutants contrasts with the situation for CO_2 where the soil flux is often assumed constant, and for water vapour where dry soil and/or a lack of energy often restrict evaporation rates. Transfer between soil or short vegetation and the atmosphere below canopies has received little attention. Shuttleworth (1979) considered 'below canopy flux' in a theoretical model of vegetation-atmosphere interaction, but there is a need for more measurements of transfer to and from the soil and vegetation below canopies, in particular to evaluate the aerodynamic resistances to transfer imposed by the relatively still layers of air close to the ground.

PREDICTION OF CANOPY FLUXES FROM RESISTANCE ANALOGUES

One of the best ways of integrating knowledge of physiological and micrometeorological mechanisms determining gas exchange and of identifying limitations in this knowledge is by developing mathematical models for simulating the interactions between vegetation and the atmosphere. The topic was thoroughly reviewed by Waggoner (1975). In the present context two types of models of canopy exchange have already been illustrated. The first is concerned with transfer to the canopy as a whole, and attributes all aspects of physiological and biochemical control of fluxes to bulk canopy resistances, which may also be influenced by the efficiency of transfer in the canopy and to the soil. From this type of model Monteith *et al.* (1964) evaluated the effects of the environment on the carbon assimilation of field and glasshouse crops, Jarvis *et al.* (1976) interpreted differences in evaporation rates from coniferous forests at several different sites, Fowler & Unsworth (1979) predicted deposition rates for SO_2 to several types of short vegetation, and Garland & Branson (1977) estimated SO_2 uptake by a pine forest.

The second type of model treats a canopy either as discrete layers of foliage (Waggoner, Furnival & Reifsnyder 1969), or as a continuous distribution (Cowan 1968) and links resistances associated with the vegetation with transfer resistances in the canopy. The most important determinants of fluxes to vegetation in such models are often the stomatal and residual resistances of leaves, and one of the challenges of this approach is the specification of the spatial distribution of such resistances and their response to the environment. A corollary of this view is that provided the distribution of leaf resistances is specified, the influence of transfer in the canopy on the gas exchange of leaves is small and may sometimes be neglected. For example, Monteith *et al.* (1965) estimated canopy evaporation from stomatal resist-

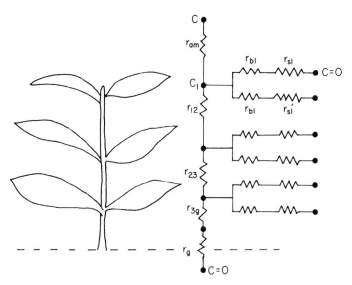

FIG. 7.10. A 'discrete layer' model for estimating ozone fluxes to a tobacco crop. The leaves are treated as three layers with bulk boundary layer resistances, r_{bl}, etc., and bulk stomatal resistances, r_{sl}, r_{sl}', etc. for adaxial and abaxial surfaces, respectively. Atmospheric resistances, r_{am}, r_{12}, etc. link the atmosphere, foliage and soil. The soil has a resistance r_g.

ances, Biscoe *et al.* (1975) estimated net photosynthesis of barley from light response curves of single leaves, and Black & Unsworth (1979b) used a similar model to that of Biscoe *et al.* for speculating on the effects of SO$_2$ on the dry matter production of a crop of *Vicia faba*.

When the partitioning of fluxes between vegetation and the soil is required, atmospheric transfer resistances become important. Waggoner (1975) described elegant matrix methods for solving the simultaneous equations which arise from stratified models, and he and his colleagues used this approach to model evaporation from a maize crop (Waggoner & Turner 1972) and removal of ozone from the atmosphere by crop canopies and by the soil below them (Turner, Waggoner & Rich 1974). Unsworth (unpublished) used a similar model to estimate fluxes of ozone to tobacco crops in Southern Ontario. Fig. 7.10 shows a 3-layer resistance model of the tobacco canopy. Each layer of leaves was characterized by the bulk stomatal resistance of

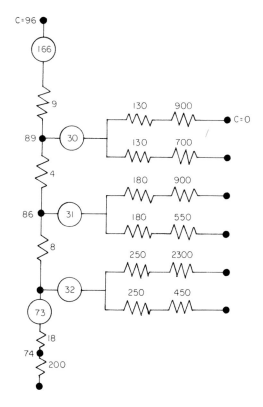

FIG. 7.11. Potentials, resistances and fluxes of ozone predicted from the model of Fig. 7.10 at noon on a day when the ozone concentration above the crop was 96 parts 10^{-9}. Units: potentials, parts 10^{-9} by volume (1 part 10^{-9} $\simeq$ 2 μg m^{-3}); resistances, s m^{-1}; fluxes, 10^{-8} g m^{-2} s^{-1}.

adaxial and abaxial surfaces, r_{sl} and r_{sl}', found by dividing values measured with a porometer by the leaf area index of each layer. A leaf boundary layer resistance, determined as a function of wind speed and leaf size (Thom 1968), was calculated for the adaxial and abaxial surfaces of each layer. Aerodynamic resistances to transfer between layers depend on eddy diffusivity and on the spacing between layers (Monteith 1973); in this model the eddy diffusivity was assumed to be constant in the upper half of the canopy (Thom 1971) and to decay exponentially below this level. A constant value of 200 s m^{-1} was assumed for the soil boundary layer resistance and soil resistance (Turner *et al.* 1973). Fig. 7.11 illustrates the potentials, resistances and fluxes in the model at noon on a day when the ozone concentration above the canopy was 96 parts 10^{-9}. It is clear from the figure that stomatal and leaf boundary layer resistances were the main factors determining ozone fluxes to vegetation; the small transfer resistances in the canopy led to little change in pollutant concentration with height except in the lowest few centimetres. The main factor determining the flux to the soil is the large and imprecisely known resistance associated with the soil boundary layer. Consequently the partitioning of the total flux into uptake by vegetation (56%) and by the soil (44%) has a large uncertainty. Better knowledge of transfer processes in the lowest layers of canopies is essential for modelling the pathways for pollutants in canopies. Such knowledge is also necessary for modelling the transpiration or productivity of vegetation below tall canopies, such as bracken on forest floors or short crops planted between rows of tall crops in tropical agricultural systems.

CONCLUSIONS

A number of developments in the measurement and analysis of gas exchange have been identified. There is increasing recognition that plants in the field do not respond to stress in the same way as plants in greenhouses or growth rooms, and this has encouraged the development of methods for studying the exchange of carbon dioxide and pollutants out-of-doors. It seems likely that cuvette methods will be more popular than micrometeorology because they allow studies of individual organs and can be used to investigate effects of different treatments or comparisons of different crop varieties within a relatively small area. With the development of more robust and portable gas analysers and with increasing use of microprocessors and computers it is likely that automatically-controlled cuvettes or enclosures will be designed in increasing numbers to monitor gas exchange over long periods.

For studying gas exchange of natural ecosystems where there is great diversity of species and habitats, micrometeorological methods are a powerful tool provided that a sufficiently large area is available. All micrometeorological methods are relatively 'blunt instruments' because they measure net fluxes and there are sometimes large uncertainties in partitioning the fluxes. Consequently such methods seem unlikely to further our understanding of biological mechanisms controlling gas exchange unless they are amalgamated with parallel programmes of physiological measurements. Nevertheless, careful analysis of fluxes measured above canopies can reveal bulk properties of the vegetation and may, within limits, be used to investigate bulk physiological responses.

Flux-gradient methods in micrometeorology require large amounts of expensive equipment and there is doubt over the validity of the aerodynamic method when used close to rough, tall vegetation. Eddy correlation methods when further developed with cheap, robust sensors and electronics will be an important tool for measuring gas exchange of canopies directly and will offer exciting opportunities for environmental physiologists to study gas exchange in many natural environments where the variety of species precludes other methods.

The measurement of pollutant fluxes to vegetation is in its infancy and there is considerable potential for further developments in instrumentation, methods and analyses. As well as evaluating the role of vegetation as a sink for pollutants, and the subsequent pathways in ecosystems, there is much scope for investigating the responses of plants to the stresses imposed by pollutant uptake.

The interpretation of gas concentration profiles within canopies to reveal source and sink strengths seems to have reached an impasse; the assumptions of one-dimensional transfer and flux-gradient analysis are a gross oversimplification of the within-canopy environment. In particular there is unlikely to be a unique distribution of the eddy diffusivity. In the long run, eddy correlation measurements of fluxes within canopies may allow micrometeorologists to make progress in identifying sources and sinks but the designing of instruments which have minimum interference with air movement will be difficult. Even with appropriate instrumentation, problems of spatial variability of fluxes in canopies pose another barrier to progress in direct measurement.

It seems unlikely that current theoretical models of canopy transfer can be refined to be of much use to the ecologist or crop physiologist interested in source and sink distributions. Progress probably requires new theories, likely to be considerably more complex than the one-dimensional methods which have been so successful above canopies.

Finally, resistance analogues, firmly established for analysing and modelling heat and water vapour exchange in canopies, are equally useful for dealing with air pollutants. The interpretation of resistances in models of CO_2 exchange is more controversial, but even here they may be useful in analysing the role of the atmosphere and stomata in controlling fluxes.

ACKNOWLEDGEMENTS

Some of the work in this review was completed during sabbatical leave at the University of Guelph, Canada; I am grateful to Professor K.M. King and P.J. Smith at Guelph for allowing me to use some of their data.

REFERENCES

Beadle C.L., Stevenson K.R., Thurtell G.W. & Dubé P.A (1974) An open system for plant gas-exchange analysis. *Canadian Journal of Plant Science*, **54**, 161–165.

Belot Y., Bourreau J.C., Dubois M.L. & Pauly C.S. (1974) Measure de la vitesse de captation du dioxide de souffre sur les feuilles des plantes au moyen du souffre—34. FAO/IAEA Isotope ratios as pollutant source and behaviour indicators. 18–22 Nov, Vienna (IAEA-SM-191-18).

Bennett J.H., Hill A.C. & Gates D.M. (1973) A model for gaseous pollutant sorption by leaves. *Journal of the Air Pollution Control Association*, **23**, 957–962.

Biscoe P.V., Scott R.K. & Monteith J.L. (1975) Barley and its environment. III. Carbon budget of the stand. *Journal of Applied Ecology*, **12**, 269–293.

Biscoe P.V., Incoll L.D., Littleton E.J. & Ollerenshaw J.H. (1977) Barley and its environment. VII. Relationships between irradiance, leaf photosynthetic rate and stomatal conductance. *Journal of Applied Ecology*, **14**, 293–302.

Biscoe P.V., Gallagher J.N., Littleton E.J., Monteith J.L. & Scott, R.K. (1975) Barley and its environment. IV. Sources of assimilate for the grain. *Journal of Applied Ecology*, **12**, 295–318.

Black V.J. & Unsworth M.H. (1979a) A system for measuring effects of sulphur dioxide on gas exchange of plants. *Journal of Experimental Botany*, **30**, 81–88.

Black V.J. & Unsworth M.H. (1979b) Effects of low concentrations of sulphur dioxide on net photosynthesis and dark respiration in *Vicia faba*. *Journal of Experimental Botany*, **30**, 473–483.

Black V.J. & Unsworth M.H. (1976c) Resistance analysis of sulphur dioxide fluxes to *Vicia faba*. *Nature*, **282**, 68–69.

Businger J.A. (1975) Aerodynamics of vegetated surfaces. *Heat and Mass Transfer in the Biosphere* (Eds. D.A. DeVries & N. Afgan), pp. 139–166. Scripta Book Company, Washington D.C.

Campbell G.S. & Unsworth M.H. (1979) An inexpensive anemometer for eddy-correlation. *Journal of Applied Meteorology*, **18**, 1072–1077.

Chamberlain A.C. (1975) Pollution in plant canopies. *Heat and Mass Transfer in the Biosphere* (Eds. D.A. DeVries & N. Afgan), pp. 561–582. Scripta Book Company, Washington D.C.

Cowan I.R. (1968) Mass, heat and momentum exchange between stands of plants and their atmospheric environment. *Quarterly Journal of the Royal Meteorological Society*, **94**, 523–544.

Den Hartog G. & Shaw R.H. (1975) A field study of atmospheric exchange processes within a vegetative canopy. *Heat and Mass Transfer in the Biosphere* (Eds. D.A. DeVries & N. Afgan), pp. 299–310. Scripta Book Company, Washington D.C.

Denmead O.T. (1976) Temperate cereals. *Vegetation and the Atmosphere*. Vol. II. (Ed. J.L. Monteith), pp. 1–32. Academic Press, London.

Denmead O.T. & McIlroy I.C. (1971) Measurement of carbon dioxide exchange in the field. *Plant Photosynthetic Production: Manual of Methods* (Eds. Z. Šesták, J. Čatský and P.G. Jarvis), pp. 467–516. Junk, The Hague.

Desjardins R.L. (1974) A technique to measure CO_2 exchange under field conditions. *International Journal of Biometeorology*, **18**, 76–83.

Eckardt F.E. (1968) Techniques de mesure de la photosynthese sur le terrain, basées sur l'emploi d'enceintes climatisées. *Functioning of Terrestrial Ecosystems at the Primary Production Level* (Ed. F.E. Eckardt), pp. 289–319. Unesco, Paris.

Fowler D. (1975) Uptake of sulphur dioxide by crops and soil. Ph.D. thesis, University of Nottingham, School of Agriculture.

Fowler D. (1978) Dry deposition of SO_2 on agricultural crops. *Atmospheric Environment*, **12**, 369–373.

Fowler D. & Unsworth, M.H. (1979), Turbulent transfer of sulphur dioxide to a wheat crop. *Quarterly Journal of the Royal Meteorological Society*, **105**, 767–783.

Gaastra P. (1959) Photosynthesis of crop plants as influenced by light, carbon dioxide, temperature and stomatal diffusion resistance. *Mededelingen van de Landbouwhogeschool te Wageningen, Nederland*, **59(13)**, 1–68.

Galbally I.E. (1971) Ozone profiles and ozone fluxes in the atmospheric surface layer. *Quarterly Journal of the Royal Meteorological Society*, **97**, 18–29.

Galbally I.E., Garland J.A. & Wilson M.J.G. (1979) Uptake of sulphur from the atmosphere by forest and farmland. *Nature*, **280**, 49–50.

Garland J.A. & Branson J.R. (1977) The deposition of sulphur dioxide to pine forest assessed by a radioactive tracer method. *Tellus*, **29**, 445–454.

Garsed S.G. & Read D.J. (1977) Sulphur dioxide metabolism in soy-bean, *Glycine max* Var. *Biloxi. New Phytologist*, **78**, 111–119.

Hill A.C. (1971) Vegetation: a sink for atmospheric pollutants. *Journal of the Air Pollution Control Association*, **21**, 341–346.

Incoll L.D. (1977) Field studies of photosynthesis: monitoring with $^{14}CO_2$. *Environmental Effects on Crop Physiology* (Eds. J.J. Landsberg & C.V. Cutting), pp. 137–155. Academic Press, London.

Inoue E., Uchijima Z., Saito T., Isobe S. & Vemura K. (1969) The 'Assimitron', a newly devised instrument for measuring CO_2 flux in the surface air layer. *Journal of Agricultural Meteorology, Tokyo*, **25**, 19–26.

James G.B. (1977) Exchanges of mass and energy in Sitka spruce. Ph.D. thesis, University of Aberdeen.

Jarvis P.G. (1971) The estimation of resistances to carbon dioxide transfer. *Plant Photosynthetic Production: Manual of Methods.* (Eds. Z. Šesták, J. Čatský & P.G. Jarvis), pp. 566–631. Junk, The Hague.

Jarvis P.G. & Čatský J. (1971) Gas exchange systems. *Plant Photosynthetic Production: Manual of Methods* (Eds. Z. Šesták, J. Čatský and P.G. Jarvis), pp. 50–56. Junk, The Hague.

Jarvis P.G., James G.B. & Landsberg J.J. (1976) Coniferous forest. *Vegetation and the Atmosphere*, Vol. II (Ed. J.L. Monteith), pp. 171–240. Academic Press, London.

Jones E.P., Ward T.V. & Zwick H.H. (1978) A fast-response atmospheric CO_2 sensor for eddy correlation flux measurements. *Atmospheric Environment*, **12**, 845–851.

Legg B.J. & Monteith J.L. (1975) Heat and mass transfer within plant canopies. *Heat and Mass Transfer in the Biosphere* (Eds. D.A. DeVries & N. Afgan), pp. 167–186. Scripta Book Company, Washington D.C.

Lemon E.R. & Wright J.L. (1969) Photosynthesis under field conditions. XI. Assessing sources and sinks of carbon dioxide in a corn crop using a momentum balance approach. *Agronomy Journal*, **61**, 408–411.

Leuning R. (1976) Ph.D. thesis, University of Melbourne, Australia.

Leuning R., Neumann H.H. & Thurtell G.W. (1979a) Ozone uptake by corn: a general approach. *Agricultural Meteorology*, 20, 115–136.

Leuning R., Unsworth M.H., Neumann H.H. & King K.M. (1979b) Ozone fluxes to tobacco and soil under field conditions. *Atmospheric Environment*, 13, 1155–1163.

Liss P.S. & Slater P.G. (1974) Flux of gases across the air–sea interface. *Nature*, 247, 181–184.

Louwerse W. & Eikhoudt J.W. (1975) A mobile laboratory for measuring photosynthesis, respiration and transpiration of field crops. *Photosynthetica*, 9, 31–34.

Macdowall F.D.H. (1974) Importance of soil in the absorption of ozone by a crop. *Canadian Journal of Soil Science*, 54, 239–240.

Marshall B. & Biscoe P.V. (1977) A mobile apparatus for measuring leaf photosynthesis in the field. *Journal of Experimental Botany*, 28, 1008–1017.

Monteith J.L. (1963) Gas exchange in plant communities. *Environmental Control of Plant Growth* (Ed. L.T. Evans), pp. 95–112. Academic Press, New York.

Monteith J.L. (1973) *Principles of Environmental Physics*. Edward Arnold, London.

Monteith J.L., Szeicz G. & Waggoner P.E. (1965) The measurement and control of stomatal resistance in the field. *Journal of Applied Ecology*, 2, 345–355.

Monteith J.L., Szeicz G. & Yabuki K. (1964) Crop photosynthesis and the flux of carbon dioxide below the canopy. *Journal of Applied Ecology*, 1, 321–337.

Mooney H.A., Dunn E.L., Harrison A.T., Morrow P.A., Bartholomew B. & Hays R.L. (1971) A mobile laboratory for gas exchange measurements. *Photosynthetica*, 5, 128–132.

National Research Council of Canada (1939) *Effects of Sulphur Dioxide on Vegetation*. Publication 815, Ottawa.

Nielson R.E. (1977) A technique for measuring photosynthesis in conifers by $^{14}CO_2$ uptake. *Photosynthetica*, 11, 241–250.

Penman H.L. & Long I.F. (1960) Weather in wheat: an essay in micrometeorology. *Quarterly Journal of the Royal Meteorological Society*, 86, 16–50.

Philip J.R. (1966) Plant water relations: some physical aspects. *Annual Review of Plant Physiology*, 17, 245–268.

Rawson H.M. & Begg J.E. (1977) The effect of atmospheric humidity on photosynthesis, transpiration and water use efficiency of leaves of several plant species. *Planta*, 134, 5–10.

Raybould C.C., Unsworth M.H. & Gregory P. (1977) Sources of sulphur in rain collected below a wheat canopy. *Nature*, 267, 146–147.

Rich S., Waggoner P.E. & Tomlinson H. (1970) Ozone uptake by bean leaves. *Science*, 169, 79–80.

Ripley E.A. & Redmann R.E. (1976) Grassland. *Vegetation and the Atmosphere*, Vol. II (Ed. J.L. Monteith), pp. 351–398. Academic Press, London.

Saugier B. & Ripley E.A. (1978) Evaluation of the aerodynamic method of determining fluxes over natural grassland. *Quarterly Journal of the Royal Meteorological Society*, 104, 257–270.

Šesták Z., Čatský J. & Jarvis P.G. (1971) *Plant Photosynthetic Production: Manual of Methods*. Junk, The Hague.

Shuttleworth W.J. (1976) A one-dimensional theoretical description of the vegetation—atmosphere interaction. *Boundary-Layer Meteorology*, 10, 273–302.

Shuttleworth W.J. (1979) Below-canopy fluxes in a simplified one-dimensional theoretical description of the vegetation—atmosphere interaction. *Boundary-Layer Meteorology*, 17, 315–331.

Spedding D.J. (1969) Uptake of sulphur dioxide by barley leaves at low sulphur dioxide concentrations. *Nature*, 224, 1229–1230.

Szeicz G., Van Bavel C.H.M. & Takami S. (1973) Stomatal factor in the water use and dry matter production by sorghum. *Agricultural Meteorology*, **12**, 361–389.

Tan C.S. & Black T.A. (1976) Factors affecting the canopy resistance of a Douglas-Fir forest. *Boundary-Layer Meteorology*, **10**, 475–488.

Thom A.S. (1968) The exchange of momentum, mass and heat between an artificial leaf and the airflow in a wind tunnel. *Quarterly Journal of the Royal Meteorological Society*, **94**, 44–55.

Thom A.S. (1971) Momentum absorption by vegetation. *Quarterly Journal of the Royal Meteorological Society*, **97**, 414–428.

Thom A.S. (1975) Momentum, mass and heat exchange in plant communities. *Vegetation and the Atmosphere*, Vol. I (Ed. J.L. Monteith), pp. 57–110. Academic Press, London.

Thom A.S., Stewart J.B., Oliver H.R. & Gash J.H.C. (1975) Comparison of aerodynamic and energy budget estimates of fluxes over a pine forest. *Quarterly Journal of the Royal Meteorological Society*, **101**, 93–105.

Thomas M.D. & Hill G.R. (1937) Relation of sulphur dioxide in the atmosphere to photosynthesis and respiration of alfalfa. *Plant Physiology*, **12**, 309–383.

Turner N.C., Rich S. & Waggoner P.E. (1973) Removal of ozone by soil. *Journal of Environmental Quality*, **2**, 259–264.

Turner N.C., Waggoner P.E. & Rich S. (1974) Removal of ozone from the atmosphere by soil and vegetation. *Nature*, **250**, 486–489.

Unsworth M.H., Biscoe P.V. & Black V.J. (1976) Analysis of gas exchange between plants and polluted atmospheres. *Effects of Air Pollutants on Plants* (Ed. T.A. Mansfield), pp. 5–16, Cambridge University Press.

Waggoner P.E. (1975) Micrometeorological models. *Vegetation and the Atmosphere*, Vol. I (Ed. J.L. Monteith), pp. 205–228. Academic Press, London.

Waggoner P.E., Furnival G.M. & Reifsnyder W.E. (1969) Simulation of the microclimate in a forest. *Forest Science*, **15**, 37–45.

Waggoner P.E. & Turner N.C. (1972) Comparison of simulated and actual evaporation from maize and soil in a lysimeter. *Agricultural Meteorology*, **10**, 113–123.

Wesely M.L., Eastman J.A., Cook D.R. & Hicks B.B. (1978) Daytime variations of ozone fluxes to maize. *Boundary-Layer Meteorology*, **15**, 361–373.

8. TURBULENT TRANSFER OF SULPHUR DIOXIDE TO CEREALS: A CASE STUDY

DAVID FOWLER *

*University of Nottingham, School of Agriculture,
Sutton Bonington, Loughborough, Leicester*

SUMMARY

Sulphur dioxide fluxes to a wheat crop were determined from vertical gradients of SO_2 concentration, windspeed and air temperature above the crop. Analysis of these results provides an example of the extent to which boundary layer measurements may be used to evaluate the importance of different sites of uptake. Sequential measurements over 24–36 hours revealed diurnal changes in the fluxes, and in properties of the crop. Using a resistance analogy the relative importance of stomata, the cuticle and the soil layer as sinks for SO_2 have been quantified. In daytime, before senescence 60% of the SO_2 flux is absorbed through stomata and most of the remaining 40% is absorbed by the cuticle (soil uptake accounting for at most 10% of the total flux). Water on plant surfaces, if pure, provides an efficient sink for SO_2 and absorption by the water short-circuits other routes of uptake until an equilibrium is reached between the various oxidized sulphur species in solution and ambient SO_2 concentration.

INTRODUCTION

Most micrometeorological measurements of turbulent mass transfer between vegetation and the atmosphere have been confined to studies of water vapour and carbon dioxide exchange. These measurements, combined with more conventional growth analysis, have provided the necessary field testing and development of methods, and enabled important relationships between energy budgets, water status and exchange rates of water vapour and carbon dioxide to be established.

By the time interest in the fate and effects of pollutant gases had generated sufficient incentive for field measurements of fluxes to the ground, a number of well tried micrometeorological methods were available. Laboratory measurements showed stomata to be important sites for uptake of sulphur dioxide (Spedding 1969), though the overall affinity of vegetation for pollutant gases was uncertain. The lack of information on removal rates of SO_2

* Present address: Institute of Terrestrial Ecology, Bush Estate, Penicuik, Midlothian EH26 0QB.

from the atmosphere by vegetation (and other natural surfaces) was a major factor limiting knowledge of the residence time of sulphur in the atmosphere and consequently, how far it may be transported before being deposited.

A number of investigations of fluxes of pollutant gases (notably SO_2 and O_3) onto vegetation have now been completed using micrometeorological methods (Garland 1977) and, in some, attempts were made to indicate whether the process was controlled by surface or atmospheric factors. This contribution is intended to show for a pollutant gas how a set of micro-meteorological flux measurements may be used to discover the major sites of absorption in vegetation and their relative importance in a range of surface and atmospheric conditions.

METHODS

Fluxes of SO_2 onto a wheat crop were estimated from average vertical gradients of SO_2 concentration, horizontal windspeed and air temperature in the turbulent boundary layer over the crop. The theory, methods of analysing data and theoretical problems with interpretation are treated in the preceding paper and in several texts (e.g. Thom 1975). Concentrations of SO_2 were determined at five levels above the crop using methods described by Fowler (1976). Air temperatures at six levels were measured using thermocouple sensors described by Biscoe *et al.* (1975). Windspeeds were measured at six levels using sensitive cup-anemometers.

In addition, measurements of the presence of dew using a dew balance, and stomatal resistance to water vapour diffusion were made.

Following analysis of the gradients to obtain fluxes of SO_2, the deposition velocity (v_g) and atmospheric resistance ($r_{am} + r_{bg}$) were estimated (see pp. 111–138 for definitions of these quantities and techniques for estimating their values). In the following results, v_g is used in place of the flux as this quantity is normalized for ambient concentration. There was no evidence from the measurements that properties of the sinks for SO_2 varied with SO_2 concentration.

RESULTS

The 90 individual one or two hour mean values of v_g ranged from 1 to 15 mm s^{-1}. To separate the effects of atmospheric processes on v_g from properties of the absorbing surface a resistance analogue of the deposition process was used. This analogy has been used to parameterize water vapour and CO_2 exchange between the atmosphere and vegetation.

The preceding paper describes techniques for applying a similar analogy to the transfer of pollutant gases from the atmosphere to vegetation, the

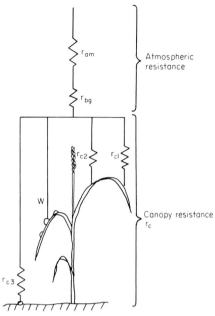

FIG. 8.1. Resistances to SO_2 deposition on wheat: r_{c1}, stomatal component of r_c; r_{c2}, cuticular component of r_c; r_{c3}, soil component of r_c; r_{am} and r_{bg} are atmospheric resistances; w refers to SO_2 uptake by water on the foliage.

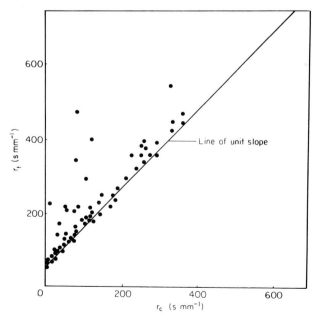

FIG. 8.2. Total resistance to transfer of SO_2, r_t ($= 1/v_g$) plotted against canopy resistance r_c (two field seasons of measurement).

components of which are summarized in Fig. 8.1.

The first step in separating the resistance components shown in Fig. 8.1 is to evaluate the atmospheric terms (r_{am} and r_{bg}) and subtract these from the total resistance r_t ($r_t = 1/v_g$). The residual component is the surface or canopy resistance, r_c, and is the result of all sites of uptake (or 'sinks') at the surface acting in parallel. Canopy resistance r_c was usually the largest component of r_t (Fig. 8.2) showing that surface factors exert most control on rates of deposition. There is seldom therefore any dependence of v_g on windspeed.

There were hourly, daily and seasonal changes in v_g largely due to changes in surface properties. These changes provide clues to the identity and importance of different components of the surface for uptake of SO_2.

For dry plant surfaces while the crop was still in the vegetative phase, diurnal changes in v_g closely followed stomatal movements, the maximum

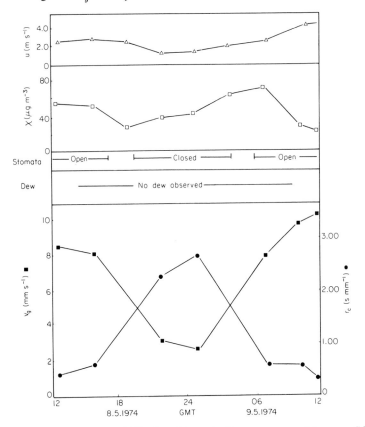

FIG. 8.3. The diurnal course of SO_2 deposition velocity v_g, canopy resistance r_c, SO_2 concentration χ (at 2 m), windspeed u (at 2 m), the presence of dew and the state of stomata for 8–9 May 1974.

values of v_g occurring when stomata were fully open (Fig. 8.3). Unlike water vapour exchange, however, the rates of SO_2 deposition on the dry crop at night (stomata closed) were still appreciable. The canopy resistance for SO_2 in these conditions was about 250 s m^{-1}, whereas the equivalent canopy resistance for water vapour transfer is $\geqslant 1000$ s m^{-1} (Demead 1975). These observations indicate the presence in the wheat canopy of a significant sink for SO_2 that is not a major source of water vapour.

The presence of dew on the foliage also had a marked effect on v_g. Typical results (Fig. 8.4) show an increase in v_g and a decrease in r_c (to approximately zero) in the presence of dew. As the crop became senescent and the green-leaf area index decreased, daytime deposition velocities became smaller, less variable and most of any diurnal change was caused by the appearance of dew. The data presented in Fig. 8.5 were obtained during this period and as the area of green foliage was small the stomata were assumed to be permanently closed. Unlike other sequences of v_g measurements in the presence of

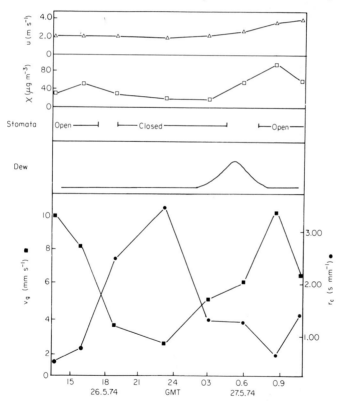

Fig. 8.4. The diurnal course of SO_2 deposition velocity v_g, canopy resistance r_c, SO_2 concentration χ (at 2 m), windspeed u (at 2 m), the presence of dew and the state of stomata for 26–27 May 1974.

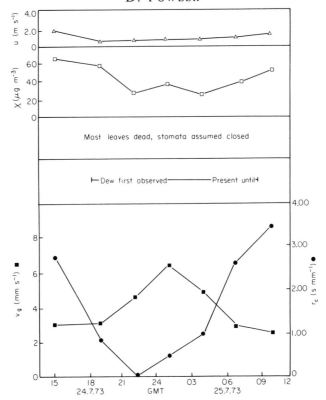

FIG. 8.5.   The diurnal course of SO₂ deposition velocity v_g, canopy resistance r_c, SO₂ concentration χ (at 2 m), windspeed u (at 2 m), the presence of dew and the state of stomata for 24–25 July 1975.

dew, on this occasion v_g increased initially but then decreased before the dew evaporated. Finally, immediately before harvest, rates of deposition onto a dry crop were smaller than the lower limit of detection, despite excellent atmospheric conditions for the measurements. In these conditions the crop is a very poor sink for SO_2.

These features may be summarized as follows. Deposition velocities on wheat occur in the range 1 to 15 mm s⁻¹ and are controlled mainly by conditions at the surface. Stomata appear to be an important site of SO_2 uptake, but not the only one. The presence of dew generally leads to large rates of deposition. The measurements shortly before harvest allow a minimum value for the soil component of canopy resistance, r_{c3}, to be estimated. At this time (early August), soil was visible from above the canopy for the first time since early May and leaf area index had declined from 4·5 in early June to 3·5. Profiles of SO_2 concentration could not be detected and the fluxes were therefore very small. If we assume that the entire flux was to the soil, then the maximum value of the flux may be estimated from the maximum gradient that could have existed without detection (*viz.* $0.5 \mu g \ m^{-3} \ m^{-1}$). This yields a minimum value for r_{c3} of 900 s m^{-1}. Using this value, the maximum flux into the soil during the months May and June was only 5 to 10% of the total flux. It follows that for most of the growing season in dry conditions, SO_2 uptake is primarily by surfaces within the sub-stomatal cavity and by foliage surfaces. Further, rates of uptake by these surfaces are the limiting steps in the deposition process. Average values of the resistances r_{c1} and r_{c2} for the period prior to senescence may now be estimated. At night the value of r_c is very much smaller than the equivalent canopy resistance for water vapour. As the stomata are closed at night r_{c1} may be assumed to be very much greater than r_{c2} so that $r_c \simeq r_{c2}$ at night in the absence of dew. With this assumption nine measurements of r_c at night in the absence of dew were averaged to obtain the value 250 ± 30 s m^{-1} for r_{c2}. Providing conditions that influence SO_2 uptake at the cuticle surface remain fairly constant throughout day and night, this value of r_{c2} can be used to estimate cuticular uptake during daylight. The value of r_{c1} can then be estimated from daytime measurements with open stomata early in the growing season and was found to be about 100 s m^{-1}. If relative humidity had decreased during the day, doubling the value of r_{c2}, then r_{c1} would be 80 s m^{-1} so the assumptions in this calculation of r_{c1} are not very sensitive to changes in r_{c2}.

The influence of dew on v_g also provides an important guide to uptake mechanisms at the surface. Although atmospheric stability poses serious problems for application of the flux-gradient principle (Thom 1975), several sets of observations were obtained on nights when strong stability was prevented by moderate windspeeds. On such occasions the initial effect of dew was to reduce r_c to zero, normal sinks for uptake in this case being 'short-circuited' by absorption into the dew. For the condition $r_c =$ zero, the surface is a perfect sink and in these conditions atmospheric processes limit rates of uptake. On one occasion the surface behaved as a perfect sink for only 2 to 3 hours (Fig. 8.5) and r_c increased with time in the presence of the dew. This has been interpreted as the result of a decrease in the pH of the dew to the value at which liquid phase resistances limit uptake because of uptake and oxidation of SO_2 (Fowler and Unsworth 1979).

DISCUSSION

In setting boundary conditions for studies of long range transport (Fisher 1975) and for regional (Dovland, Joranger & Semb 1976) and global sulphur budgets (Robinson & Robbins 1970) it has been customary to apply a single value of v_g (usually 8 mm s^{-1}) with little regard to differences between surfaces over which the original measurements were made and those to which the value is being applied. The generalized description of SO_2 deposition and its limiting steps obtained using methods presented in this paper provides a framework within which the most important steps may be identified and studied separately. In this way rates of deposition may be more reliably predicted. The use of mean values for surface resistance, together with appropriate values of atmospheric resistance estimated from known conditions of windspeed and surface roughness, provide values of v_g that agree well with independent measurements (Fowler & Unsworth 1979), and enable the quantity of SO_2 absorbed by the crop over the growing season to be estimated. However, use of such mean values does conceal large, short-term changes in surface properties, especially of stomatal resistance. Also, cuticular surfaces are unlikely to retain a constant affinity for SO_2 as chemical and physical reactions between epicuticular wax and the atmosphere change the structure and composition of the waxes. The application of micrometeorological techniques to study mechanisms limiting the exchange of pollutant gases between vegetation and the atmosphere is however an indication of the promise that such methods hold for future work. Further, the recent development in gas analytical and data logging equipment will make future measurements more discriminating and applicable for a larger range of gases (Unsworth, this volume).

REFERENCES

Biscoe P.V., Clark J.A., Gregson K., McGowan M., Monteith J.L. & Scott R.K. (1975) Barley and its environment. 1. Theory and practice. *Journal of Applied Ecology*, **12**, 227–247.

Denmead O.T. (1975) Temperate cereals. *Vegetation and the Atmosphere*, Vol. 2 (Ed. J.L. Monteith), pp. 1–31. Academic Press, London.

Dovland H., Joranger E. & Semb A. (1976) Deposition of air pollutants in Norway. Impact of acid precipitation on forest and freshwater ecosystems in Norway. *Report 6/76* (Ed. Braekke, F.H.).

Fisher B.E.A. (1975) The long range transport of sulphur dioxide. *Atmospheric Environment*, **9**, 1063–1070.

Fowler D. (1976) Uptake of SO_2 by crops and soil Ph.D. Thesis., University of Nottingham.

Fowler D. & Unsworth M.H. (1979) Turbulent transfer of sulphur dioxide to a wheat crop. *Quarterly Journal of the Royal Meteorological Society*, **105**, 767–784.

Garland J.A. (1977) The dry deposition of SO_2 to land and water surfaces. *Proceedings of the Royal Society London*, A.**354**, 254–268.

Robinson E. & Robbins R.C. (1970) Gaseous sulphur pollutants from urban and natural sources. *Journal of the Air Pollution Control Association*, **20**, 233–235.

Spedding D.J. (1969) Uptake of sulphur dioxide by barley leaves at low sulphur dioxide concentrations. *Nature*, **224**, 1229–1230.

Thom A.S. (1975) Momentum, mass and heat exchange of plant communities. *Vegetation and the Atmosphere*, Vol. 1 (Ed. J.L. Monteith), pp. 57–109. Academic Press, London.

9. TRANSPORT AND CAPTURE OF PARTICLES BY VEGETATION

A.C. CHAMBERLAIN AND P. LITTLE

Environmental and Medical Sciences Division,
Atomic Energy Research Establishment,
Harwell, Oxfordshire, OX11 0RA

SUMMARY

1. Particles are deposited on vegetation by several processes, *viz.* sedimentation, impaction, interception and Brownian motion.

2. It has been shown experimentally that impaction rate depends on particle size, wind speed, the nature of the surface and the stickiness of the particle. In some cases, particles are observed to bounce off the surface.

3. Very small particles (< 1 μm) are deposited by Brownian diffusion, analogous to the diffusion of gases. Rates of deposition are much less than that of spores and pollen and also less than those of gases.

4. Results of wind tunnel studies have been applied to model the deposition of particles to forest canopies, and the deposition of fungal spores within a cereal crop.

5. In assessing the accumulation of pollutants on foliage allowance must be made for the dilution effect caused by new growth, and other effects caused by loss of leaf material and uptake from the roots.

INTRODUCTION

The airborne dispersion and capture of particles and droplets on vegetation has interested plant pathologists for many years. The review by Gregory (1945) marks the first attempt to apply modern theories of atmospheric dispersion and particle physics to the subject. Gregory's (1973) book *Microbiology of the Atmosphere* is mainly about spores and pollen but also gives useful general information on particle-plant interactions. Theoretical work and experiments in the field and in wind tunnels have been done at Hanford, USA, with particular reference to deposition and resuspension of particles from the ground in arid climates (Energy Research and Development Administration 1976). Chamberlain (1975) has also reviewed the subject. The aims of this paper are to describe some more recent work.

In principle, the theory of travel and deposition of particles above and within plant canopies can be synthesized from data on interactions with individual plant elements together with well-established theory of eddy

147

diffusion. The theory of vapour transfer is analogous, but there is one simplification and several complications where particles are considered. There are no short term physiological modifications of leaf surfaces which affect particle deposition as drastically as stomatal opening and closure affect exchange of water vapour or CO_2. However leaf and stem structure and the stickiness and wetness of surfaces, partly morphological and partly adventitious in nature, greatly influence capture efficiency for particles. Also wind speed within the canopy is more critical for particle capture than for gaseous transfer, and the same is true for leaf posture which is partly determined by wind. The size and physico-chemical nature of the particles are very important, quite different mechanisms of deposition applying over the range of sizes of natural and man-made particles in the atmosphere.

There is great variation in the size and nature of particles in the atmosphere. A broad four-way classification is:

1. Spores and pollen
2. Dusts, formed by fragmentation of matter
3. Smokes and fumes, formed by condensation of metallic vapours or volatile organic substances, or by gas-phase reactions in the atmosphere
4. Mists formed by condensation of water into nuclei.

The classification is not rigid. For example ammonium sulphate particles, which cause haze in country air, are formed by gas-phase reactions, and may be called smoke, but when relative humidity exceeds 82% they deliquesce and become mist.

In Table 9.1, approximate ranges of particle diameters are given, but there are exceptions in almost every case. For example, the particle size of salt nuclei (or droplets if relative humidity exceeds 75%) depends on wind speed offshore and proximity to the coast. The modes of deposition in Table 9.1 refer to dry weather. Most particles can be incorporated in fog or cloud droplets, be captured by raindrops, or coagulate with other particles, and when this happens their physical properties are completely altered. It is believed that bacteria and viruses in the atmosphere are usually attached to other particles.

TABLE 9.1. Particles in the atmosphere.

Particles	Range of diameter (μm)		Main modes of deposition (other than in rain)
Soil and road dust	10	to 100	Sedimentation
Spores and pollen	5	to 50	Sedimentation, impaction
Fly ash	1	to 20	Sedimentation, impaction
Bacteria	1	to 10	Uncertain
Sea salt nuclei	0·5	to 5	Impaction, sedimentation
Ammonium sulphate	0·1	to 1	Uncertain
Lead from road vehicles	0·04	to 1	Brownian diffusion
Diesel and domestic smoke	0·01	to 1	Brownian diffusion

MECHANISM OF DEPOSITION

Particles are deposited on surfaces by sedimentation, impaction, interception and Brownian diffusion. Deposition by electrical or thermal forces is sometimes effective in the laboratory but rarely in the field. The following terminology is used:

χ Number or mass of particles per unit volume of air

$Q(x)$ Number or mass of particles airborne at distance x downwind

N   Number or mass of particles deposited per second per unit plan area of leaves, per unit maximum projection area of petioles, stems or twigs or per unit area of ground

$u(z)$ Wind speed at height z

$u(h)$ Wind speed at height h of top of canopy

u_* Friction velocity

d_p Diameter of particle

ρ Density of particle

ρ_0 Density of air

v_s Sedimentation velocity: for particles with $0\cdot1 < d_p < 50\ \mu$m, Stokes Law gives $v_s = (d_p{}^2\, g\rho)/(18\, \rho_0 \nu)$

v_g Velocity of deposition, equal to N/χ

S Stopping distance, equal to horizontal distance travelled by particle in still air when given an initial horizontal velocity u. To an approximation, $S = v_s u g^{-1}$

L Effective diameter of obstacle, equal to diameter of sphere or cylinder. For leaves, L is usually taken as diameter of circle of equivalent area

Re   Reynolds number of flow round obstacle, equal to $uL\nu^{-1}$

St   Stokes number equal to S/L, a parameter which determines impaction efficiency of particles on obstacles

C_I Impaction efficiency, defined as number of particles striking an obstacle divided by number that would have passed through the space occupied by it if it had not been there

C_P Capture efficiency, equal to C_I if all impacts are captures. C_P is also capture efficiency of ground surface or canopy, equal to v_g/u

D Diffusion coefficient. For gases this is the molecular diffusion coefficient, for particles it is the Brownian diffusion coefficient

ν Kinematic viscosity of air

Sc   Schmidt number, equal to D/ν, a non-dimensional measure of the diffusivity

Sh   Sherwood number, equal to $v_g L D^{-1}$, a non-dimensional mass transfer coefficient. By convention, in deriving Sh, v_g is calculated with reference to the total area of the surface, not the plan or projection area.

The physical processes governing deposition on surfaces are described in textbooks on aerosol physics, for example, Fuchs (1964), Green & Lane (1964), Friedlander (1977), and the application to deposition on vegetation by Gregory (1973) and Chamberlain (1975). The main factors affecting the efficiency of the mechanisms of deposition are as follows.

Sedimentation

As v_s depends on the product of ρ and $d_p{}^2$ it depends strongly on particle size and density. It is independent of wind speed, but if sedimentation is the only mechanism of deposition, C_p, equal to v_s/u, is inversely proportional to wind speed.

A sedimentation velocity of 2 cm s^{-1}, typical of spores and pollen with diameters about 30–40 μm, is small in comparison with normal wind speeds above the canopy, and the particle deviates only slightly from the streamlines of air flow. Within dense canopies, v_s/u is much greater, and particles sediment out of the air flow to much greater effect.

Impaction

Impaction occurs where streamlines of air flow bend round obstacles. Particles deviate from the streamlines by virtue of their inertia. If a particle carried by wind with velocity u approaches an obstacle, it will deviate from the streamlines by a distance related to its stopping distance, S, but depending also on the geometry of the obstacle. Since streamlines bend more sharply round small than round large obstacles, the chance of the particle impacting is a function of the Stokes number, St, equal to S/L. The dependence of C_I on St for certain shapes of obstacle (cylinders, spheres, discs, strips) has been worked out theoretically (Fuchs 1964) and measured experimentally with droplets (May & Clifford 1967).

Gregory & Stedman (1953) showed that the impaction of spores on cylinders followed the same rules, except that at very high u (about 10 m s^{-1}) there was a reduction in efficiency of capture (C_P no longer equal to C_I even though the cylinders were made sticky).

Because S increases with u, C_I increases with u, and v_g, which equals uC_I, increases more rapidly. This is an important distinction between impaction and sedimentation. Impaction is most important when particles are large (since S is proportional to v_s), wind speeds are high, and obstacles are small (since St $= S/L$).

Interception

Interception occurs when the streamlines carrying a particle approach so near an obstacle that the particle touches it without deviation. It is an important mechanism of filtration by fine fibrous media, but is probably only of minor importance in nature. The hairs on leaves may catch particles by interception.

Brownian motion

In sedimentation and impaction, particles cross the sublayer of air flow and deposit on surfaces because their mass and momentum cause them to react to gravity and to impulsive forces. In Brownian motion, which is important only for particles less than 1 μm diameter, particles diffuse to surfaces in the same way that SO_2 or any other gas diffuses, except that the diffusivity of the particle is many orders of magnitude less. The Schmidt number, Sc, is the ratio of the kinematic viscosity of air, ν, to the diffusivity, D, of the particle or gas. For a 0·1 μm particle, Sc $= 2\cdot2 \times 10^4$, and is independent of its density. This does not mean that the mass transfer coefficient is $2\cdot2 \times 10^4$ times less than the mass transfer coefficient of a gas, because the gradient of concentration of particles near the surface, set up in response to the deposition at the surface, is steeper than the gradient of concentration of gases. It turns out that C_P is approximately proportional to $Sc^{-2/3}$. Particles small enough to have appreciable Brownian motion adhere very strongly to surfaces and there is no bounce-off and little or no blow-off of these particles.

DEPOSITION BY IMPACTION ON SHOOTS

Wind tunnel experiments

Spores and pollen grains can be counted under the microscope when they are deposited on artificial surfaces, but it is more difficult to do this when plant shoots are used. Carter (1965) placed apricot wood infected with the fungus *Eutypa armeniacae* in a wind tunnel and measured the deposition of the ascospore octads on fresh shoots downwind. The ascospores were resuspended from the shoots into water, aliquots were plated out onto agar, and colonies counted after incubation.

Belot (1975) and Belot & Gauthier (1975) made particles by dissolving uranine and methylene blue dye in methanol and spraying this from a spinning top aerosol generator. The methanol rapidly evaporated, leaving dye particles, of which the size could be varied. The particles were impacted on shoots of pine (*Pinus sylvestris*) and oak (*Quercus sessiliflora*) and after the experiment the particles were washed off the foliage and the amount of dye measured with a fluorimeter. Little (1977) and Little & Wiffen (1977) generated polystyrene particles tagged with the radioactive tracer [99m]Te, and measured the airborne concentration and the deposition by radioactive counting methods on shoots of nettle (*Urtica dioica*), beech (*Fagus sylvatica*) and white poplar (*Populus alba*). Recently, Little (unpublished) has repeated the measurements with shoots of pine (*Pinus sylvestris*) to obtain a comparison with the results of Belot and Gauthier. Comparisons were made of the catch on sticky (vaselined) and untreated pine needles and stems. As it was found impracticable to treat needles with vaseline in the position of growth, a number of individual sticky (S) and non-sticky (NS) needles and stems were exposed in the wind tunnel, normal to the airstream, during the release of the polystyrene particles.

Stems and twigs of the various plants used in these studies, and also individual pine needles, could be arranged perpendicular to the flow, and the

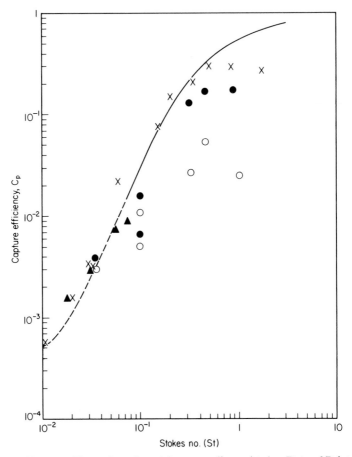

Fig. 9.1.   Efficiency of impaction of particles on needles and twigs. Data of Belot (1975) using dye particles on pine needles (×), Little (1979) using polystyrene particles on sticky (●) and non sticky (○) pine needles and Carter (1965) using ascospore octads on apricot twigs (▲).

collection efficiency C_P compared with theoretical values. The results for the three series of experiments are assembled in Fig. 9.1. The full curve is the theoretical efficiency for impaction on cylinders (May & Clifford 1967) and the dashed curve is an extrapolation by Belot (1975), who found that his results correlated well when C_P was plotted against St, even for values of St less than 0·1, at which theoretically C_P should be zero. Carter's results with ascospores, and Little's with polystyrene particles on sticky pine needles, fall near to Belot's curve, but the points for polystyrene particles on untreated needles fall below the line. Little's (unpublished) results are shown in Table 9.2 in terms of v_g to isolated needles and stems and also to needles, stems and whole shoots in the position of growth.

TABLE 9.2. Velocity of deposition (cm s^{-1}) of polystyrene particles on pine shoots and on isolated needles and stems.

Particle size (μm)	2·75		5			8·5	
Wind speed (m s^{-1})	2·5	5	1·5	2·5	5	2·5	5
Pine shoots:							
needles	1·0	5·6	1·6	2·6	15	9·6	11·7
stems	0·63	3·2	0·59	0·98	16	6·2	15·5
whole shoots	0·93	5·1	1·5	2·5	15	9·4	12·0
Isolated needles:							
non-sticky	0·74	2·5	1·9	—	13	13	12
sticky	0·94	3·1	2·4	—	64	43	110
ratio NS/S	0·79	0·81	0·79	—	0·20	0·30	0·11
Isolated stems:							
non-sticky	0·73	2·4	1·2	—	14	5·1	13
sticky	0·41	2·5	0·93	—	23	4·8	25
ratio NS/S	1·8	1·0	1·3	—	0·61	1·1	0·52

Clearly, with polystyrene particles of 5 or 8·5 μm diameter, the capture by sticky needles and stems exceeds that of non-sticky needles and stems, particularly at the higher wind speeds. Belot's dye particles, and Carter's ascospore octads, gave impaction efficiencies close to the theoretical values, even though the surfaces were non-sticky. This is explicable since the ascospores (certainly) and the dye particles (probably) were rather moist and sticky themselves whereas the polystyrene particles were not.

In Fig. 9.2 (middle and right hand panels) the collection efficiency of the whole shoots is shown as a function of wind speed. As the leaves and petioles trailed downwind, the area normal to the wind flow, and the characteristic lengths, are not known, and St cannot be calculated. C_P is calculated as $N\chi^{-1} u^{-1}$ from the particles deposited per unit plan area of leaf, or per unit maximum projection area of petiole or stem, without allowing for the inclination to the air flow.

In the left hand panel of Fig. 9.2 results are shown of Gregory and his colleagues (Gregory 1973) obtained in the wind tunnel by impaction of *Lycopodium* spores on potato (*Solanum tuberosum*), and broad bean (*Vicia faba*) leaves, and on inclined sticky microscope slides. This again shows the wide divergence of results between natural and artificial (sticky) surfaces for impaction of the larger pollen particles at moderate and high wind speeds.

Even with one type of particle (polystyrene) there is considerable variation, in Fig. 9.2, in the results for nettle, white poplar and beech. This can probably

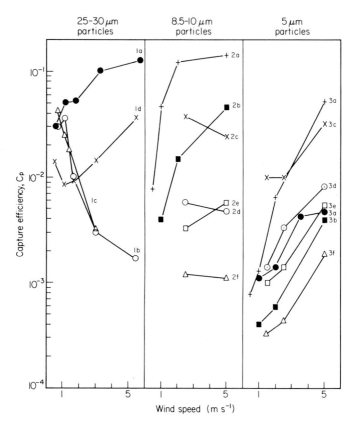

FIG. 9.2. Efficiency of deposition of particles on shoots, leaves and slides. Data of Gregory (1973) using *Lycopodium* spores (30 μm) depositing on microscope slides at 45° (1a), potato leaves (1b), broad bean leaves (1c), *Erysiphe graminis* spores (25 × 12 μm) on microscope slides at 45° (1d); Belot (1975) using 10 μm dye particles on pine shoots (2a), oak shoots (2b), 5 μm dye particles on pine shoots (3a), oak shoots (3b); Little (1979) using 8·5 μm polystyrene particles on pine shoots (2c), 5 μm polystyrene particles on pine shoots (3c); Little (1977) using 8·5 μm polystyrene particles on nettle (2d), white poplar (2e), beech shoots (2f), 5 μm polystyrene particles on nettle (3d), white poplar (3e), beech shoots (3f); Carter (1965) using *Eutypa armeniacae* ascospore octads (8 × 4 μm) on apricot shoots (3g).

be explained by the nature of the leaf surface, the rough and coarsely hairy leaves of nettle and the densely tomentose leaves of white poplar being more efficient at collection (or retention) of particles than the smooth shiny leaves of mature beech.

Other authors have reported the effect of leaf surface structure on collection efficiency. Wedding *et al.* (1975) in wind tunnel experiments with 6·8 μm diameter uranine dye particles found $C_P = 0·05$ for the pubescent leaves of sunflower (*Helianthus annuus*) but $C_P = 0·008$ for the glabrous leaves of tulip poplar (*Liriodendron tulipifera*) at $u = 2·7$ m s^{-1}. Forster (1977) found that the presence or absence of wax on leaves of spruce (*Picea sitchensis*) and sycamore (*Acer pseudoplatanus*) considerably affected the deposition of *Lycopodium* spores, and Tauber (1967) found many more pollen grains deposited naturally in a forest on hairy twigs of willow than on smooth twigs of birch. Roberts (1975) found 5 times higher lead concentrations in pubescent birch than in glabrous maple leaves exposed to lead in air and dust fall near a smelter. However White & Turner (1970) found the opposite when they sprayed shoots of various tree species with droplets of salt solution. Hazel (*Corylus avellana*), with hairs on both surfaces, was least efficient, ash (*Fraxinus excelsior*) and birch (*Betula spp.*) with no hairs, most efficient. Possibly hairs reduce the impaction efficiency for droplets, by increasing the boundary layer thickness, but increase the collection efficiency for dry particles, by cushioning the impact and reducing bounce-off.

The evidence that particles in the size range 5 to 30 μm bounce off leaves and stems is indirect, but in the laboratory particles have been observed bouncing off surfaces by microcinematography (Löffler & Umhauer 1971; Hiller & Löffler 1978; Esmen, Ziegler & Whitfield 1978). The probability of bounce-off depends on the coefficient of restitution of the particle and on the energy of interaction between particle and surface. When a particle nears a surface it falls into the particle-surface potential well, and if it rebounds with insufficient energy to escape the well, it will be captured. The depth of the potential well depends on particle and surface deformation, which is the probable reason why even a very thin film of moisture increases capture efficiency.

Experiments with particles of glass, quartz and paraffin wax impinging on artificial (polyamide) fibres of 20 μm diameter (Hiller & Löffler 1978) show that collection efficiency falls below 20% when the approach velocity exceeds about 0·5 m s^{-1} for 10 μm particles and 1 m s^{-1} for 5 μm particles. Bounce-off is particularly effective when the obstacle is very small, for two reasons:

(a) the boundary layer is small, so the particle is not slowed down much before it strikes the obstacle

(b) if rebound occurs, the particle is unlikely to strike the obstacle a second time.

This effect was shown clearly in Gregory's (1951) experiments on the impaction of *Lycopodium* spores on cylinders. Although the cylinders were made sticky, collection efficiency decreased with increasing wind speed when the cylinder diameter was reduced to 0·18 mm.

Recently Kyaw Tha Paw U & Reifsnyder (1979) at Yale, have observed the bounce-off of *Lycopodium* spores from leaves of American elm (*Ulmus americana*). To obtain reproducible geometry, the leaves were wrapped round 3·74 mm diameter cylinders. At wind speeds of 5·5 and 9·7 m s^{-1}, the spores were observed to bounce off the leaves with rebound coefficients (ratios of rebound velocity to incident normal velocity) of 0·28. The authors imply that rebound would also take place at typical canopy wind speeds of about 1·5 m s^{-1}. If the leaf was at a 45° angle, a particle impinging at this wind speed, and bouncing off, would strike the leaf again less than 4 mm from the point of original contact, and as its velocity would be much reduced, would have a good chance of capture. If the spore struck a petiole, and rebounded similarly, it would be unlikely to impact a second time.

DEPOSITION BY BROWNIAN DIFFUSION

Impaction efficiency is negligible for particles of diameter less than about 1 μm. Some interception may occur, for example on leaf hairs, but the efficiency of this is low, since the particles are caught only if the streamlines of their travel pass within a distance from the surface less than the radius of the particle. The diffusivity of a particle is independent of its density, and depends only on its size and shape. For 0·1 μm and 0·01 μm spherical particles, $D = 7 \times 10^{-6}$ and 5×10^{-4} cm^2 s^{-1} respectively, so the diffusivity of even the smallest particle is much less than that of gases, for which D is of order 0·1 cm^2 s^{-1}. Chemical engineers have found that mass transfer of gases and particles with varying diffusivity can be correlated if the product ShSc$^{-1/3}$ is correlated with the Reynolds number of the flow over the surface. The Sherwood number Sh is a deposition velocity made non-dimensional by multiplying by the characteristic length of the surface and dividing by D, and the Schmidt number Sc is the ratio of the kinematic viscosity of air to the diffusivity of the gas or particle.

Fig. 9.3 shows the correlation of ShSc$^{-1/3}$ with Re for the transport of both gases and particles to leaves, petioles and stems. The results of Grace & Wilson (1976) were obtained by measuring the rate of evaporation of water ($D = 0·25$ cm^2 s^{-1}) from a flat filter paper replica of a leaf of poplar (*Populus* x *euramericana*). Those of Chamberlain (1974) refer to transport of ^{212}Pb vapour ($D = 0·054$ cm^2 s^{-1}) to leaves of bean (*Vicia faba*).

The bean leaves were in their normal posture in a canopy in the wind tunnel, whereas the model leaves of Grace & Wilson were glued to aluminium templates and placed parallel to the mean wind flow. The lines drawn in

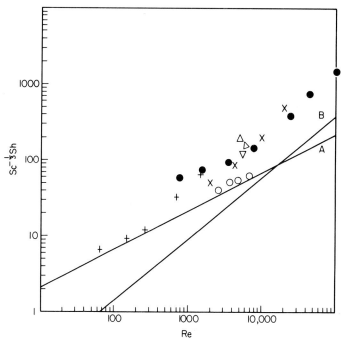

FIG. 9.3. Transport of particles and gases to and from leaves by Brownian diffusion.
Particles: 0·17 μm on pine (+) and oak (×) (Belot 1975); 0·03 μm on nettle (△), white
poplar (▷) and beech (▽) (Little & Wiffen 1977). Gases: water vapour from poplar (●)
(Grace & Wilson 1976); ^{212}Pb on bean (○) (Chamberlain 1974). The lines show the
theoretical result for laminar flow (A) and turbulent flow (B).

Fig. 9.3 are theoretical correlations of mass transfer to smooth surfaces in
laminar (A) or turbulent (B) flow. The paper of Grace and Wilson gives
references to several other correlations of heat or water vapour transport
to model leaves, which show similar results.

Also shown in Fig. 9.3 are the results of experiments by Belot (1975) on
transport of 0·17 μm particles ($D = 3\cdot2 \times 10^{-6}$ cm^2 s^{-1}, Sc $= 4\cdot7 \times 10^4$)
to leaves of oak (*Quercus sessiliflora*) and pine (*Pinus sylvestris*), and experiments by Little & Wiffen (1977) with 0·03 μm particles ($D = 7 \times 10^{-5}$
cm^2 s^{-1}, Sc $= 2 \times 10^3$) to leaves of nettle (*Urtica dioica*), beech (*Fagus
sylvatica*) and white poplar (*Populus alba*). The values of ShSc$^{-1/3}$ for particles
are greater by a factor of up to 2·5 than the theoretical smooth surface values,
but this also appears in most heat and vapour transfer measurements, and is
attributed to the effects of surface roughness, leaf posture, turbulence and
aspect ratio (limited cross-wind dimension of leaf surface). Considering that
the range of diffusivity is five orders of magnitude, the correlation is good.
It follows from Fig. 9.3 that the rate of deposition by Brownian motion varies

with u raised to an exponent less than unity, so C_P decreases with increasing wind speed.

Belot (1975) and Belot, Baille & Delmas (1976) have used their results to predict the deposition of sub-micron particles in coniferous forests (see below). As would be expected from theory and from previous wind tunnel experiments (Chamberlain 1966), the rate of deposition of sub-micron particles is much less than that of spores and pollen (5 to 30 μm diameter) and also less than that of gases such as SO_2.

DISTRIBUTION OF CATCH BETWEEN LEAVES, PETIOLES AND STEMS

Table 9.3 shows the percentages of the total catch of particles on plant shoots that were found on leaves, petioles and stems in the wind tunnel experiments of Little (1977), Little & Wiffen (1977) and Carter (1965). Although the specific surface and impaction efficiency generally increase in the sequence

TABLE 9.3. Mean percentage of total catch intercepted by leaf laminas, needles, petioles and stems.

Particle size	Plant part	Beech	White poplar	Nettle	Pine	Apricot
Little (1977); Little & Wiffen (1977); Little (unpublished)						
0·3 μm	leaf laminas	80	75	83	—	—
	petioles	15	12	7	—	—
	stems	6	14	10	—	—
0·2 μm	leaf laminas	88	77	86	—	—
	petioles	7	—	6	—	—
	stems	5	11	8	—	—
2·75 μm	leaf laminas	63	74	68	—	—
	petioles	30	11	16	—	—
	needles				97	
	stems	7	15	16	3	—
5·0 μm	leaf laminas	71	57	82	—	—
	petioles	23	27	7	—	—
	needles	—	—	—	98	—
	stems	6	16	11	2	—
8·5 μm	leaf laminas	63	45	68	—	—
	petioles	17	29	11	—	—
	needles	—	—	—	97	—
	stems	20	26	20	3	—
Carter (1965)						
8·1 × 4·4 μm	leaf laminas	—	—	—	—	81
(ascospore octads)	petioles	—	—	—	—	13
	stems					6

leaves, stems, petioles, this is more than compensated by the greater area of leaves. Over a wide range of particle sizes leaves account for 60 to 80% of the total catch.

DEPOSITION IN CANOPIES

Coniferous forests

Given the shape and size of leaves and stems in a canopy, the wind speed versus height, and the impaction efficiency of each crop element at the appropriate wind speed, the rate of deposition to the canopy as a whole can be calculated. Belot and his colleagues (Belot 1975; Belot *et al.* 1976) have done this for a canopy of pine (*Pinus sylvestris*). The extent to which the predictions of the model are borne out in practice gives some indication of the accuracy of the premises, in particular the assumption that the rate of capture of particles is equal to the calculated rate of impaction or sedimentation.

Belot *et al.* (1976) considered deposition in a pine canopy of aerosol particles advected from outside, as for example from a nearby town. The forest was modelled on that at Thetford, with canopy height h, zero displacement d and roughness length z_0, equal respectively to 15, 11 and 1 m (for definitions of the last two terms see Thom 1975). The wind speed was characterized either by the wind at canopy height, $u(h)$, or by the friction velocity, u_*. In neutral conditions, to which calculations applied, the usual logarithmic law gave

$$u(h)/u_* = 5{\cdot}75 \log_{10} (h - d)/z_0$$
$$= 3{\cdot}45$$

Belot *et al.* calculated the 'vitesse integrales de depot' (velocity of deposition) to the canopy for a range of particle sizes from 0·2 to 10 μm diameter. Fig. 9.4 shows the velocity of deposition of 10, 5 and 2 μm particles as a function of the friction velocity. The dashed line in Fig. 9.4 corresponds to $u_*^2/u(h)$, which may be considered the velocity of deposition for momentum, since it is equal to the ratio of the shearing stress to the momentum of the air at height h. The theory of Belot *et al.* predicts that v_g for particles of 10 μm diameter and friction velocities exceeding 0·5 m s^{-1} actually exceeds the corresponding velocity for momentum (calculated with h as reference height).

There is no direct experimental evidence of the velocity of deposition of particles to forests, but measurements have been made in Japan of the deposition of fog droplets on coniferous forests. The measurements were done in a study of the efficacy of coastal forests in diminishing the advection over the land of coastal mists driven by wind. The mean diameter of the

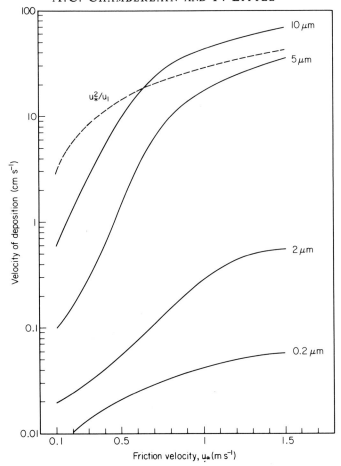

Fig. 9.4.   Velocity of deposition of particles to a coniferous forest (Belot *et al.* 1976).

droplets was about 17 μm. The deposition of water was estimated by a variety of methods, and it was concluded (Yosida 1953) that 'the forests . . . captured on the average 0·5 litres of fog water in one hour for each 1 square meter of their areas, when the wind containing fog water of 800 mg/m³ was blowing over them with velocity 4 m/s.' This is equivalent to a velocity of deposition of 17 cm s^{-1}. From the details of the forests and height of measurement of wind speed it can be estimated that the friction velocity was about 1 m s^{-1}. Thus the rate of deposition of fog droplets was comparable with, though rather less than, the value calculated by Belot *et al.*

In experiments on deposition of *Lycopodium* spores to a canopy of pine seedlings on the floor of a wind tunnel ($h = 8$ cm, $z_0 = 0·6$ cm, $u(h) = 2·5$ m s^{-1}, $u_* = 0·76$ m s^{-1}) Chamberlain found v_g equal to 9·0 cm s^{-1} when the seedlings were wetted, but only 2·9 cm s^{-1} when they were dry. This and other wind tunnel work implies that particles can readily bounce off pine needles unless either particle or needle is wet or sticky.

The theoretical curves for deposition of 10 and 5 μm particles to the model forest in Fig. 9.4 show v_g varying with u_* raised to a power of nearly 2. Consequently the canopy collection efficiency C_P, equal to $v_g/u(h)$, increases with increasing wind speed.

Distance of travel of particles

The distance of travel of particles in or above a canopy is related to C_P. If Q_0 is the source strength of particles emitted from a point, and $Q(x)$ is the number still airborne at distance x downwind, then $Q(x)/Q_0$ is known as the suspension ratio. Elementary theories of particle travel and deposition, such as that of Chamberlain (1953), predict that $Q(x)/Q_0$ is strongly dependent on C_P. For example, if the height of release is 10 m, the elementary theory predicts that, at 1 km downwind, $Q(x)/Q_0$ equals 0·73 if $C_P = 0·01$ but only 0·20 if $C_P = 0·05$ (see Gregory 1973, Fig. 33).

This dependence of $Q(x)/Q_0$ on C_P also emerges from the more exact theory of Belot and his co-workers. Fig. 9.5 shows the depletion of the plume in travel over coniferous forest for three particle sizes in moderate and light winds (u_* equal to 0·6 and 0·2 ms^{-1}). The theory predicts that fewer particles travel a given distance in strong than in light winds, especially if the particle diameter is equal to or greater than 10 μm.

The limited experimental evidence on transport of pollen grains tends to the opposite conclusion. Raynor, Hayes & Ogden (1974) measured the concentration of the ragweed pollen (*Ambrosia trifida*) at distances up to 60 m into a forest of pine trees, and found that the horizontal mass flux of pollen was depleted faster with a low wind speed than with a higher one. The authors concluded 'These results reinforce earlier evidence that the major mechanism of particle loss is deposition within the forest due to low wind speeds and slow transport below the canopy rather than impaction on the foliage. If the latter mechanism were predominant, the trend with wind speed should be reversed.' Tampieri, Mandrioli & Puppi (1977) measured the airborne concentration of pollen of chestnut (*Castanea sativa*) in the Po valley. Chestnut is common in the foothills of the Apennines but is absent from the plain. Measurements at three stations, one in the foothills and two others 20 and 100 km into the plain enabled the relaxation distance (the distance in which the airborne concentration fell by a factor e) to be estimated. In winds of 10 to 20 km h^{-1} this was about 20 km but in winds of 50 km h^{-1} it was 100 km.

Another possible test of the theory would be to measure the vertical gradient $\chi(z)$ of concentration of particles in and above the forest. Belot *et al.*

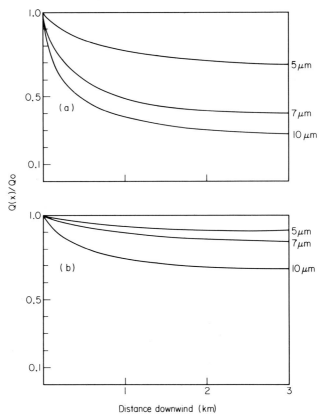

Q(x)/Qo

Distance downwind (km)

Fig. 9.5. Fractions of particles remaining airborne versus distance of travel over forest (Belot *et al.* 1976). (a) $u_* = 0.6$ m s^{-1}, (b) $u_* = 0.2$ m s^{-1}.

(1976) predict that in moderate winds ($u_* = 0.6$ m s^{-1}), χ would increase by about a factor two from just above the canopy to a height of 30 m, in the case where particles are advected from a source 1 km upwind. This increase of $\chi(z)$ with z is a consequence of the predicted high efficiency of impaction in the pine canopy, which depletes χ to the extent that deposition is limited by the rate of eddy diffusion from above. In the study of deposition of fog droplets on forest in Japan, the profile of χ above the forest was found to match the profile of u, in accordance with the hypothesis that the rate of eddy diffusion was limiting. Measurements of the profile of grass or herbaceous pollen above a forest would be of great interest but have not so far been reported.

It has been suggested that the dispersion of motor exhaust lead across the adjoining countryside could be mitigated by planting belts of trees alongside motorways. The deposition of small particles by Brownian diffusion is generally less effective than the impaction of large particles (Fig. 9.4), although there is no question of the small particles bouncing off surfaces. Making use of measurements in the wind tunnel of the deposition of exhaust lead to leaves, Little & Wiffen (1978) have calculated that a 10 m wide belt of white poplar alongside a motorway would intercept no more than about 1·5 % of the lead emitted.

Cereal crops

Legg & Powell (1979) have developed a mathematical model for deposition of *Lycopodium* spores (30 μm diameter) and *Erysiphe graminis* spores (elliptical 28 $\times$ 10 μm) in a ripe barley crop ($h = 1$ m). They calculated the concentration $\chi(xz)$ at various distances x downwind and heights z in and above the canopy. The results were compared with measured concentrations (Bainbridge & Stedman 1979) of *Lycopodium* spores downwind of a line of artificial point sources level with the top of the barley crop, and also with concentrations of *Erysiphe graminis* spores downwind of a narrow strip of infected barley. Good agreement between calculated and measured concentrations of *Lycopodium* were found when sedimentation to leaves (contributing 27 % of total deposition) and impaction on leaves (18 %), stems (36 %) and heads (16 %) of the barley were included. However a discrepancy appeared if impaction on the awns of the barley was included. Theoretically, the awns should trap spores very efficiently, and this would give a deficiency of spores at the height of the awns which was not observed experimentally. The awns may have failed to trap the spores because they became aligned parallel to the wind or because spores bounced off them. The observed profiles of *Erysiphe graminis* were not inconsistent with impaction on the awns, though the agreement was not good enough to be conclusive. Legg and Powell point out that *Erysiphe graminis* is a foliar pathogen and has sticky spores.

In Fig. 9.6 is shown the ratio $\chi(xz)/\chi(oz)$ of concentrations at some reference height z and distance x into the crop to the concentration at the same height at the upwind edge of the crop. The experimental results of Bainbridge & Stedman (1979) and calculations of Legg & Powell (1979) refer to *Lycopodium* spores dispersed from a line source at the height of the top of the canopy and 2 m upwind of the crop. Also shown in Fig. 9.6 are the results of measurements by Hirst & Stedman (1971) of penetration into a wheat crop of pollen from an adjacent plot of sugar beet. The reference height is 0·75 m for Hirst and Stedman's results and 1 m for Legg and Powell's. The fall off of χ with x was less rapid in Hirst and Stedman's experiment, probably because concentrations were reinforced more effectively by downwards diffusion from above. In the absence of such reinforcement, penetration of spores and pollen into a canopy from sources below canopy height is very limited.

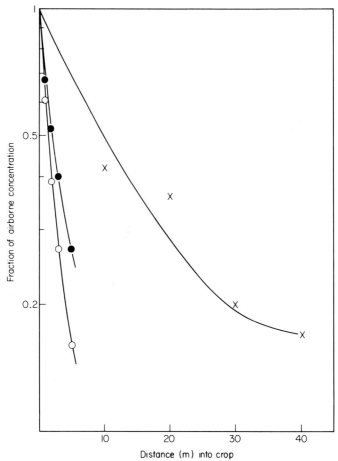

FIG. 9.6. Penetration of spore and pollen clouds into crops according to experiments by Legg & Powell (1979) (○) and Hirst & Stedman (1971) (×), and the theory of Legg & Powell (1979) (●).

Legg and Powell's calculations show that a spore at height 0·5 m within the barley crop has a mean airborne life of about 8 seconds before being deposited on leaves or stems, and as the wind speed at this height is only 0·2 m s⁻¹ it only travels about 1·6 m before becoming trapped. At height 0·8 m, the wind speed is 1 m s⁻¹, but the crop density is higher at this height, owing to the presence of heads, and the mean life before impaction is only about 2·5 seconds. Also, spores at this height are rapidly lost upwards by dispersion out of the crop.

The spores and pollen which have the best chance of long range travel are those which are carried out of the crop in the first few metres of travel. Fig. 9.7a shows Legg & Powell's calculated distribution of *Lycopodium* spores downwind of a line source running cross wind and releasing spores equally at all heights within the barley crop. Neutral stability conditions with $u(h) = 1\cdot7$ m s^{-1} were assumed. The proportion A airborne below crop height diminishes rapidly and virtually vanishes beyond 10 m. At this distance a proportion $0\cdot2$ of the original source remains airborne above the crop, and this constitutes the effective source for long range transport. Fig. 9.7b shows the same calculation for very unstable conditions with $u(h) = 0\cdot75$ m s^{-1}, as might apply on a sunny day in summer. In these conditions a rather higher proportion of the source remains airborne 10 m downwind because impaction on the foliage at the top of the crop is less efficient and also because eddy diffusion in unstable conditions carries the spores out of the crop more readily. These are the conditions under which flowering plants normally release airborne pollen.

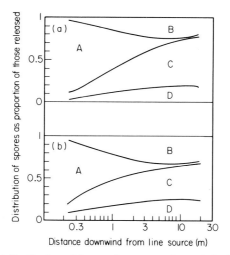

Fig. 9.7. Calculated distribution of *Lycopodium* spores downwind of line source releasing spores at all heights within a barley crop (Legg & Powell 1979): (a) neutral stability, $u(h) = 1\cdot7$ m s^{-1}, (b) unstable, $u(h) = 0\cdot75$ m s^{-1}. Figure gives proportion of spores airborne within the crop (A), airborne above the crop (B), deposited on foliage (C), and deposited on soil (D).

Even discounting capture by the awns, impaction contributed 73% of the total deposition in Legg and Powell's calculations of deposition of *Lycopodium* spores in a barley crop. For the case $\chi(z)$ constant $(0 < z < h)$, the calculation predicts a velocity of deposition of $0\cdot14$ m s^{-1}, corresponding to a canopy resistance to uptake of spores of 7 m^{-1} s. Since $u(h)$ was $1\cdot7$ m s^{-1} and u_* was $0\cdot35$ m s^{-1}, the aerodynamic resistance $u(h)/u_*^2$ was 14 m^{-1} s. In practice $\chi(z) =$ constant does not apply, since a profile develops within the crop, but the comparison of the resistances shows that aerodynamic resistance would be limiting when spores are carried from a distance and deposited on a crop.

This should result in a gradient of spore concentration above the crop similar to the gradient of wind speed. The development of such a gradient can be seen in the measurements by Hirst & Stedman (1971) of pollen carried from a plot of sugar beet over an adjacent plot of wheat. Bainbridge & Stedman (1979) give profiles of the concentration of *Erysiphe graminis* carried from a distance over a crop of mildew-resistant barley which do not appear to show the expected vertical gradient. It is to be hoped that further measurements of this type will be made.

Impaction, sedimentation and Brownian diffusion

The question of whether impaction constitutes an important or even the predominant mode of deposition on canopies of particles in the 10 to 40 μm size range, first raised by Gregory (1945) more than 30 years ago, still remains largely unanswered. Theory indicates that impaction should be predominant at moderate and high wind speeds provided most impacts are captures. For droplets this is true, and it is probably true also for those spores and other particles which are moist or sticky or easily deformed on impact. If impaction is effective and predominant over sedimentation then this should reveal itself in several ways:

(a) Canopy efficiency C_P increasing with increasing wind speed.

(b) Fraction of particles travelling a given distance decreasing with increasing wind speed.

(c) A noticeable gradient of concentration, with χ increasing with z, when particles are carried over a canopy from a distant source.

(d) Particles caught primarily on petioles and stems, needles and other slender foliage components, rather than on leaves.

The evidence, such as it is, for (a), (b) and (c) has already been discussed. The location of capture of natural spores and pollen grains is difficult to determine. Aylor (1975) released stained pollen of ragweed (*Ambrosia elatior*) upwind of a field of corn (*Zea mays*), and subsequently sprayed a thin coating of acrylic plastic on the foliage. When this was stripped, the pollen grains were removed quantitatively onto the plastic, and could be counted under the microscope. The relative contributions of sedimentation and impaction were assessed by comparing deposition on leaf surfaces at various angles to the horizontal. In an experiment in which the wind speed $u(h)$ at canopy height was 0.5 m s^{-1} the number of pollen grains on leaves nearly equalled that expected from sedimentation alone. In another experiment with $u(h) = 2$ m s^{-1}, deposition on leaves within the canopy again corresponded to sedimentation alone, but upper leaves collected pollen at three times the gravitational rate. In both experiments, more pollen was collected on the wider, nearly horizontal, surface near the leaf midsection than on the narrower, nearly vertical surface near the leaf top, indicating that overall sedimentation was more important than impaction.

It would be satisfactory to be able to conclude that spores of foliar pathogens, being mostly moist and sticky, are adapted to be caught by impaction on foliage, whereas pollen grains are adapted to bounce off leaves and stems, and to be caught only by the specially adapted stigmas. This however remains to be proved.

The calculated deposition of sub-micron particles by Brownian motion seems to accord well with experiments. Deposition by this mechanism is an important mode of contamination of foliage by industrial fumes and smokes, for example lead particles from motor vehicle exhaust. The velocity of deposition is much less than for particles of the size of most pollen and spores, and canopies are not so effective in limiting the spread of the smaller particles.

ACCUMULATION OF POLLUTANTS ON FOLIAGE

The phytotoxicity of pollutants usually depends more on the concentration in the foliage, in μg per g or mg per kg dry matter, than on the rate of deposition. Also the amounts of the pollutant taken up by grazing animals, and present in their meat or milk, depends on the concentration in the foliage. This is particularly important in the evaluation of hazards from nuclear fission products such as the radioisotopes of iodine, strontium and caesium, for which the pathways of uptake via forage plants and herbivorous animals are particularly important (Scott Russell 1966; Chamberlain 1970).

To assess such effects, it is necessary to monitor the input of contaminant to the crop over the growing season and also to evaluate the contribution of uptake from the soil of the cumulative fallout. Very often there are insufficient long term measurements of the airborne concentration χ, but more probably the rate of fallout N in μg m^{-2} d^{-1} or μCi m^{-2} d^{-1} is measured by means of deposit gauges of some type. Whether or not the deposit gauge gives an accurate measure of the input to the ground surface is sometimes difficult to determine.

The relation between the concentration in the foliage and the rate of fallout can be calculated if the following parameters can be estimated:

μ = filtration efficiency of vegetation in units m^2 kg^{-1}.
   This is the fraction of the fallout per m^2 ground surface which is intercepted by the foliage per kg dry weight.

λ = field loss factor, units day^{-1}.
   This is the fraction of the deposited material lost from the foliage per day, by re-suspension, wash-off, die-back of leaves, loss of cuticle and other mechanisms.

Clearly any calculation is subject to many uncertainties, for example the dependence of μ on wind speed and particle size and on whether the pollutant is deposited in rainfall or by dry deposition. Chamberlain (1970) derived a value of μ equal to $3 \text{ m}^2 \text{ kg}^{-1}$ from data published in a variety of experiments in which particles were deposited on crops. Subsequently, Miller (1979) analysing more recent data confirmed this value of μ for grasses but not for other canopies.

It might be thought that λ would vary greatly according to the physical and chemical nature of the pollutant, but it seems that it is usually determined by inherent crop factors rather than by the type of pollutant or the frequency of rain. Milbourn & Taylor (1956) found that 50% of ^{89}Sr applied in a spray to pastures was lost from the foliage in 14 days when the activity was measured per unit area of ground. Per unit dry weight the half period was 9 days, owing to dilution by growth. Assuming an exponential loss curve, the field loss half life time T_B is equal to $0.693 \lambda^{-1}$. Chadwick & Chamberlain (1970), in similar experiments, found T_B for ^{89}Sr to be 19 days in summer and 49 days in winter. Results with ^{51}Cr, applied either in solution or incorporated in 1 μm diameter polystyrene particles, were similar to those with ^{89}Sr. The finding that T_B is longer in winter is supported by other evidence (Chamberlain 1970). It also appears to be longer for foliage which is slow growing for reasons of climate or soil type. Moorby & Squire (1963) found that loss of radioactivity from the leaves of plants, in the absence of rain, was most rapid when the plants were growing rapidly, and surmised that this might be related to the loss of waxy cuticle, which is continuously produced and removed during growth. On the micro-scale this might help to explain the very non-uniform distribution of particulate deposit on leaves which has been observed.

Normalized specific concentration

The normalized specific concentration, F, on foliage, is defined by Chamberlain (1970) as:

$$F = \frac{\text{Amount of pollutant per kg dry matter in foliage}}{\text{Amount deposited per m}^2 \text{ of ground per day}}$$

The units of F are $\text{m}^2\text{d kg}^{-1}$. The use of F presupposes that foliar uptake is the main route of contamination.

Chamberlain calculated values of F as shown in Table 9.4. To allow for dilution by new growth, a constant growth rate (in $\text{kg m}^{-2} \text{ d}^{-1}$) was assumed. Two values of λ, 0·054 and 0·037 d^{-1}, corresponding to the experimental results of Milbourn & Taylor (1956) and Chadwick & Chamberlain (1970) were assumed. F was found to increase only slowly as the period of growth

TABLE 9.4. Normalized specific concentrations of herbage exposed to constant daily fallout (Chamberlain 1970).

Period of growth (days)	30	50	80
F (m² d kg⁻¹)			
$\lambda = 0.054$ days⁻¹	27	35	41
$\lambda = 0.037$ days⁻¹	32	44	55

increased from 30 to 80 d. This was because the T_B values assumed were comparable with, or shorter than, the growth period, so that a quasi-equilibrium between deposition and field loss was achieved.

A variety of measured values, derived from data on radioactive species, and also lead and fluorine, collected by Chamberlain (1970) showed F in the range 30–60 m²d kg⁻¹ for grass and cereal crops in good growth conditions. In winter, and also in summer on upland pastures, values in the range 60–150 m²d kg⁻¹ were found. Measurements in the literature of fission products in arctic vegetation (birch, willow, cowberry) gave F values in the range 100–700 m²d kg⁻¹.

These results must not be interpreted as some sort of universal law, but rather as a norm against which other measurements can be compared. To

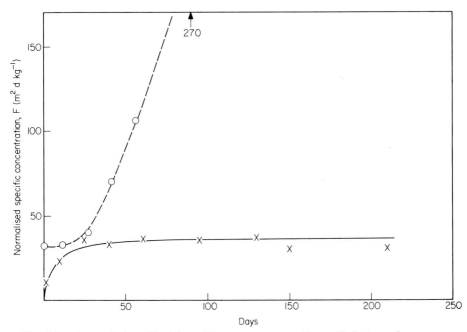

FIG. 9.8. Accumulation of lead from fallout on grass according to Roberts *et al.* (1974) (×), and Tjell *et al.* (1979) (○).

illustrate this, Fig. 9.8 shows measurements of lead in grass exposed to fallout.

Roberts *et al.* (1974) transplanted uncontaminated turf (grass species not stated) to an area near a smelter in Toronto where the fallout of lead averaged 250 mg m^{-2} per 30 days. Within 30 days, the accumulation of lead in leaves (and also in humus) reached an apparent equilibrium value, corresponding to $F = 35$ m^{2}d kg^{-1}, which persisted from July through to the following April. Tjell, Hovmand & Mosbaek (1979) grew Italian rye grass (*Lolium multiflorum* Lam.) in pots at Taastrup, Denmark, where the fallout of lead, mainly derived from automobiles, was 0·9 mg m^{-2} per 30 days. Cuttings were taken periodically and analysed for lead content. The measurements on 29th June, 12th July and 27th July showed about 1 to 1·2 mg kg^{-1} of lead in the leaves, giving F equal to 30 to 40 m^{2}d kg^{-1}, but thereafter the concentration increased to 3·2 mg kg^{-1} on 24th August ($F = 107$ m^{2}d kg^{-1}) and 7 mg kg^{-1} on 30th September ($F = 270$ m^{2}d kg^{-1}). On 2nd May of the following year, at the beginning of the new growth season, the concentration was 15 mg kg^{-1}, giving $F = 500$ m^{2}d kg^{-1} (not shown in Fig. 9.8). Chamberlain *et al.* (1979) in measurements of fallout and lead in grass near the M4 motorway, found F about 50 m^{2}d kg^{-1} in normal growing conditions but about 6 times higher during the very hot and dry summer of 1976, when the grass was brown and parched. It is unlikely that the increase was due simply to lack of rain to wash the vegetation, since Rains (1971) found a big increase in the lead content of wild oats (*Avena fatua*) in California in winter, although that is the rainy season. Mitchell & Reith (1966) had earlier noted an increase in the lead content of pasture and herbage in late autumn, persisting through to spring. The increase was sometimes by an order of magnitude. The same authors mentioned unpublished results with various crop species and deciduous shrubs showing the same effect. They thought that uptake from the soil by the dormant crop in winter was unlikely. In experiments by Tjell *et al.*, this is known not to be the explanation, since an independent estimate of the uptake was available from the concentration in the foliage of the isotope ^{210}Pb which had been added to the soil. Possible explanations are

(a) The reduction in rate of field loss in water and increase in T_B as found in the field experiments of Chadwick and Chamberlain (1970) with ^{89}Sr.

(b) The reduction in dry weight of the foliage at senescence.

(c) Strong binding of metals to exposed cell walls (Rains 1971).

(d) Remobilization from the roots.

High F values are found to apply to measurements of both radioactive and non-radioactive pollutants in lichens. Hasanen and Miettinan (1966) found that ^{137}Cs fallout is retained almost quantitatively in lichens over periods of years. This results in high concentrations of ^{137}Cs in the flesh of reindeer and thence of Eskimos. Andersen, Hovmand & Johnson (1978) have reported concentrations of trace metals (Cn, Pb, Zn, V, Fe) in lichens and bryophytes from the Copenhagen district and have also given the rates of fallout in deposit gauges. The F values for lichens range from 300 to 1200 and for bryophytes from 400 to 6000 $m^2 d\ kg^{-1}$. These high values can be related to the slow growth rates and high specific surface of these plants, and also possibly to affinity for metals.

The accumulation of lead from fallout in crops is a possible route of entry to humans, either directly or via milk.

If, as an alternative to F, we define

$$G = \frac{\text{Amount per g dry weight in foliage}}{\text{Amount per } m^3 \text{ of air}}$$

then Tjell, Hovmand & Mosbaek's results indicate a G value of about 15 $m^3 g^{-1}$ in the growing season (2 mg kg^{-1} in grass with airborne lead 0·13 $\mu g\ m^{-3}$).

Measurements by Davies & Holmes (1972) and Chamberlain *et al.* (1979) of Pb in grass near roads, and of Motto *et al.* (1970) of Pb in corn leaves, when referred to measured or estimated concentrations in air, give G values of 10 to 40 $m^3 g^{-1}$.

Leaves of grasses and cereals are effective traps of pollutants because their surface to weight ratios are high. High G values are found when the accumulation of reactive gases by grasses is measured. Thus Less *et al.* (1975) exposed rye grass (*Lolium perenne* L.) to HF gas in a glasshouse and found G values of 100 to 150 $m^3 g^{-1}$.

REFERENCES

Anderson A., Hovmand M.F. & Johnson I. (1978) Atmospheric heavy metal deposition in the Copenhagen area. *Environmental Pollution*, **17**, 133–151.

Aylor D. (1975) Deposition of particles in a plant canopy. *Journal of Applied Meteorology*, **14**, 52–57.

Bainbridge A. & Stedman O.J. (1979) Dispersal of *Erysiphe graminis* and *Lycopodium clavatum* spores near to the source in a barley crop. *Annals of Applied Biology*, **91**, 187–198.

Belot Y. (1975) Etude de la captation des polluants atmospheriques par les vegetaux. Commissariat a l'Energie Atomique, Fontenay-aux-Roses, Paris.

Belot Y. & Gauthier D. (1975) Transport of micronic particles from atmosphere to foliar surfaces. *Heat and Mass Transfer in the Biosphere* (Ed. by D.A. de Vries & N.H. Afgan), pp. 583–591. Scripta Book Co., Washington D.C.

Belot Y., Baille A. & Delmas J.L. (1976) Modele numerique de dispersion des polluants atmospheriques en presence des couverts vegetaux. *Atmospheric Environment*, **10**, 88–98.

Carter M.V. (1965) Ascospore deposition in *Eutypa armeniacae*. *Australian Journal of Agricultural Research*, **16**, 825–836.

Chadwick R.C. & Chamberlain A.C. (1970) Field loss of radionuclides from grass. *Atmospheric Environment*, **4**, 51–56.

Chamberlain A.C. (1953) *Aspects of Travel and Deposition of Aerosol and Vapour Clouds.* Atomic Energy Research Establishment, Harwell, Report 1261, H.M.S.O., London.

Chamberlain A.C. (1966) Transport of *Lycopodium* spores and other small particles to rough surfaces. *Proceedings of the Royal Society*, A **296**, 45–70.

Chamberlain A.C. (1970) Interception and retention of radioactive aerosols by vegetation. *Atmospheric Environment*, **4**, 57–78.

Chamberlain A.C. (1974) Mass transfer to bean leaves. *Boundary-Layer Meteorology*, **6**, 477–486.

Chamberlain A.C. (1975) The movement of particles in plant communities. *Vegetation and the Atmosphere* (Ed. J.L. Monteith), Vol. 1, pp. 155–203. Academic Press, London.

Chamberlain A.C., Heard M.J., Little P. & Wiffen R.D. (1979) The dispersion of lead from motor exhausts. *Philosophical Transactions of the Royal Society*, A **290**, 577–589.

Davies B.E. & Holmes P.L. (1972) Lead contamination of roadside soil and grass in Birmingham, England, in relation to naturally occurring levels. *Journal of Agricultural Science, Cambridge*, **79**, 479–484.

Energy Research and Development Administration (1974) *Atmosphere-Surface Exchange of Particulate and Gaseous Pollutants* (CONF-740921). N.T.I.S., Springfield, Va.

Esman N.A., Ziegler P. & Whitfield R. (1978) The adhesion of particles upon impaction. *Journal of Aerosol Science*, **9**, 547–556.

Forster G.F. (1977) Effect of leaf surface wax on the deposition of airborne propagules. *Transactions of the British Mycological Society*, **68**, 245–250.

Friedlander S.K. (1977) *Smoke, Dust and Haze.* John Wiley & Sons, London.

Fuchs N.A. (1964) *The Mechanics of Aerosols.* Pergamon Press, Oxford.

Grace J. & Wilson J. (1976) The boundary layer of a *Populus* leaf. *Journal of Experimental Botany*, **27**, 231–241.

Green H.L. & Lane W.R. (1964) *Particulate Clouds: Dusts, Smokes and Mists*, 2nd Edn. E. & F.N. Spon, London.

Gregory P.H. (1945) The dispersion of air-borne spores. *Transactions of the British Mycological Society*, **28**, 26–71.

Gregory P.H. (1951) Deposition of airborne *Lycopodium* spores on cylinders. *Annals of Applied Biology*, **38**, 357–376.

Gregory P.H. (1973) *The Microbiology of the Atmosphere*, 2nd Edn. Leonard Hill, Aylesbury.

Gregory P.H. & Stedman O.J. (1953) Deposition of airborne *Lycopodium* spores on plane surfaces. *Annals of Applied Biology*, **40**, 651–674.

Hasanen E. & Miettinan J.K. (1966) Gamma-emitting radionuclides in subarctic vegetation during 1962–4. *Nature*, **212**, 379–382.

Hiller R. & Löffler F. (1978) The influence of bouncing of solid particles and oil drops on the filtration efficiency in fibre filters. *Deposition and Filtration of Particles from Gases and Liquids*, Society of Chemical Industry.

Hirst J.M. & Stedman O.J. (1971) Patterns of spore dispersal in crops. *Ecology of Leaf Surface Microorganisms.* (Ed. T.F. Preece & C.H. Dickinson), pp. 229–237. Academic Press, London.

Kyaw Tha Paw U & Reifsnyder W.E. (1979) The physics of pollen and spore rebound from vegetative surfaces. *14th Conference on Agricultural and Forest Meteorology, Minneapolis*, pp. 244–246. American Meteorological Society.

Legg B.J. & Powell F.A. (1979) Spore dispersal in a barley crop: a mathematical model. *Agricultural Meteorology*, **20**, 47–67.

Less L.N., McGregor A., Jones L.H.P., Cowling D.W. & Leafe E.L. (1975) Fluorine uptake by grass from aluminium smelter fume. *International Journal of Environmental Studies*, **7**, 153–160.

Little P. (1977) Deposition of 2·75, 5·0 and 8·5 μm particles to plant and soil surfaces. *Environmental Pollution*, **12**, 293–305.

Little P. & Wiffen R.D. (1977) Emission and deposition of petrol engine exhaust Pb. I. Deposition of exhaust Pb to plant and soil surfaces. *Atmospheric Environment*, **11**, 437–447.

Little P. & Wiffen R.D. (1978) Emission and deposition of lead from motor exhausts. II. Airborne concentration, particle size and deposition of lead near motorways. *Atmospheric Environment*, **12**, 1331–1341.

Löffler F. & Umhauer H. (1971) An optical method for the determination of particle separation on filter fibers. *Staub-Reinhalt. Luft* (in English), **31**, 9–16.

May K.R. & Clifford R. (1967) The impaction of aerosol particles on cylinders, spheres, ribbons and discs. *Annals of Occupational Hygiene*, **10**, 83–95.

Milbourn G.M. & Taylor R. (1956) The contamination of grassland with radioactive strontium—I, Initial retention and loss. *Radiation Botany*, **5**, 337–347.

Miller C.W. Validation of a model to predict aerosol interception by vegetation. *International Symposium on Biological Implications of Radionuclides released from Nuclear Industries*. International Atomic Energy Agency, Vienna (in press).

Mitchell R.L. & Reith J.W.S. (1966) The lead content of pasture and herbage. *Journal of Science in Food & Agriculture*, **17**, 437–440.

Moorby J. & Squire H.M. (1963) The loss of radioisotopes from the leaves of plants in dry conditions. *Radiation Botany*, **3**, 163–167.

Motto H.L., Daines R.H., Chilko D.M. & Motto C.K. (1970) Lead in soils and plants: its relationship to traffic volume and proximity to highways. *Environmental Science & Technology*, **4**, 231–238.

Rains D.W. (1971) Lead accumulation by wild oats in a contaminated area. *Nature*, **233**, 210–211.

Raynor G.S., Hayes J.V. & Ogden E.C. (1974) Particulate dispersion into and within a forest. *Boundary-Layer Meteorology*, **7**, 429–456.

Roberts T.M., Paciga J.J., Jervis R.E., Van Loon J.C., Hutchinson T.C., Chattopadhyay A. & Hahn F. (1974) *Lead Contamination around two Secondary Smelters in Downtown Toronto*. Institute for Environmental Studies, University of Toronto.

Roberts T.M. (1975) A review of some biological effects of lead emissions from primary and secondary smelters. *International Conference on Heavy Metals in the Environment*, Vol. II, Part 2, pp. 503–532. University of Toronto.

Scott Russell R. (Ed.) (1966) *Radioisotopes and Human Diet*, Pergamon Press, Oxford.

Tampieri F., Mandrioli P. & Puppi G.L. (1977) Medium range transport of airborne pollen. *Agricultural Meteorology*, **18**, 9–20.

Thom A.S. (1975) Momentum, mass and heat exchange. *Vegetation and the Atmosphere* (Ed. J.L. Monteith), pp. 57–109. Academic Press, London.

Tjell J.C., Hovmand M.F. & Mosbaek H. (1979) Atmospheric lead pollution of grass grown in a background area in Denmark. *Nature*, **280**, 425–426.

Wedding J.B., Carlson R.W., Stubel J.J. & Bazzaz F.A. (1975) Aerosol deposition on plant leaves. *Environmental Science and Technology*, **9**, 151–153.

White E.J. & Turner F. (1970) A method of estimating income of nutrients in catch of airborne particles by a woodland canopy. *Journal of Applied Ecology*, **7**, 441–461.

Yosida Z. (1953) General survey of the studies on fog-preventing forest. *Studies on Fogs in relation to Fog-Resisting Forests* (Ed. T. Hori), pp. 1–23. Tanne Trading Co. Ltd., Sapporo, Japan.

10. STOMATAL CONDUCTANCE, GASEOUS EXCHANGE AND TRANSPIRATION

P.G. JARVIS

Department of Forestry & Natural Resources, University of Edinburgh, King's Buildings, Mayfield Road, Edinburgh EH9 3JU

SUMMARY

1. The history of the measurement of leaf conductance is outlined and current applica‹ tions of the measurement are considered.

2. The role of stomatal conductance in the control or the exchange of carbon dioxide and pollutant gases is briefly considered.

3. The physical and physiological controls of transpiration are considered. A derivation of the Penman-Monteith equation is given in an Appendix and the implications of the equation considered. A simplification for rough canopies of large boundary layer conductance is given.

4. An example is given of the estimation of transpiration rate from two Scots pine stands of different densities using the Penman-Monteith equation, weather station records and measurements of stomatal conductance made with a null-balance diffusion porometer. The transpiration rate was consistently higher on the plot with the largest leaf area index.

5. The observations are discussed in relation to the main variables which determine the rate of transpiration: available energy, saturation deficit, boundary layer conductance and canopy conductance.

6. Some problems of graphical presentation arising out of the association between available energy and saturation deficit are discussed.

7. The implications of the size of the boundary layer conductance and canopy conductance for transpiration from different kinds of canopies are considered. Canopy conductance is proportional to leaf area index which can therefore have a major influence on the rate of transpiration from rough canopies.

8. The rate of transpiration from the Scots pine canopies is less than expected from the weather conditions because of an interaction between stomatal conductance and saturation deficit.

9. It is concluded that the diffusion porometer together with the Penman-Monteith equation can provide a practical means of estimating transpiration rate from vegetation and of analysing the variations in transpiration rate.

LEAF CONDUCTANCE

A number of papers in this volume are concerned with leaf resistance or conductance to the transfer of water vapour, carbon dioxide and pollutants, such as sulphur dioxide, between the leaf and the atmosphere. I propose, therefore, to begin with a short discussion of leaf or stomatal conductance before going on to discuss the interrelationship between stomatal conduct-ance and the weather in controlling evaporation.

I define leaf conductance as the proportionality parameter relating the flux of a property in the gas phase in or out of a leaf to the driving force existing between the leaf and the bulk air outside the leaf boundary layer. The property might be water vapour, carbon dioxide or a pollutant such as gaseous sulphur dioxide. The flux of water in transpiration, F, for example, is proportional to the difference in specific humidity, Δq (kg kg^{-1}) between leaf and air:

$$F = g \, \Delta q \rho \qquad\qquad 10.1$$

where ρ is the density of air (kg m^{-3}) and g is the leaf conductance in units of m s^{-1} if F is in kg m^{-2} s^{-1}. The leaf conductance can be partitioned into components of the cuticular and stomatal pathways, i.e. boundary layer, cuticle, stomatal ante-chamber, stomatal pore, substomatal cavity etc. (Milthorpe 1961; Jarvis 1971). However, for most leaves by far and away the largest part of the gaseous flux is through the stomatal pore, and conse-quently leaf conductance, after allowance for the boundary layer, can generally be regarded as almost synonymous with stomatal conductance, g_s, although of course not in the cases of the mosses and algae discussed elsewhere in this volume by Proctor, and by Norton & Jones. The stomatal conductance can also be identified with anatomical features of the leaf and derived from appropriate measurements of the dimensions of the water vapour transport pathway (e.g. Jarvis, Rose & Begg 1967; Milthorpe & Penman 1967).

Leaf or stomatal conductance of plants in the field is nowadays measured with a diffusion porometer. All diffusion porometers are based on the solution of equation 10.1 for g. Leaf conductance is derived from a rapid measurement of the rate of transpiration of water from a source of known specific humidity into an atmosphere of known specific humidity (Jarvis 1971; Slavik 1974; Kanemasu 1975). Today there are essentially three basic designs of porometer in use: the unventilated 'transit-time' type (Kanemasu, Thurtell & Tanner 1969; Stiles 1970), the ventilated 'transit-time' type (Turner & Parlange 1970; Körner & Cernusca 1976) and the ventilated 'steady-state' type (Parkinson & Legg 1972; Beardsell, Jarvis & Davidson 1972). They all embody a small chamber which is attached to part of a leaf, or which encloses one or more leaves, for the time required to make a measurement of about one minute. With the 'transit-time' porometers the rate of transpiration is derived from the time taken for the humidity in the chamber to rise by a certain amount. With the 'steady-state' porometers, the transpiration rate is derived either from the increase in humidity of an air-stream flowing through the chamber at a constant rate, or, in the null-balance version, from the rate of inflow of dry air necessary to maintain the humidity in the chamber constant. In the unventilated porometers, water vapour diffuses from the leaf across a short distance of still air to the humidity sensor. In the ventilated porometers, the thickness of the boundary layer is minimized by a small fan or pump which also circulates the air to the humidity sensor which is some distance away from the leaf. All incorporate measurement of leaf and/or air temperature and most have a quantum flux sensor associated with them.

With the recent widespread adoption of the diffusion porometer by ecologists and crop physiologists, measurements of stomatal conductance in the field have become commonplace over the last five years. However, between about 1900 and 1950, virtually all estimates of stomatal conductance were made either by calculation from stomatal numbers and dimensions or by derivation from measurements made with mass flow porometers (for reviews of these methods see Jarvis 1971; Slavik 1974; Kanemasu 1975). A number of useful, portable mass flow porometers were developed (Alvim 1965; Bierhuizen, Slatyer & Rose 1965; Raschke 1965) but there were considerable uncertainties in the derivation from both methods of quantitative values of stomatal conductance which could be used in models. Some attempts were made to 'calibrate' the mass flow stomatal resistance in terms of diffusive resistance for stomata of particular species (Jarvis, Rose & Begg 1967) but, because of the difficulties in obtaining quantitative information on stomatal conductance, widespread use of mass flow porometers never became a reality. Nonetheless, mass flow porometers are still today one of the most successful ways of detecting changes in the aperture of the stomatal *pore* itself (Meidner, in press).

During the 1960s the significance of quantitative estimates of diffusive conductance to the interpretations of gaseous exchange became widely recognized and large numbers of measured values began to appear in the literature. At this time nearly all values of stomatal conductance were obtained using cuvettes or assimilation chambers (Bierhuizen & Slatyer 1964; Holmgren, Jarvis & Jarvis 1965). Whilst yielding accurate values of particular usefulness in experiments in controlled environments these techniques certainly did not meet the requirements of ecologists, whether crop ecologists or ecologists working on wild plants in the field. Consequently the 1960s saw the emergence of portable diffusion porometers, initially as instruments which were rather clumsy and difficult to use in the field (Grieve & Went 1965; Van Bavel, Nakayama & Ehrler 1965; Stiles 1970) but later of increasing sophistication and ease of use.

The main problems in the early development of diffusion porometers were in finding suitable humidity sensors, temperature sensitivity and calibration of the sensor, and the water retention characteristics of the structural materials. Because of these difficulties diffusion porometers were still not established as useful practical tools by 1970. In a review of techniques for the determination of stomatal resistance, they were given only 3 pages out of 65 (Jarvis 1971). However, during the early 1970s, there were considerable improvements in design and several of the designs have subsequently become available on a commercial or semi-commercial basis (Kanemasu, Thurtell & Tanner 1969; Stiles 1970; Turner & Parlange 1970; Beardsell, Jarvis & Davidson 1972; Körner & Cernusca 1976). None the less there are still some problems today with respect to the temperature sensitivity, stability and hysteresis of the sensor and water adsorption properties of the construction materials used in some of the available instruments, especially when used in the tropics. There is also a need for a convenient, universally acceptable calibration device which can be used to check any porometer in the field.

Nonetheless as a result of the widespread availability of practical, portable diffusion porometers, large numbers of measurements of leaf or stomatal conductance have been made over the last six or seven years and these have recently been summarized by Körner, Scheel & Baur (1979). There has been an explosion of information, comparable with that which resulted a few years earlier as a consequence of the widespread adoption of the pressure chamber for measurements of xylem pressure potential. New practical instruments always result in the acquisition of large amounts of new data, especially new field instruments in ecology. However, in the enthusiasm of exploiting a new technique, the scientific purpose for which the data are acquired is not always immediately apparent! I shall therefore discuss some of the applications of measurements of stomatal conductance in general terms.

As a number of papers in this volume show, a knowledge of leaf conductance is currently of importance in the following areas of ecology:

(a) Assessment of adaptation and acclimation to environment, e.g. Björkman (this volume).

(b) Interpretation of measurements and building of models of photosynthesis, pollutant exchange and transpiration by leaves and canopies, e.g. Landsberg, Unsworth, Körner & Mayr, Cernusca & Seeber, Grace, Monteith (this volume).

Considering the role of stomatal conductance in the exchange of gases and water by leaves and canopies, I shall make use of examples most of which have been obtained using techniques suitable for sampling leaves in an extensive, rather than intensive, manner in the field.

EXCHANGE OF GASES

Carbon dioxide

Gaastra pointed out in 1959 that CO_2 uptake depends upon both stomatal conductance and a number of biochemical and biophysical variables which, for purposes of comparison with the stomatal resistance, he treated as a mesophyll resistance. Fig. 10.1 shows an example of the dependence of net CO_2 uptake F_c, on stomatal conductance, g_s. As g_s approaches infinity, F_c is limited entirely by the mesophyll component. While comparisons of stomatal and mesophyll resistance are usually made to determine their relative importance, consideration of the rate of photosynthesis at infinite g_s conveniently enables the demonstration of the magnitude of the biochemical and biophysical limitations to photosynthesis. The rate of photosynthesis if g_s were infinite can readily be found from the response curve of F_c to CO_2 concentration, by equating the mean intercellular space CO_2 concentration, C_i, with the ambient concentration, C_a (Björkman *et al.* 1972).

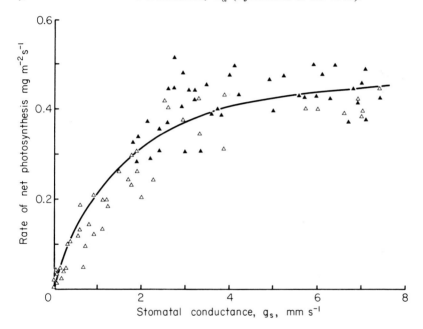

FIG. 10.1. The relation between rate of net photosynthesis and stomatal conductance for water vapour, g_s. The points ▲ are from an experiment in which vapour pressure deficit was the variable and the points △ are from an experiment in which shoot water potential was the variable. The line is the hyperbolic curve of best fit to the data and represents a CO_2 mesophyll conductance of 0·92 mm s^{-1} (redrawn from Watts and Neilson 1978).

The line in Fig. 10.1 is the hyperbolic relationship which results from the assumption of a constant mesophyll resistance (and rate of photorespiration) in series with the stomatal resistance. The data are uniformly distributed about the curve at the high values but fall below at the lower values suggesting that the mesophyll resistance is underestimated in water stressed leaves.

There is some argument as to whether the stomata do in fact control the rate of photosynthesis or whether photosynthesis controls stomatal conductance (Jarvis, in press). Since, in many species, stomata are sensitive to the CO_2 concentration in their immediate environment, it has been argued that the stomatal conductance attained is the result of feedback from the particular rate of photosynthesis, and hence that the rate of photosynthesis controls stomatal conductance (Farquhar, Dubbe & Raschke 1978). Furthermore, on the basis of a narrow range of values of C_i of around 200 cm³ m⁻³ over a wide range of conditions and species, Cowan and Farquhar (1977) have argued that evolution has resulted in the optimization of the feedback between photosynthesis and stomatal conductance so that the maximum amount of CO_2 is fixed per unit of water transpired.

However, the stomata of some species show little or no response to CO_2 and may indeed open in response to light in the presence or absence of CO_2 (Ng & Jarvis 1980). Furthermore when the ambient CO_2 concentration is changed, the same stomatal conductance occurs at widely different values of C_i. Thus if photosynthesis is controlling g_s in such cases, rather than vice versa, it is not through the medium of C_i (Jarvis 1980).

At the present time, the mechanism by which stomatal conductance responds to light is far from clear. Although the relationship between photosynthesis and stomatal conductance may be a finely controlled system with perhaps both dependent upon the same electron transport pathways and photophosphorylation, Gaastra's supposition that the stomata control the entry of CO_2 into the leaf, as well as the loss of water, seems still to be valid today, at least for some species (Jarvis, in press). Similarly, comparison between the stomatal resistance and the mesophyll resistance is still a useful

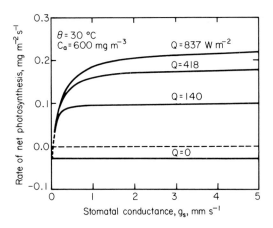

FIG. 10.2. The predicted rate of net photosynthesis of yellow poplar (*Liriodendron tulipifera* L.) in relation to stomatal conductance at four different irradiances (redrawn from Reed *et al.* 1976).

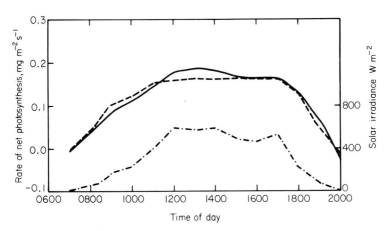

F<small>IG</small>. 10.3. Predicted (– – –) and measured (——) rates of net photosynthesis of yellow poplar throughout the course of a day and solar irradiance (– · – ·) (redrawn from Reed *et al.* 1976).

way of assessing the relative magnitude of the stomatal and non-stomatal components of limitation. Comparison between the rates of photosynthesis at the measured stomatal conductance and at infinite conductance, i.e. zero resistance, is also a very useful and demonstrative way of making the same assessment (see Björkman, this volume).

The explicit inclusion of measured stomatal conductance in models of photosynthesis can result in models with a high predictive ability. Fig. 10.2 shows response curves, relating net CO_2 uptake to irradiance Q and stomatal conductance, the parameters for which were derived from fitting a model to field data (Reed *et al.* 1976). The main variables in the model were irradiance, temperature, ambient CO_2 concentration and stomatal conductance. The predictive ability of the model is shown by comparison between predicted rates of net photosynthesis with rates measured on another occasion in Fig. 10.3.

Pollutants

Sulphur dioxide follows a similar entry pathway to CO_2, becoming dissolved in the mesophyll cell walls which appear to act as a substantial sink. Fig. 10.4 shows that leaf conductance for SO_2, directly determined from $^{35}SO_2$ uptake, is closely related to the stomatal conductance for water vapour measured on adjacent shoots with a diffusion porometer. Although there is a lot of scatter in the data, the slope of the line of best fit is very close to the ratio of diffusivities of SO_2 and water vapour in air. Thus these data suggest that there is no mesophyll resistance to SO_2 comparable to that for CO_2.

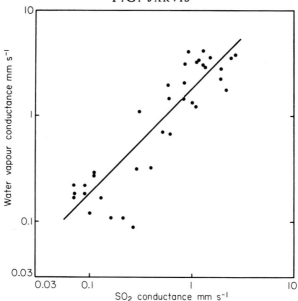

FIG. 10.4. The relation between stomatal conductances for water vapour and SO_2 of shoots of Scots pine (*Pinus sylvestris* L.) (redrawn from Garland & Branson 1977).

Conclusions

From these examples, it is clear that measurements of stomatal conductance are crucial to the interpretation of rates of uptake of gases and of photo-synthesis: stomatal conductance is a variable which should be included in models of gaseous exchange. Because stomatal conductance is itself dependent upon a set of environmental and plant variables, such models usually require sub-models for stomatal conductance (Jarvis 1976), since measured values cannot be provided for every occasion for which an estimate of gaseous flux is desired. At present, models of stomatal conductance as a function of the weather are very empirical because we do not understand the stomatal mechanism properly. It is to be hoped that models of stomatal conductance which are more mechanistic may emerge in the near future so that models of pollutant uptake, photosynthesis and growth can be more soundly based mechanistically and made more informative.

TRANSPIRATION

Physical and physiological control

Since molecules of water leave the leaf through the stomata, stomatal conductance has an appreciable influence on the rate of transpiration and up until the mid 1930s virtually all plant physiologists and ecologists regarded transpiration as a physiological property of the plant.

In 1948 Penman derived the first physically based equations for evapotranspiration from crop canopies by the application of conservation of energy to the heat and water vapour transport equations. Penman's treatment included some empirical constants related to windspeed and daylength which had to be known for different crops but the overall emphasis was of transpiration as an essentially physical process, largely determined by environmental variables. During the 1950s, the pendulum swung further in that direction and the first leaf energy budgets which explicitly included transpiration in relation to environmental variables, including leaf temperature, appeared (Raschke 1956, 1958), to be followed by papers such as 'Transpiration by sudan grass as an externally controlled process' (Van Bavel, Fritschen & Reeves 1963). During this period transpiration was regarded by crop ecologists as a particular case of the physical process of evaporation and as depending almost exclusively on physical variables which fluctuated in the environment with the weather and season (Fritschen & Van Bavel 1964).

However in the early 1960's developments of Penman's equation appeared which began to contain vegetation parameters in a more explicit form (Slatyer & McIlroy 1961) and in 1964 Monteith (1965) gave the first wholly explicit presentation in terms of stomatal and boundary layer resistances. Applications of this equation to crops were not slow to follow and at the International Botanical Congress in Edinburgh in 1964, Professor Rutter demonstrated the first application of what is now known as the Penman-Monteith equation to the determination of evaporation from coniferous forest (Rutter 1967). However, ecologists have been slow to make use of the equation although it allows the determination of the rate of transpiration from individual species in mixed communities.

Many ecologists find the derivation of the equation in Monteith's (1965) paper difficult to follow so a short simple derivation is given in the Appendix to this chapter.

Monteith's initial water vapour transfer equation was the familiar expression for transpiration $\mathbf{E}_T$, as written in terms of the conductance of the water vapour pathway, g_W, as:

$$\mathbf{E}_T = \frac{c_p \rho}{\gamma \lambda} (e_s(T_l) - e_a) g_W \tag{10.2}$$

where the constants (see Appendix), $c_p/(\gamma \lambda) = k$, convert the leaf-air vapour pressure difference, $(e_s(T_l) - e_a)$, to specific humidity. His final equation was

$$\mathbf{E}_T = \frac{s\mathbf{A} + c_p \rho D g_a}{\lambda(s + \gamma(1 + g_a/g_c))} \tag{10.3}$$

where $\mathbf{A}$ is available energy, D is vapour pressure deficit of the air, s is the slope of the curve relating saturated vapour pressure to temperature and g_c and g_a are the canopy stomatal and boundary layer conductances (see Appendix).

Both equations are correct and equivalent statements within the framework of the assumptions made (see Appendix). The first requires a knowledge of leaf temperature whereas this is replaced in the second by the available energy or the net radiation. A number of methods are available for obtaining transpiration given the leaf temperature (Jarvis 1971; Butler 1976) but that information is often not easily available. The elimination of leaf temperature was an essential part of Penman's original contribution (1948). The second equation utilizes available energy in place of leaf temperature, the leaf temperature being treated essentially as the result of the energy balance.

Equation 10.3 can be applied to single leaves or to extensive canopies. When applied to canopies an approximation is involved since the derivation assumes identical location of the sources and sinks of water vapour and heat, whereas this is unlikely to be the case in a canopy. However comparisons with more complex models have shown that this approximation does not lead to serious error in the estimation of $\mathbf{E}_T$ (Shuttleworth 1976; Sinclair, Murphy & Knoerr 1976).

Equation 10.3 can be applied successfully over periods of minutes, hours, days or weeks provided that appropriate values of the variables are known. Application over short periods of time, such as an hour, is usually limited by the availability of hourly meteorological data and hourly measurements, or reliable estimates, of leaf and boundary layer conductance. The boundary layer conductance can usually be obtained with adequate precision from measurements of windspeed and equations derived from studies in a wind tunnel (e.g. Thom 1968, Landsberg & Thom 1971; Landsberg & Powell 1973) or from wind profiles (Thom 1975). Estimation of the leaf or canopy conductance requires regular, extensive measurements or an adequate submodel (Jarvis 1976).

The conductance of a leaf varies with its age, position on the plant and position in the canopy, as well as upon the season, the time of day and recent and current weather. The conductance of a canopy is the arithmetical sum of the conductances of all the individual leaves in the canopy. Thus it depends upon the number and area of leaves present (i.e. the leaf area index) and the conductance of the individual leaves. Because of the variation in conductance amongst the leaves in a canopy, the canopy conductance depends upon the proportion of leaves of different properties which are present, i.e. it depends on the partial leaf area indices of leaves of different age, position, branching hierarchy, etc. For practical purposes, the location of samples can be stratified with respect to these, or other, variables, Then

$$g_c = \sum (g_{s,i} \cdot L_i) \tag{10.4}$$

where L_i is the leaf area index of a sample of leaves with average leaf conductance $g_{s,i}$.

There are several problems in making an integration of this kind which is suitably precise but only now are they being tackled systematically. Up to the present, the sample size has been chosen largely on the basis of the number of measurements of g_s it is feasible to make in say an hour with available porometers, without regard to variation amongst the measurements. Secondly, stratification of the samples has been based largely on intuition. Stratification certainly seems to be necessary as there is a large systematic variation in g_s amongst leaves at different positions in the canopy (Watts, Neilson & Jarvis 1976). Thirdly, analysis of the errors in deriving E_T from equation 10.3 shows that by far the largest source of error lies in the determination of the partial leaf area indices, L_i, for use in equation 10.4 (J. Wallace, personal communication). Nonetheless canopy conductance determined in this way from measurements made with the diffusion porometer on samples of leaves has been shown to agree reasonably well with measurements made by other methods (Tan & Black 1976; James & Jarvis, in press).

In the last 10 years equation 10.3 has become an extremely valuable tool for the calculation of transpiration from single leaves, plants and canopies. It is interesting to note that when in 1960 Eckardt reviewed the techniques available for the measurement or estimation of transpiration this approach did not receive even a mention: at that time it was not regarded as a practical proposition. Today it is used widely for the estimation of transpiration from a wide range of crops, though to a much lesser extent in natural vegetation. Transpiration calculated in this way has been found to agree quite well with other methods (see for example Szeicz, Van Bavel & Takami 1973). In Scots pine (*Pinus sylvestris* L.) we found that the cumulative transpiration of excised trees standing in water estimated from equation 10.3 agreed within 5% with the volume flux of isotopically labelled water in the stem (Waring, Whitehead & Jarvis 1980).

However, some problems do remain in the estimation of transpiration rates from the equations. Firstly, there are the problems mentioned earlier in the accurate estimation of canopy conductance. Secondly, it is difficult to define the net amount of energy available to isolated plants and plants in rows. In such cases novel techniques of measuring **A** such as integration over the surface of a sphere or cylinder containing the foliage are required (e.g. Thorpe 1978). The extent to which this is actually a problem of real significance depends on the accuracy with which **A** is required to be known, and hence on the relative size of the two terms in the numerator.

A simplification for rough canopies

Equation 10.3 describes how E_T increases with available energy and vapour pressure deficit. The relative importance of available energy and vapour pressure deficit in driving E_T can be assessed as the ratio of the two terms in the numerator of equation 10.3:

$$J = c_p \rho D g_a / (s\mathbf{A}) \tag{10.5}$$

Clearly the larger the boundary layer conductance, the larger will be the effect of vapour pressure deficit on E_T. If both g_a and D are large, the deficit term may greatly exceed the radiation term. This is commonly the case for tall, rough vegetation like forest plantations for which J typically lies in the range 5–20, whereas for shorter vegetation like heathland and grassland g_a, and hence J, is at least an order of magnitude smaller. When J is large, say greater than 10, substantial error in the estimation of A is unimportant and the radiation term can be neglected.

A further feature of the equation is the appearance of g_a in both numerator and denominator. As pointed out by Monteith (1965) and Lowry (1969) this makes the effect of windspeed on transpiration difficult to assess intuitively (see Grace, this volume). However, this leads to a further simplification of equation 10.3 if g_a is much larger than g_c; this is usually the case for forest because the canopy is aerodynamically rough and the canopy conductance not especially large. For such vegetation g_a/g_c is in the range 5–20 so that the small terms in the denominator of equation 10.3 can be ignored.

Thus for forest canopies, for which both J and g_a/g_c are large, the rate of transpiration can be approximated by:

$$\mathbf{E}_T = \frac{c_p \rho D g_c}{\lambda \gamma} = k\rho D g_c \tag{10.6}$$

where k is the factor for converting partial pressure of water vapour into specific humidity. This equation is identical with equation 10.2 when leaf and air temperatures are equal. For small leaves or rough canopies leaf temperatures are closely coupled to air temperatures (see Monteith, this volume) and are usually within $0.5\,^{\circ}\text{C}$ of air temperature (Yamaoka 1958) so that this condition is closely met. On the other hand, this approximation should not be used when leaf temperatures depart from air temperature by more than about $0.5\,^{\circ}\text{C}$.

Partitioning of energy

Equations 10.2, 10.3 and 10.6 provide a means of determining the rate of transpiration in particular conditions. For the purposes of discussing the relative importance of the variables which control transpiration it is also useful to express transpiration in terms of the energy available to drive it:

$$\mathbf{A} + \mathbf{C} + \lambda\mathbf{E}_T = 0 \tag{10.7}$$

where **C** is the convective exchange of sensible heat to or from the foliage. The numerical value of all three terms on the left may be positive or negative, the fluxes usually changing in sign over 24 hours. In the daytime, if the stomata are wide open and leaf temperature lower than air temperature, heat is added to the leaves by convection and **C** is positive: when leaf and air temperatures are equal, **C** is zero. Two alternative parameters are commonly used to describe the partitioning of available energy by leaves and vegetation:

1. The fraction of the available energy used in transpiration

$$\alpha = \lambda \mathbf{E}_T / \mathbf{A} \tag{10.8}$$

2. The Bowen ratio

$$\beta = \mathbf{C} / \lambda \mathbf{E}_T \tag{10.9}$$

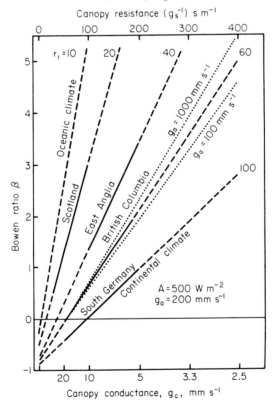

FIG. 10.5. The relationship between Bowen ratio, β, and canopy conductance, g_c (equation 10.12) in different climatic regions as defined by the climatological resistance r_i (equation 10.11). The figures by the lines are the values of r_i in s m^{-1}. The solid lines indicate ranges of β and g_c for coniferous stands in the areas indicated. The dotted lines show the effect of variation in g_a between 100–1000 mm s^{-1} (redrawn from Jarvis *et al.* 1976).

From equations 10.7 and 10.9, E_T can also be written as:

$$E_T = \frac{A}{\lambda(1 + \beta)}$$

(10.10)

Following Monteith (1965) and Stewart and Thom (1973), a climatological resistance (with dimensions of s m^{-1}) can be defined as:

$$r_i = \frac{c_p \rho}{\gamma} \cdot \frac{D}{A}$$

(10.11)

From equations 10.3, 10.7, 10.9 and 10.11 the Bowen ratio can be written as:

$$\beta = \frac{1 + g_a/g_c - g_a \cdot r_i}{s/\gamma + g_a \cdot r_i}$$

(10.12)

Equation 10.12 demonstrates that the partitioning of the available energy depends not only on the boundary layer and leaf conductances, but also depends on the general nature of the climate and weather as described by r_i. This equation is solved for a particular set of conditions in Fig. 10.5. Fig. 10.5 shows that β increases with decreasing canopy conductance but that in addition, for particular leaf and boundary layer conductances, β decreases as r_i increases, i.e. as D/A goes up. Thus the general level of the transpiration rate depends not only on the canopy conductance but also on the continentality of the site.

A case-study

Estimates have been made of the transpiration rates of mature Scots pine plantations managed to different densities near the Moray Firth in north-east Scotland (Waring, Whitehead & Jarvis 1979). Some properties of the two stands with most extreme densities are given in Table 10.1. The null balance diffusion porometer (Beardsell et al. 1972) was used to provide estimates of g_c and a portable weather station (Strangeways 1972) to provide estimates of A, D and g_a.

TABLE 10.1. Characteristics of the two stands of Scots pine at Roseisle Forest in 1977 (from Waring, Whitehead and Jarvis 1979).

Plot	Trees ha^{-1}	Basal area m^2 ha^{-1}	Volume m^3 ha^{-1}	Height m $\pm$ SE	Leaf area index
1	608	26·6	168	15·0 $\pm$ 0·3	2·4
2	3281	57·7	365	15·0 $\pm$ 0·2	3·1

Figure 10.6 shows the diurnal course of transpiration in these stands. Transpiration rate varied from day to day and between the two stands. The transpiration rate of the dense stand was consistently higher than that of the open stand, the difference being closely related to the difference in leaf area index. Plots of λE_T against both $\mathbf{A}$ and D suggest that transpiration rate is strongly dependent on both variables but that g_c declines and β increases with increasing $\mathbf{A}$ and D (Fig. 10.7).

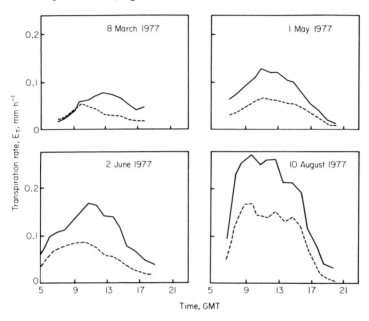

Fig. 10.6.   The diurnal course of transpiration on four days from two stands of Scots pine (*P. sylvestris* L.) differing in density of trees and leaf area (see Table 10.1) at Roseisle Forest by the Moray Firth, N.E. Scotland: ------ plot 1, ——— plot 2 (from Whitehead, Waring & Jarvis, unpublished).

The rate of transpiration

In principle the rate of transpiration could vary widely and reach very high rates depending upon the value of the variables $\mathbf{A}$, D, g_a and g_c in equation 10.3. For example, a transpiration rate of ca 1 mm h^{-1} is to be expected from $\mathbf{A} = 600$ W m^{-2}, $D = 15$ mb, $g_a = 200$ mm s^{-1} and $g_c = 20$ mm s^{-1} at 15 °C. This is over four times the highest rates shown in Fig. 10.7. In practice, therefore, the range of rates of transpiration encountered was restricted by the *combinations* of values of these variables which occurred. Thus it is useful to consider the ranges of values of the variables which normally occur together and their effects on the rate of transpiration.

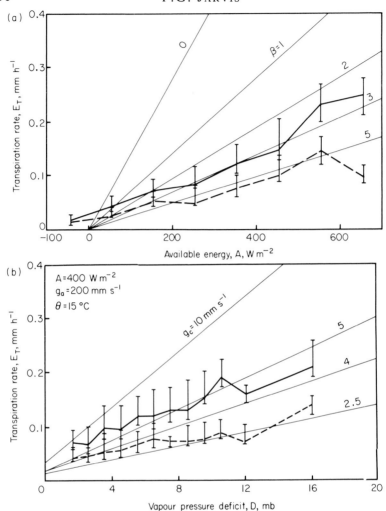

Fɪɢ. 10.7. The relation between transpiration rate and (a) available energy, A, (b) vapour pressure deficit, D, for two stands of Scots pine (*P. sylvestris* L.) (see Fig. 10.6 and Table 10.1 for details). The faint lines on (a) are lines of transpiration rate at equal Bowen ratio. The faint lines on (b) are lines of transpiration rate at equal canopy conductance for the conditions indicated (from Whitehead, Waring & Jarvis, unpublished).

Available energy, A

In dense canopies which intercept most of the incident radiation the available energy in the U.K. reaches about 600 W m^{-2} on a ground area basis. The amount per unit leaf area is substantially less depending upon the leaf area index. For more open canopies more of the radiant energy reaches the ground and less is absorbed but the amount available per unit leaf area is still small (Landsberg *et al.* 1975; Thorpe 1978).

The influence of available energy and vapour pressure deficit appear in the numerator of equation 10.3 as additive. In the past this has led to the description of transpiration as having a radiation component and a wind-speed, or vapour pressure deficit, component. This is a convenient but some-what artificial distinction since, at the large-scale level of climate, maintenance of the atmospheric vapour pressure deficit depends upon the input of energy into the system.

Equation 10.3 can be written in terms of **A** and D as follows:

$$\mathbf{E}_T = a\mathbf{A} + bD \qquad (10.13)$$

where a is $s/(\lambda(s + \gamma(1 + g_a/g_c)))$

and b is $c_p\rho g_a/(\lambda(s + \gamma(1 + g_a/g_c)))$

Thus if bD is regarded as constant for a particular set of conditions (i.e. constant values of g_a, g_c, D and temperature, θ) $\mathbf{E}_T$ increases linearly with **A**, as shown in Fig. 10.8a, with slope g_a and intercept bD. When g_a is large changes in the income of available energy have only small effects on the overall rate of transpiration. This is usually the case for forest. On the other hand shorter, less rough vegetation has much smaller values of g_a, so that the overall rate of transpiration responds appreciably to changes in **A** (Fig. 10.8a).

In practice, D tends to increase with **A** so that bD is not constant but J does not vary much. As a result a plot of $\lambda\mathbf{E}_T$ against **A** may have quite a steep slope, especially when J is large (Fig. 10.8b). This can easily lead to the misapprehension that the transpiration rate is strongly dependent upon **A**, in a situation when $\lambda\mathbf{E}_T$ is in fact largely dependent on D and hardly depen-dent on **A** at all (Fig. 10.8b). From Fig. 10.7a it is not therefore clear how dependent $\lambda\mathbf{E}_T$ is on **A** without a calculation of J.

Vapour pressure deficit, D

The normal range of vapour pressure deficits in temperate maritime climates is small. Over most of the U.K., for example, D seldom exceeds 15 mb whereas in continental climates, vapour pressure deficits of twice that value are not uncommon. If $a\mathbf{A}$ is regarded as constant in equation 10.13, $\mathbf{E}_T$ increases linearly with D with slope b and intercept $a\mathbf{A}$. When g_a is large the contribution which $a\mathbf{A}$ makes to the overall rate of transpiration is small and the overall rate increases in proportion to D, as emphasized by equation 10.6. Fig. 10.9a shows that at quite moderate values of D, very high rates of transpiration would be expected, several times the measured rates, provided

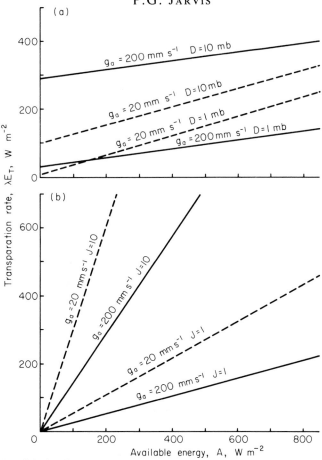

FIG. 10.8. Calculated relationships between transpiration rate and available energy with either (a) constant values of vapour pressure deficit, D, and variable J (eqn. 10.5), or (b) constant values of J (i.e. $D \propto A$) and variable D ($\theta = 15$ °C; $g_c = 20$ mm s^{-1}).

that g_c remains constant. That such high rates do not occur in practice (cf. Fig. 10.7b) is the result of stomatal closure and a consequent fall in g_c. In contrast when g_a is smaller, for example in the case of a less rough canopy, the intercept aA is a much larger proportion of the overall transpiration rate and the rate of increase of transpiration with increasing D is also lower (Fig. 10.9a).

Again, however, because of the association between A and D in practice, plots of transpiration rate against D are open to misinterpretation. A steeper slope is obtained at lower values of J than at higher values, suggesting a strong dependence of λE_T on D in circumstances when the dependence of λE_T on D is in fact small (Fig. 10.9b). Consequently it is not clear from a plot such as Fig. 10.7b alone how dependent λE_T is on D.

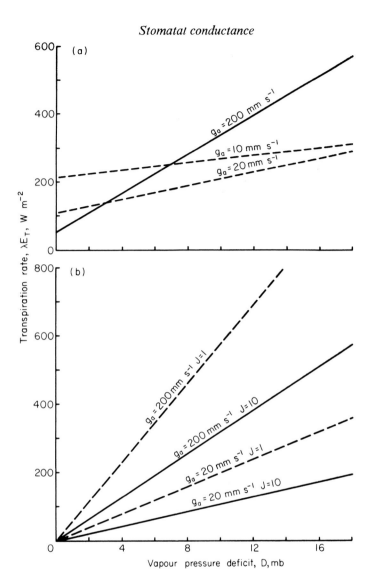

FIG. 10.9 Calculated relationships between transpiration rate and vapour pressure deficit with either (a) constant values of available energy, **A**, and variable *J*, or (b) constant values of *J* (i.e. *D* ∝ **A**) and variable **A** ($\theta = 15\ ^{\circ}\text{C}$, $g_c = 20\ \text{mm s}^{-1}$).

Boundary layer conductance

The boundary layer conductance of leaves depends on the size of the leaf and on the windspeed (Grace, this volume). The boundary layer conductance of a canopy depends on the surface roughness of the canopy and on the windspeed. For canopies of grassland and heathland, boundary layer conductances are typically less than 20 mm s^{-1} (Table 10.2, Fig. 10.10). On the other hand, the boundary layer conductance of forest canopies is typically in the range 100 to 300 mm s^{-1} because of the height of the trees and the much rougher surface (Rutter 1968) and because of the higher windspeeds, which are usually more than 0·5 m s^{-1}, at the top of the canopy.

The boundary layer conductance of grassland and heathland is in the same range as the stomatal conductance (Table 10.2) and consequently has a considerable influence on the rate of transpiration (Fig. 10.11). In contrast g_a of forest is very large compared with values of g_c (Table 10.2) with the

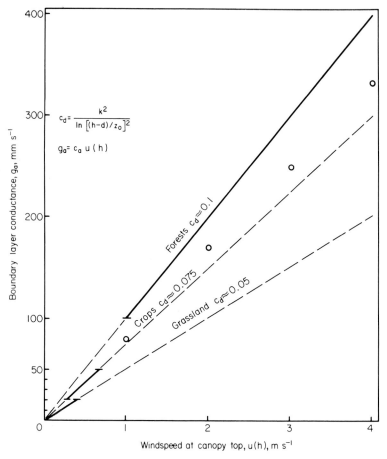

FIG. 10.10.   The relation between boundary layer conductance and windspeed at the top of the canopy for plantation forests, cereal crops and grassland. Crop heights of 16, 1 and 0·3 m were assumed. C_d is a nondimensional canopy drag coefficient which is dependent upon the aerodynamic parameters z_0 and d (Jarvis et al. 1976). Heavy lines indicate typical ranges of values in the field. Data points are based on 66 mean hourly wind profiles measured over Sitka spruce forest (redrawn from Jarvis & Stewart 1979).

TABLE 10.2. Canopy (g_c) and boundary layer (g_a) conductances for different kinds of vegetation. Based on figures given in *Vegetation and the Atmosphere*, Vol. II, ed. J. L. Monteith, Academic Press, 1976.

	Maximal g_c, m s^{-1}		Range of g_a, m s^{-1}
	per unit leaf area	per unit ground area	per unit ground area
Grassland/heathland	0·01	0·02	0.05–0·02
Agricultural crops	0·02	0·05	0·02–0·05
Plantation forest	0·006	0·02	0·1–0·3

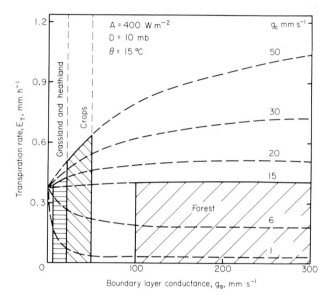

FIG. 10.11. The calculated relationships between transpiration rate and boundary layer conductance at different canopy conductances for the conditions $\mathbf{A} = 400$ W m^{-2}, $D = 10$ mb and $\theta = 15°$C. Observations for particular kinds of vegetation are expected to lie within the areas shaded.

consequence that g_a can vary widely with little or no effect on transpiration (Fig. 10.11). Thus the different transpiration rates of the two stands and the changes in transpiration rate with time of day and weather shown in Figs. 10.6 and 10.7 are unlikely to be the result of variation in g_a. Nonetheless the low transpiration rates of coniferous forest result in part from the large boundary layer conductance because this facilitates removal of sensible heat from the foliage so that the leaf temperatures of conifers seldom differ from air temperature by more than 0·5 °C (Jarvis *et al.* 1976). As a result the transpiration rate is lower, with smaller α and larger β, than it would be for crops and grassland with a lower boundary layer conductance (Table 10.2).

Leaf and canopy conductance

The largest value of canopy conductance is probably about 50 mm s^{-1} in some field crops, but many crops have smaller maximal values. We know little about the canopy conductance of other kinds of vegetation, especially natural vegetation. The maximal value for grassland and heathland is probably about 20 mm s^{-1} (McKerron 1971; Ripley & Saugier 1978). The maximal canopy conductance of Sitka and Norway spruce forest canopies is also about 20 mm s^{-1} and for Scots pine, lodgepole pine and Douglas fir, somewhat less than 10 mm s^{-1} (Jarvis *et al.* 1976).

The influence of canopy conductance on the partitioning of the available energy into transpiration was shown in Fig. 10.5: the Bowen ratio, β, increases linearly with increasing canopy resistance ($1/g_c$). Fig. 10.12 shows that transpiration increases with canopy conductance, particularly for canopies with large boundary layer conductances, such as forest. In grassland and heathland with much smaller g_a, transpiration increases with increasing g_c over only a very much smaller range before becoming saturated with respect to g_c (Fig. 10.12).

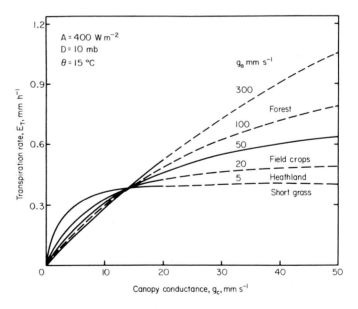

FIG. 10.12.   The calculated relationships between transpiration rate and canopy conductance at different boundary layer conductances for the conditions shown. The values of boundary layer conductance have been chosen to be appropriate for different kinds of vegetation the canopy conductances of which are generally within the solid lines.

The large consistent differences in transpiration rate of the two plots of Scots pine shown in Figs 10.6 and 10.7 result from consistent differences in g_c. The transpiration rate is strongly dependent on g_c because the canopy is rough. Consequently, J is large and g_a/g_c is large, so that equation 10.6 is applicable. The canopy conductance is consistently different between the plots because g_c depends on leaf area index as illustrated by equation 10.4.

Multiple effects of the variables

The final overall rate of transpiration depends on the values of g_c and g_a which occur in conjunction with particular weather conditions. If the influence of either of the driving variables **A** or **D** is to reduce g_s then transpiration will not increase as anticipated. This is evident in Fig. 10.7 where the rate of transpiration crosses the lines of equal g_c. At Roseisle Forest, even at deficits of 16 mb, $\mathbf{E}_T$ was only 0·2 mm h^{-1} because g_c declined as vapour pressure deficit increased (Fig. 10.13). This observation in the forest is completely consistent with laboratory studies which have shown that stomatal conductance in Scots pine is very sensitive to vapour pressure deficit (Jarvis 1980).

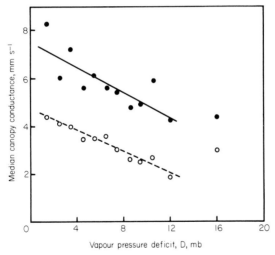

FIG. 10.13. The correlation between canopy conductance and vapour pressure deficit for the two stands of Scots pine (*P. sylvestris* L.) at Roseisle Forest (for details see Table 10.1 and Figs 10.6 and 10.7).

CONCLUSIONS

The rate of transpiration depends on the effects of a number of variables some of which are environmental and some of which are characteristics of the plants and vegetation. These variables can be interrelated through the Penman–Monteith equation which provides a suitable means of estimating transpiration rate and of analysing its variations. Whilst this approach has been familiar to ecologists of agricultural crops for some years, it is equally useful to plant ecologists.

In an analysis of this kind it is not possible to cover all combinations of the main variables. The only answer to a particular question is to solve the equation for the appropriate circumstances.

The examples discussed here demonstrate that stomatal conductance is a variable which is invaluable in the interpretation of gaseous exchange by leaves and canopies and indispensable for modelling the processes of gaseous exchange. In the field stomatal conductance may be measured extensively with the porometer, but this is somewhat laborious and tedious. It may also be measured with cuvettes and chambers but these measurements are expensive and intensive rather than extensive. It is to be hoped that the great surge in knowledge about the stomatal mechanism of the past decade will lead in the next one to mechanistic models of stomatal action which will enable useful predictions of stomatal conductance in the field. In the meantime diffusion porometers are a useful tool for the ecologist willing to spend long hours in the field.

APPENDIX

Derivation of the Penman–Monteith equation
Consider a slice of canopy 1 m² in cross-section.

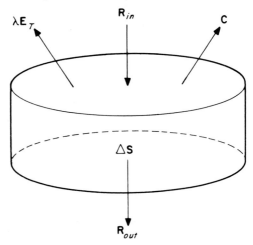

Conservation of energy requires that:

$$\mathbf{R}_{in} + \mathbf{R}_{out} + \Delta \mathbf{S} + \lambda \mathbf{E}_T + \mathbf{C} = 0 \tag{A1}$$

here $\mathbf{R}$ is the flux of net radiation per unit area of slice cross-section

$\Delta \mathbf{S}$ is the net change in energy stored in the slice and includes changes of sensible heat and metabolic energy in the biomass and changes in the sensible heat and latent heat content of the air contained in the slice.

$\lambda \mathbf{E}_T$ is the net flux of latent heat in or out of the slice.

$\mathbf{C}$ is the net flux of sensible heat in or out of the slice.

Gains of energy by the slice are regarded as positive and losses as negative. Thus the energy 'available' for the exchanges of sensible and latent heat is:

$$\mathbf{A} = \mathbf{R}_{in} - \mathbf{R}_{out} - \Delta \mathbf{S} = \lambda \mathbf{E}_T + \mathbf{C} \tag{A2}$$

For daytime conditions, apart from dawn and dusk, $\Delta \mathbf{S}$ is small relative to $(\mathbf{R}_{in} - \mathbf{R}_{out})$ and can often be neglected.

Following Thom (1975) the equations for transfer across the boundary layers of the leaves in the slice are:

$$\mathbf{C} = c_p \rho (T_l - T_a) g_H = c_p \rho \Delta T g_H \tag{A3}$$

and

$$\lambda \mathbf{E}_T = \frac{c_p \rho}{\gamma} (e_s(T_l) - e_a) g_W \tag{A4}$$

where c_p is the specific heat of dry air (1006 J kg^{-1} °C^{-1})

ρ is the density of dry air (1·22 kg m^{-3} at 15 °C)

γ is the psychrometric constant (0·655 mg °C^{-1} at 15 °C)

T_l is the leaf temperature (°C)

T_a is air temperature (°C)

$e_s(T_l)$ is the saturated water vapour pressure of air at leaf temperature (mb)

e_a is the water vapour pressure of ambient air (mb)

g is the transfer conductance for leaves referred to unit area of slice cross-section (m s^{-1}); subscript H for heat, W for water vapour.

Following Monteith (1965), we define s as the slope of the curve relating saturated water vapour pressure to temperature at the temperature $(T_l + T_a)/2$. Then we may write

$$e_s(T_l) - e_a = e_s(T_l) - e_s(T_a) + e_s(T_a) - e_a = s\Delta T + D \tag{A5}$$

where D is vapour pressure deficit $(e_s(T_a) - e_a)$ (mb).

Combining equations A2, A3, A4 and A5 with the elimination of $\mathbf{C}$ and ΔT, the following expression for the latent heat flux is obtained:

$$\lambda \mathbf{E}_T = \frac{s\mathbf{A} + c_p \rho D g_H}{s + \gamma g_H / g_W} \tag{A6}$$

For foliage the surface of which is wet, the pathways of heat and water vapour transport may be regarded as similar so that $g_H \simeq g_W$, giving the familiar equation for potential transpiration.

For dry foliage, the water vapour pathway includes transfer from the sources within the leaves across the substomatal cavities and out through the stomatal pore. The diffusive resistance of this pathway is in series with the boundary layer so

$$\frac{1}{g_W} = \frac{1}{g_c} + \frac{1}{g_a} \tag{A7}$$

where g_a is the total boundary layer conductance for water vapour and g_c is the total leaf conductance for water vapour. Substituting for $1/g_W$ from equation A7 in equation A6 gives

$$\lambda E_T = \frac{s\mathbf{A} + c_p\rho Dg_H}{s + \gamma(g_H/g_a + g_H/g_c)} \tag{A8}$$

The conductance of the boundary layer is proportional to the (molecular diffusivity in air)$^{2/3}$ of the property being considered (Thom 1968). The diffusivities of heat and water vapour in air at 15 °C are 20·8 and 23·4 mm^2 s^{-1}, respectively (Monteith 1973). Assuming that the pathway across the boundary layer is identical for both heat and water vapour,

$$g_H/g_a = (20\cdot8/23\cdot4)^{2/3} = 0\cdot93 \tag{A9}$$

Substituting for g_H from equation A9 in equation A8 leads to

$$\lambda E_T = \frac{s\mathbf{A} + c_p\rho D\, 0\cdot93g_a}{s + 0\cdot93\gamma(1 + g_a/g_c)} \tag{A10}$$

However if the assumption is made that $g_H = g_a$, equation A8 reduces to equation 10.3 in the body of the paper, viz.

$$\lambda E_T = \frac{s\mathbf{A} + c_p\rho Dg_a}{s + \gamma(1 + g_a/g_c)}$$

This approximation is sufficiently precise for most purposes.

The derivation for an individual leaf is similar. In that case, ΔS can certainly be neglected; $\mathbf{R}$ and g are referred to unit leaf area rather than unit area of slice cross-section. It is desirable to ensure that both g_H and g_W are expressed on the same area basis initially (i.e. plan or total for both) or confusing complexities arise later.

REFERENCES

Alvim P.deT. (1965) A new type of porometer for measuring stomatal opening and its use in irrigation studies. *Arid Zone Research*, **25**, 325–329.

Beardsell M.F., Jarvis P.G. & Davidson B. (1972) A null-balance porometer suitable for use with leaves of many shapes. *Journal of Applied Ecology*, **9**, 677–690.

Bierhuizen J.F. & Slatyer R.O. (1964) Photosynthesis of cotton leaves under a range of environmental conditions in relation to internal and external diffusive resistances. *Australian Journal of Biological Science*, **17**, 348–360.

Bierhuizen J.F., Slatyer R.O. & Rose C.W. (1965) A porometer for laboratory and field operation. *Journal of Experimental Botany*, **16**, 182–191.

Björkman O., Boardman N.K., Anderson J.M., Thorne S.W., Goodchild D.J. & Pyliotis N.A. (1972) Effect of light intensity during growth of *Atriplex patula* on the capacity of photosynthetic reactions, chloroplast components and structure. *Carnegie Institution of Washington Year Book*, **71**, 115–135.

Butler D.R. (1976) Estimation of transpiration rate in an apple orchard from net radiation and vapour pressure deficit measurements. *Agricultural Meteorology*, **16**, 277–290.

Cowan I.R. & Farquhar G.D. (1977) Stomatal function in relation to leaf metabolism and environment. *Integration of Activity in the Higher Plant*. Symposium of the Society for Experimental Biology 31. (Ed. by D.H. Jennings) pp 491–505. Cambridge University Press.

Eckardt F.E. (1960) Eco-physiological measuring techniques applied to research on water relations of plants in arid and semi-arid regions. *Arid Zone Research*, **15**, 139–171.

Farquhar G.D., Dubbe D.R. & Raschke K. (1978) Gain of the feedback loop involving carbon dioxide and stomata. *Plant Physiology*, **62**, 406–412.

Fritschen L.J. & Van Bavel C.H.M. (1964) Energy balance as affected by height and maturity of Sudan grass. *Agronomy Journal*, **56**, 201–204.

Gaastra P. (1959) Photosynthesis of crop plants as influenced by light, carbon dioxide, temperature and stomatal diffusion resistance. *Mededelingen van de landbouwhogeschool te Wageningen, Nederland*, **59** (13), 1–68.

Garland J.A. & Branson J.R. (1977) The deposition of sulphur dioxide to pine forest assessed by a radioactive tracer method. *Tellus*, **29**, 445–454.

Grieve B.J. & Went F.W. (1965) An electric hygrometer apparatus for measuring water vapour loss from plants in the field. *Arid Zone Research*, **25**, 247–258.

Holmgren P., Jarvis P.G. & Jarvis M.S. (1965) Resistances to carbon dioxide and water vapour transfer in leaves of different plant species. *Physiologia Plantarum*, **18**, 557–573.

James G.B. & Jarvis P.G. Carbon dioxide and water vapour exchanges by a Sitka spruce forest canopy. II. Evaporation and surface resistance. In press.

Jarvis P.G. (1971) The estimation of resistances to carbon dioxide transfer. *Plant Photosynthetic Production: Manual of Methods* (Ed. by Z. Šesták, J. Čatský and P.G. Jarvis), pp. 566–631. Junk, The Hague.

Jarvis P.G. (1976) The interpretation of the variations in leaf water potential and stomatal conductance found in canopies in the field. *Philosophical Transactions of the Royal Society London B*. **273**, 593–610.

Jarvis P.G. (1980) Stomatal response to water stress in conifers. *Adaptation of Plants to Water and High Temperature Stress* (Ed. by N.C. Turner and P.J. Kramer), pp. 105–122. Wiley-Interscience, New York.

Jarvis P.G. Stomatal control of transpiration and photosynthesis. *Stomatal Physiology*, (Ed. by P.G. Jarvis and T.A. Mansfield). S.E.B. Seminar Series, Cambridge University Press. In press.

Jarvis P.G., James G.B. & Landsberg J.J. (1976) Coniferous forest. *Vegetation and the*

Atmosphere, Vol. 2 (Ed. by J.L. Monteith). pp. 171–240. Academic Press, London.

Jarvis P.G., Rose C.W. & Begg J.G. (1967) An experimental and theoretical comparison of viscous and diffusive resistances to gas flow through amphistomatous leaves. *Agricultural Meteorology*, **4**, 103–117.

Jarvis P.G. & Stewart J.B. (1979) Evaporation of water from plantation forest. *The Ecology of Even-Aged Forest Plantations*. (Ed. by E.D. Ford, D.C. Malcolm and J. Atterson) pp. 327–350. Institute of Terrestrial Ecology, Cambridge.

Kanemasu E.T., Thurtell G.W. & Tanner C.B. (1969) Design, calibration and field use of a stomatal diffusion porometer. *Plant Physiology*, **44**, 881–885.

Kanemasu E.T., Ed. (1975) *Measurement of Stomatal Aperture and Diffusive Resistance.* College of Agriculture Research Center, Washington State University, Bulletin 809.

Körner C. & Cernusca A. (1976) A semi-automatic recording diffusion porometer and its performance under alpine field conditions. *Photosynthetica*, **10**, 172–181.

Körner C., Scheel J.A. & Bauer H. (1979) Maximum leaf diffusive conductance in vascular plants. *Photosynthetica*, **13** (1), 45–82.

Landsberg J.J., Beadle C.L., Biscoe P.V., Butler D.R., Davidson B., Incoll L.D., James G.B., Jarvis P.G., Martin P.J., Neilson R.E., Powell D.B.B., Slack W.M., Thorpe M.R., Turner N.C., Warrit B. & Watts W.R. (1975) Diurnal energy, water and CO_2 exchanges in an apple (*Malus pumila*) orchard. *Journal of Applied Ecology*, **12**, 659–684.

Landsberg J.J. & Powell D.B.B. (1973) Surface exchange characteristics of leaves subject to mutual interference. *Agricultural Meteorology*, **12**, 169–184.

Landsberg J.J. & Thom A.S. (1971) Aerodynamic properties of a plant of complex structure. *Quarterly Journal of the Royal Meteorological Society*, **97**, 565–570.

Lowry W.P. (1969) *Weather and Life*, pp. 305. Academic Press, London.

McKerron D.K.L. (1971) Energy, carbon dioxide, and water vapour exchange growth of *Calluna vulgaris* (L.) Hull in relation to environment. *Ph.D. Thesis, University of Aberdeen.*

Meidner H. Measurements of stomatal aperture and responses to stimuli. *Stomatal Physiology*. S.E.B. Seminar Series (Ed. by P.G. Jarvis & T.A. Mansfield). Cambridge University Press. In press.

Milthorpe F.L. (1961) Plant factors involved in transpiration. *Arid Zone Research*, **16**, 107–115.

Milthorpe F.L. & Penman H.L. (1967) The diffusive conductivity of the stomata of wheat leaves. *Journal of Experimental Botany*, **18**, 422–457.

Monteith J.L. (1965) Evaporation and environment *The State and Movement of Water in Living Organisms*. Symposium of the Society for Experimental Biology 19, pp. 205–224. Cambridge University Press.

Monteith J.L. (1973) *Principles of Environmental Physics*. Edward Arnold, London.

Ng P.A.P. & Jarvis P.G. (1980) Hysteresis in the response of stomatal conductance in *Pinus sylvestris* L. needles to light: observations and a hypothesis. *Plant, Cell and Environment*, **3**, 207–216.

Parkinson K.J. & Legg B.J. (1972) A continuous flow porometer. *Journal of Applied Ecology*, **9**, 669–675.

Penman H.L. (1948) Natural evaporation from open water, bare soil and grass. *Proceedings of the Royal Society, London (A)*, **193**, 120–145.

Raschke K. (1956) Mikrometeorologisch gemessene Energieumsätze eines Alocasiablattes. *Archiv für Meteorologie, Geophysik und Bioklimatologie*, **7**, 240–268.

Raschke K. (1958) Über den Einfluss der Diffusionswiderstände auf die Transpiration und die Temperatur eines Blättes. *Flora*, **146**, 546–578.

Raschke K. (1965) Der seifenblasen Porometer—(Zur Messung der Stomaweite an amphistomatischen Blätter). *Planta*, **66**, 113–120.

Reed K.L., Hamerly E.R., Dinger B.E. & Jarvis P.G. (1976) An analytical model for field measurements of photosynthesis. *Journal of Applied Ecology*, **13**, 925–942.

Ripley E.A. & Saugier B. (1978) Biophysics of a natural grassland. *Journal of Applied Ecology*, **15**, 459–479.

Rutter A.J. (1967) An analysis of evaporation from a stand of Scots pine. *International Symposium on Forest Hydrology* (Ed. by W.E.S. Sopper & H.W. Lull), pp. 403–417. Pergamon Press, Oxford.

Rutter A.J. (1968) Water consumption by forests. *Water Deficits and Plant Growth* (Ed. by T.T. Kozlowski), pp. 23–84. Academic Press, London.

Slatyer R.O. & McIlroy I.C. (1961) *Practical Microclimatology*. UNESCO, Paris.

Shuttleworth J.W. (1976) A one-dimensional theoretical description of the vegetation atmosphere interaction. *Boundary-Layer Meteorology*, **10**, 273–302.

Sinclair T.R., Murphy Jr. C.E. & Knoerr K.R. (1976) Development and evaluation of simplified models for simulating canopy photosynthesis and transpiration. *Journal of Applied Ecology*, **13**, 813–830.

Slavik B. (1974) Methods of studying water plant relations. *Ecological Studies*, Vol. 9. Springer-Verlag, Berlin

Stewart J.B. & Thom A.S. (1973) Energy budgets in pine forest. *Quarterly Journal of the Royal Meteorological Society*, **99**, 154–170.

Stiles W. (1970) A diffusive resistance porometer for field use. I. Construction. *Journal of Applied Ecology*, **7**, 617–622.

Strangeways I.C. (1972) Automatic weather stations for network operation. *Weather*, **27**, 403–408.

Szeicz G., Van Bavel C.H.M. & Takami S. (1973) Stomatal factor in the water use and dry matter production by sorghum. *Agricultural Meteorology*, **12**, 361–381.

Tan C.S. & Black T.A. (1976) Factors affecting the canopy resistance of a Douglas-fir forest. *Boundary-Layer Meteorology*, **10**, 475–488.

Thom A.S. (1968) The exchange of momentum, mass and heat between an artificial leaf and the airflow in a wind-tunnel. *Quarterly Journal of the Royal Meteorological Society*, **94**, 44–55.

Thom A.S. (1975) Momentum, mass and heat exchange of plant communities. *Vegetation and the Atmosphere*, Vol. 1 (Ed. by J.L. Monteith), pp. 57–109. Academic Press, London.

Thorpe M.R. (1978) Net radiation and transpiration of apple trees in rows. *Agricultural Meteorology*, **19**, 41–57.

Turner N.C. & Parlange J.Y. (1970) Analysis of operation and calibration of a ventilated diffusion porometer. *Plant Physiology*, **46**, 175–177.

Van Bavel C.H.M., Fritschen L.J. & Reeves W.E. (1963) Transpiration by sudan grass as an externally controlled process. *Science*, **14**, 269–270.

Van Bavel C.H.M., Nakyama F.S. & Ehrler W.L. (1965) Measuring transpiration resistance of leaves. *Plant Physiology*, **40** (3), 525–540.

Waring R.H. & Roberts J.M. (1979) Estimating water flux through stems of Scots pine with tritiated water and phosphorus-32. *Journal of Experimental Botany*, **30**, 459–471.

Waring R.H., Whitehead D. & Jarvis P.G. (1979) The contribution of stored water to transpiration in Scots pine. *Plant, Cell and Environment*, **2**, 309–319.

Waring R.H., Whitehead D., & Jarvis P.G. Comparison of an isotopic method and the Penman-Monteith equation for estimation of transpiration from Scots pine. *Canadian Journal of Forest Research*. In press.

Watts W.R. & Neilson R.E. (1978) Photosynthesis in Sitka spruce (*Picea˙sitchensis* (Bong.) (Carr.). IX Measurements of stomatal conductance and $^{14}CO_2$ uptake in controlled environments. *Journal of Applied Ecology*, **15**, 245–255.

Watts W.R., Neilson R.E. & Jarvis P.G. (1976) Photosynthesis in Sitka spruce (*Picea sitchensis* (Bong.) Carr.) VIII Forest canopy measurements of stomatal conductance and $^{14}CO_2$ uptake. *Journal of Applied Ecology*, **13**, 623–638.

Yamaoka Y. (1958) Experimental studies on the relation between transpiration rate and meteorological elements. *Transactions of the American Geophysics Union*, **39**, 239–265.

11. STOMATAL BEHAVIOUR IN ALPINE PLANT COMMUNITIES BETWEEN 600 AND 2600 METRES ABOVE SEA LEVEL

CH. KÖRNER AND R. MAYR

Institut für Botanik der Universität, Sternwartestraße 15,
A-6020 Innsbruck, Austria

SUMMARY

Leaf diffusive conductance, water potential and anatomy were investigated in 37 species at 15 different sites between 600 and 2600 m above sea level in the Austrian Central Alps.

1. With increasing altitude the maximum leaf diffusive conductance increased, diurnal variation of conductance decreased and the daily sum (integral) of leaf conductance increased. Only very exposed communities in summit habitats showed lower values.

2. In the valley, periods with reduced leaf conductance under sunny weather condition occurred in the early afternoon, whereas in the high alpine sites morning depressions were observed, indicating after-effects of low night temperatures.

3. Leaf water potentials were relatively high at all sites. Graminoids exhibited the lowest potentials (minimum -1.7 MPa) with no distinct altitudinal variation. In herbaceous and woody species, with minimum values between -0.5 and -1.5 MPa, potentials increased slightly with altitude.

4. Seventy per cent of the species investigated were amphistomatous with stomatal frequency increasing with altitude.

5. Consequently, in most of the alpine species investigated, water is a factor of minor importance in controlling gas diffusion processes in the field. The results also explain the relatively high rates of evapotranspiration measured in high-alpine plant communities during sunny days in summer.

INTRODUCTION

Altitudinal variation of climate in mountain regions (Table 11.1) causes many visible changes in the composition of the vegetation and the growth habit of individual plants. Surprisingly these marked differences between valley and mountain habitats hardly find expression in the observed diurnal rate of evapotranspiration of canopies under bright weather conditions in summer. Fig. 11.1 shows the altitudinal variation of evapotranspiration in low vegetation in the Central Alps measured by weighing lysimeters. Average values range from 4·5 mm d^{-1} in grassland in the valley floor (600 m above sea level, LAI 6–8) to 3·8 mm d^{-1} in alpine sedge mats at 2600 m (LAI 2–3).

TABLE 11.1. Macroclimate in the Central Alps. An estimate of average values of different climatic factors for 600 and 2600 m above sea level, compiled from Hader 1954; Lauscher 1954; Winkler & Moser 1967; Fliri 1975; Dobesch, Neuwirth & Weiss 1978. The information given by these authors is based on measurements at different meteorological stations in the Austrian and Swiss Central Alps.

Height above sea level	600 m	2600 m
Global radiation on clear summer days, %	100	120
overcast summer days, %	100	260
Number of clear days in summer (June, July, August)	10	5
Number of clear days in July	4	2
Number of hours of sunshine in July	200	160
Annual average of air temperature, °C	+8	−3
Average air temperature in July, °C	+18	+5
Maximum air temperature in July	+32	+14
Number of days with frost (per year)	120	260
Annual average of vapour pressure, 10^2 Pa	9·2	4·3
Average vapour pressure in summer, 10^2 Pa	14·7	6·9
Maximum vapour pressure deficit in July, 10^2 Pa	ca 20	ca 8
Number of days with fog (per year)	0–10	80
Annual sum of precipitation (±30%), mm	900	1800
Number of days with snow cover	80	280
Annual sum of evapotranspiration, mm	500	200
Average windspeed in summer, m s^{-1}	1	4
Atmospheric pressure, 10^4 Pa	9·46	7·40

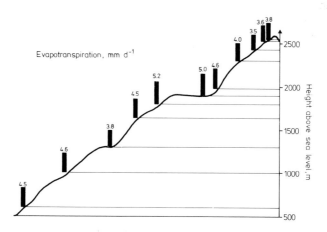

FIG. 11.1. Altitudinal variation of evapotranspiration on bright summer days obtained by weighing lysimeters (Körner 1977a; Ch. Körner & G. Wieser, unpublished).

In the alpine zone evapotranspiration from graminoid canopies with higher LAI than in sedge mats like meadows with *Deschampsia caespitosa* can be as much as 5·4 mm d^{-1} (Körner 1977a). Observed maximum values of daily evapotranspiration are 9 mm d^{-1} in wet grassland in the valley (G. Wieser, unpublished) and 8·2 mm d^{-1} in wet *Polytrichum* stands at 2300 m (Körner 1977a). For a causal analysis of these results it is necessary to study both the driving forces within the canopies and the resistances to water transfer in each compartment of the soil–plant–atmosphere continuum in different habitats at different altitudes.

In this short report we shall try to summarize results obtained on the stomatal control of water vapour diffusion in leaves. The results presented here are preliminary in some details, but we believe that some general trends are recognizable.

MATERIALS AND METHODS

The sites and plant species investigated are listed in Table 11.2. Most of the measurements have been conducted in the Hohen Tauern range and on Mt Patscherkofel near Innsbruck (47°N 11–13°E). Along the altitudinal transect covered by these investigations the duration of the period without permanent snowcover varies from 6–10 weeks at a snow-bed site at 2528 m to 36–44 weeks at a meadow site on the valley floor at 600 m. Exceptionally arid sites which cover only small areas in this part of the Central Alps are not included in this comparison.

Leaf diffusive conductance for water vapour (g_l) was measured with ventilated diffusion porometers (Körner & Cernusca 1976; Körner 1977b). The instruments used since 1977 were supplied by H. Walz, Effeltrich, F.R.G. Calibration was done as described by Körner and Cernusca (1976) using microevaporimeters (0·1 ml pipettes with sintered glass heads of different size). From the measured water flux into the porometer chamber and the dynamic response of the humidity sensor, calibration factors were calculated for different temperatures and evaporation rates. The series of polygons obtained was entered in the computer program for the calculation of diffusive conductance. The dynamic response of the sensor does not change with changing atmospheric pressure (see p. 213).

Leaf conductance was calculated for the total leaf surface area (upper plus lower leaf side in flat leaves). The use of the projected leaf area in such comparative studies which include species with non-flat leaves has two serious disadvantages:

1. Projected leaf area is a term which is not precisely defined by itself, if one does not supply detailed information about the chosen direction of the projection for each species (what is the projected leaf area of *Calluna vulgaris*?).

TABLE 11.2. Characteristics of the sites and plant species investigated along an altitudinal transect in the Central Alps.

a.s.l.	Site	Veg.	LAI	Species	s_d	s_v	ψ_i	$\int g_i$	g_1	g_2	Ref.
620	Meadow	36–44	(6·8)	Arrhenatherum elatius	+	+	.	86	0·67	0·03	[1]
				Trifolium pratense	+	+	.	194	0·65	0·35	[1]
1050	Meadow on ski-slope	28–36	(—)	Dactylis glomerata	+	+	(−1·6)	166	0·77	0·22	[1]
				Taraxacum officinale	+	+		191	0·65	0·45	[1]
1050	Pine wood	28–36	(—)	Pinus sylvestris	+	+	(−1·4)	47	0·20	0·05	[1]
				Vaccinium myrtillus	−	+		45	0·20	0·04	[1]
1630	Meadow	20–26	(8·0)	Dactylis glomerata	+	+	−1·7	360	1·25	0·62	[1]
				Taraxacum officinale	+	+	−0·9	284	1·00	0·42	[1]
				Alchemilla gracilis	+	+	−1·1	289	0·80	0·30	[1]
1800	Alpine pasture	18–24	(4·2)	Luzula sylvatica	+	+	.	117	0·63	0·12	[2]
				Geum montanum	+	+	.	227	1·00	0·40	[2]
				Gentiana punctata	−	+	.	61	0·25	0·05	[2]
				Veratrum album	−	+	.	157	0·56	0·35	[2]
1800	Green alder stand	18–24	(2·7)	Alnus viridis	−	+	−1·2	104	0·46	0·13	[3]
1910	Alpine pasture	16–22	(6·9)	Poa alpina	+	−	(−1·4)	248	1·00	0·40	[1]
				Geum montanum	+	+	.	432	1·45	0·61	[1]
				Alchemilla subcrenata			(−0·7)	360	1·40	0·56	[1]
				Arnica montana	+	+		418	1·15	1·00	[1]
1960	Dwarf shrubs	18–25	(5·3)	Vaccinium myrtillus	+	+	.	90	0·30	0·15	[4]
2180	Dwarf shrubs	20–30	(2·6)	Loiseleuria procumbens	−	+	−1·3	76	0·24	0·12	[5]
				Calluna vulgaris	+	+	−1·5	68	0·29	0·09	[5]
2290	Alpine meadow	11–16	(—)	Sesleria varia	+	−	.	180	0·72	0·28	[1]
				Aster alpinus	+	+	.	371	1·10	0·85	[1]
				Anthyllis vulneraria	+	+		405	1·25	0·90	[1]
2310	Sedge mat	10–15	(2·3)	Carex curvula	−	+	−1·4	163	0·55	0·31	[.]
				Festuca halleri	+	−	−1·7	122	0·40	0·15	[1]
				Deschampsia caespitosa	+	+	(−1·3)	488	1·40	1·00	[1]
				Geum montanum	+	+	−1·5	292	1·20	0·40	[1]
				Primula minima	+	−		245	0·75	0·50	[6]

a.s.l.	Veg.	LAI	Species	s_a	s_v	ψ_t	$\int g_l$	g_1	g_2	Ref
			Hieracium cf. alpinum	+	+	.	256	0·70	0·60	[6]
			Leontodon helveticus	+	+	.	377	1·40	0·40	[1]
			Alchemilla colorata	+	+	.	437	1·40	0·60	[1]
			Arnica montana	+	+		247	0·95	0·60	[1]
2530	Snowbed vegetation 6–10	(ca 1)	Salix herbacea	+	+	−0·9	151	0·52	0·28	[1]
			Salix retusa	+	+	−0·9	158	0·54	0·29	[1]
			Silene exscapa	+	+	−0·6	137	0·35	0·32	[7]
			Polygonum viviparum	+	+	−0·7	205	1·04	0·30	[1]
2540	Sedge fragments 8–12	(ca 2)	Carex curvula	−	+	−1·5	220	0·62	0·42	[1]
			Poa alpina	−	+	.	425	1·35	0·65	[1]
			Geum montanum	+	+	−1·3	396	1·45	0·70	[1]
			Alchemilla sp.	+	+	.	355	1·45	0·30	[1]
2610 (peak)	Sedge mat 6–12	(—)	Elyna myosuroides	−	+	−1·7	151	0·63	0·15	[1]
			Poa alpina	−	+	−1·5	148	0·55	0·20	[1]
2620 (peak)	Open vegetation 6–12	(ca 0·5)	Ranunculus alpestris	−	+	−0·5	263	0·90	0·40	[1]
			Silene exscapa	+	+	−0·8	81	0·36	0·27	[7]
			Saxifraga oppositifolia	+	−	−0·9	206	0·71	0·48	[7]
			Salix serpyllifolia	+	+	.	35	0·12	0·08	[7]

Symbols:
a.s.l. height above sea level, m.
Veg. duration of the period without permanent snowcover, weeks.
LAI leaf area index, projected area of living parts of phanerograms, m² m⁻².
Species nomenclature follows Ehrendorfer (1973); *Alchemilla* species have been identified by S. Fröhner, Nossen, G.D.R.
s_a, s_v distribution of stomata on the adaxial (s_a) and abaxial (s_v) leaf side.
ψ_t lowest hourly mean of water potential under sunny weather conditions, brackets indicate that only few values at noon have been measured, MPa.
$\int g_l$ daily sum of leaf conductance for sunny weather conditions, integral 07.00–17.00 h, 10^4 cm s⁻¹ s = 10^2 m.
g_1, g_2 highest and lowest hourly mean of leaf conductance observed in the field, cm s⁻¹

References:

[1] Unpublished
[2] Körner & Schubert 1978
[3] Körner, Jussel & Schiffer 1978
[4] Scheel 1979
[5] Körner 1976
[6] Körner 1977b
[7] Körner & Moraes 1979

2. In leaves with a strongly developed third dimension, diffusive conductance would be overestimated as compared with that of thin, flat leaves (cf. Körner, Scheel & Bauer 1979).

In large leaves, the porometer has been used on each side of the leaf separately. Small and arbitrary shaped leaves or shoots have been inserted in the chamber. For high alpine cushion plants with very small leaves the porometer has been equipped with a funnel for *in situ* measurements on

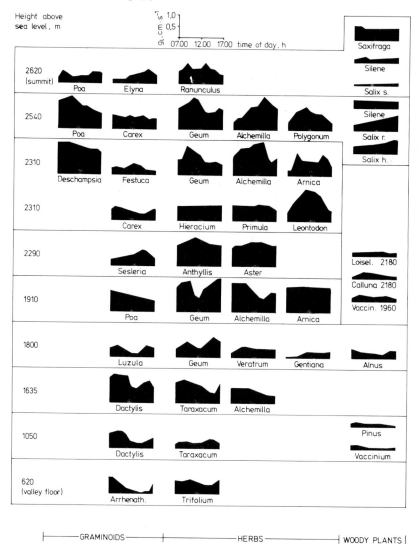

Fig. 11.2. Average bright-day courses of leaf diffusive conductance for water vapour in different sites and species in the Central Alps. For species names see Table 11.2.

the cushion's surface (Körner & Moraes 1979). For this mode of using the porometer a recalibration and the estimation of leaf area index were necessary. The aerodynamic resistance of the canopy enclosed is assumed to be relatively small as a result of ventilation. Evaporation from non-green surfaces can also be assumed to be small because of the very dense green leaf layer.

In this paper, we shall consider only the diurnal course of leaf conductance obtained under sunny weather conditions at the time of the maximum development of the canopies. This is, for instance, at the time just before the first mowing in low altitude grassland or when current year shoots have reached maturity in alpine dwarf shrubs. The average sunny day courses of leaf conductance presented in Figs 11.2 and 11.3 have been calculated as means of one to six (mostly three) individual diurnal courses. The amount of data incorporated in these curves varies considerably. For instance, at the 2310 m station the curves calculated for *Carex curvula* are based on more

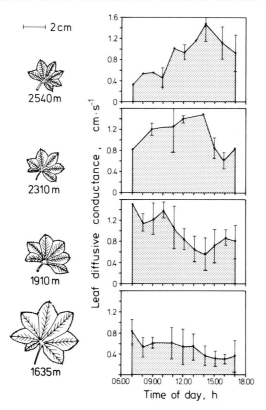

FIG. 11.3. Leaf diffusive conductance in *Alchemilla* spp. at 4 different sites. 1630 m: *A. gracilis* Opiz and *A. monticola* Opiz; 1910 m: *A. subcrenata* Buser; 2310 m: *A. colorata* Buser; 2540 m: *A.* sp. Note: With increasing altitude the peak of leaf conductance shifts from morning to afternoon.

than 200 individual measurements whereas for *Alchemilla colorata* only 20 measurements for bright weather conditions have been available. On average about 40 values for each site and species could be used. The standard deviations given in Fig. 11.3 are representative for those observed in the other species. Where these results are already published in more detail, the reader is referred to these papers for more information (Table 11.2).

Additionally, diurnal courses of xylem water potential have been measured by the pressure chamber technique (2 to 20 samples per hour). In this report only minimum potentials for sunny summer days are presented (the average of the lowest hourly means of each of 1 to 6 days). Absolute minimum potentials (single measurements) usually do not deviate from these means by more than 0·2 MPa.

RESULTS AND DISCUSSION

Leaf conductances

The average sunny day courses of leaf conductance for 37 dominant species between 600 and 2600 m are presented in Fig. 11.2. The highest and lowest hourly means of leaf conductance observed in the field are listed in Table 11.2 (columns g_1 and g_2). From these data the following trends can be seen:

In closed plant canopies between 600 and 2600 m above sea level:

1. Maximum leaf diffusive conductance increased with altitude in both herbaceous and woody plants. However the conductance values are lower for vegetation exposed to the wind at the highest sites investigated.

2. On average, the diurnal course of leaf conductance observed at the low elevation sites shows peaks in the morning and depressions at midday or in the afternoon. These are most pronounced in graminoids.

3. Plants at the high altitude sites

(a) show a less pronounced afternoon decline (e.g. *Deschampsia*) or

(b) do not show any distinct changes of leaf conductance (e.g. *Hieracium*) or

(c) even tend to exhibit peaks at midday or in the afternoon (e.g. *Elyna*).

In species which show midday or afternoon depressions of leaf conductance in the high alpine habitats like *Deschampsia* and *Geum* the lowest conductances observed between 07.00 and 17.00 h are still higher than the maximum conductances in species from valley sites. Morning depressions of leaf conductance in high alpine sites can possibly be attributed to preceding low temperatures (the after-effect of night frost) which can slow down both the rise of photosynthesis and the stomatal opening process (Larcher 1969; Drake & Salisbury 1972; Neilson & Jarvis 1975; Körner 1977b; Koh, Kumura & Murata 1978; Körner & Moraes 1979).

On average, the variation of conductance during the day decreases and, especially in herbs and woody plants, the daily sum of leaf conductance (integral 07.00–17.00 h) increases with increasing altitude (Table 11.2). Only in the open vegetation at the summit is this trend reversed. Thus, as far as leaf diffusive conductance is concerned, the alpine plant communities investigated (except the summit habitat) show a higher potential water loss than the plant communities in the valley; especially as in many alpine species (*Sesleria, Anthyllis, Aster, Loiseleuria, Festuca, Leontodon, Alchemilla, Polygonum, Salix, Elyna*) leaf conductance reaches its highest values at the time of the highest evaporative demand at midday or early afternoon when heat accumulation can exceed even that observed in valley canopies (Cernusca 1976; Larcher & Wagner 1976; Körner & Moraes 1979; Cernusca & Seeber, this volume) and the vapour pressure gradients are very steep. At the valley sites conductances are lowest during this period of the day. In seven instances it was possible to observe this altitudinal variation of leaf conductance within the same species or genus (e.g. Fig. 11.3).

The observed higher values of leaf diffusive conductance at alpine sites correspond well with the results of studies of leaf anatomy. With increasing altitude the frequency of stomata increases with a tendency to increase more in the adaxial (dorsal) leaf surface. Hypostomatous leaves are very rare in the alpine vegetation. Above 2000 m four of the species investigated have the stomata only on the adaxial, wind and radiation exposed leaf surface (Table 11.2). This trend was first discovered by Wagner (1892) and has been observed also in later investigations (Berger-Landefeldt 1936; Au 1969; Scheel 1979).

Atmospheric pressure and diffusive conductance

The measured altitudinal variation of diffusive conductance raises the question of the extent to which atmospheric pressure can influence the diffusion process between the site of evaporation in the mesophyll and the free atmosphere.

The diffusion coefficient D for the molecular diffusion of water vapour in air is directly related to the mean free path of molecular movement, λ, and the mean molecular velocity, $\bar{c}$. λ is a function of density and hence atmospheric pressure and $\bar{c}$ is a function of temperature (Moore 1972). Consequently D and hence the molecular diffusive conductance should increase by 21·8% between 9·46 × 10⁴ Pa at about 600 m and 7·40 × 10⁴ Pa at about 2600 m above sea level (+ 14·5% per 10⁴ Pa).

The diffusion from a ventilated wet surface (forced diffusion) is less affected by atmospheric pressure and the dependence varies with factors like wind speed and supply of radiant energy and cannot easily be predicted.

Using the porometer, the increase of evaporation causes a proportional decrease of the transit time (no change of the calibration factor).

The extent to which diffusive conductance of a ventilated leaf is influenced by atmospheric pressure depends on the proportions of molecular and forced diffusion in the total diffusion process. Usually it is assumed that there is free molecular diffusion inside the leaf and a transition from molecular to turbulent diffusion between stomatal pore and fully turbulent air. The development of this transition zone depends on the rate of ventilation and on the shape and dimensions of the stomatal antechamber and on the formation of wax structures and hairs. The proportion of the process in the molecular phase mainly depends on the length of the diffusion path and this varies with species, site, water status and evaporative demand.

Thus we cannot expect a unique relation between leaf conductance and decreasing atmospheric pressure, but we may consider an increase of conductance proportional to the pressure decrease as the upper limit of the possible range (e.g. $+14.5\%$ per 10^4 Pa pressure reduction or change in altitude from 620 to 1590 m). To test this hypothesis we started experiments with plant material transported from Innsbruck (620 m) to Mt Patscherkofel (1945 m) and vice versa. Three to five leaves from detached and watered shoots were investigated repeatedly over four hours after preconditioning for one hour in humid and CO_2-poor air under a glass bell jar in light at each site to obtain full stomatal opening. The preliminary results obtained with *Pinus cembra*, *Alnus viridis* and *Epilobium angustifolium* indicate an increase of leaf conductance between $+11.6$ and $+18.6\%$ (average 14.7%) for a 10^4 Pa reduction of atmospheric pressure. *Pinus cembra*, which keeps the stomata completely open for six to eight minutes after clipping off needles (tested under constant pressure), was used in another experiment: needles were cut near the top station of Mt Patscherkofel cable car, immediately inserted in the porometer, and continuous measurements were conducted in the car while it was going down 800 m in seven minutes. This was repeated 4 times. The observed decrease of leaf conductance calculated for a 10^4 Pa increase of atmospheric pressure was -9.2 to -28.0%. In both types of experiments the higher values possibly include some reduction in stomatal aperture in spite of the precautions taken. This is the main problem during such investigations (Gale 1973). Therefore these studies will be continued with potted plant material.

We may conclude that the observed increase of leaf conductance with increasing altitude (Fig. 11.2, Table 11.2) is the result of both increased frequency of stomata and increased diffusivity.

Under field conditions the influence of atmospheric pressure on leaf conductance can partly be compensated by low temperatures at high alpine sites especially on overcast days. Between 25 and 5 °C the decrease of $\bar{c}$ causes a decrease of D from 0.249 to 0.220 cm^2 s^{-1} (Monteith 1973) and thus a possible decrease of leaf conductance by 12% without a change of stomatal aperture.

The values of absolute saturation humidity and saturation vapour pressure used for the calculation of diffusive conductance are independent of atmospheric pressure (Dalton's law) and require no correction (cf. Smithsonian Institution, 1966). This is also true for absolute ambient humidity or saturation vapour pressure deficit if saturation values and relative humidity measured by a pressure independent sensor are used for the calculation. However, if humidity is derived from psychrometric measurements it must be noted that the so-called 'psychrometer constant' in reality contains a variable pressure (Monteith 1973).

To avoid misunderstandings in discussing the effect of altitude on plant transpiration one should clearly distinguish between the pure physical effects on the diffusion process discussed above and eco-meteorological effects ('mountain climate'). This has frequently been overlooked in the past.

Leaf water potential

The water potentials observed throughout the day are high, particularly in the alpine canopies, and correspond with the high osmotic potentials reported for these types of vegetation. Osmotic potentials in summer are rarely below -1.5 MPa and frequently around -1.2 MPa (Walter 1960). The lowest water potentials measured on bright summer days in graminoids at different habitats do not indicate a distinct altitudinal variation (-1.5 ± 0.2 MPa). In herbs and woody plants, potentials tended to increase with height above sea level (Table 11.2).

Nachuzrischwili (1971) investigated plant water relations in 32 species between 1700 and 3500 m in the central Caucasus. He too found minimum water potentials to be most frequently between -0.8 and -1.4 MPa (extremes -0.4 and -1.6 MPa) and he also observed generally high osmotic potentials and leaf water contents. Therefore we may conclude that the situation in the central Caucasus is not much different from that in the central Alps in this respect.

Stomatal response to ambient humidity

It can be concluded that in most species investigated above 2000 m (above the alpine timberline) internal water status may be considered as a factor of minor importance in controlling gas diffusion processes under natural field conditions, especially as sunny days, only considered in this paper, are not very frequent at this altitude (Table 11.1). However, under sunny weather conditions ambient humidity may directly influence leaf conductance in some species. This was demonstrated in field experiments with *Loiseleuria procumbens*, a prostrate dwarf shrub, the stomata of which close if the humidity near the leaf is reduced artificially below a vapour pressure deficit of 10^3 Pa whilst water potential remains high (-0.7 MPa, Körner 1977b). Undisturbed shrubs have a very high aerodynamic resistance causing a permanently high humidity in the leaf canopy (Cernusca 1976). Under those conditions leaf water potential can decrease to -1.7 MPa causing no change in leaf conductance. Similar effects of humidity were observed by Schulze *et al.* (1974) in *Prunus armeniaca*.

Loiseleuria is considered to be one of the most wind-resistant species ('windedge' plant), though from these experiments we must conclude that this is only true for the intact canopy as a whole. Curiously, the isolated individual shoot is extremely sensitive to wind and low humidity. This explains why this species is so susceptible to mechanical disturbances of the canopy structure by grazing cattle, trampling and skiing (Körner, in press).

CONCLUSIONS

The most important conclusions from this study of stomatal behaviour along an altitudinal transect in the Central Alps are:

1. In controlling the diurnal course of leaf diffusive conductance in summer the importance of the water factor decreases with increasing altitude.

2. The observed altitudinal increase of both maximum and mean leaf diffusive conductance for water vapour partly explains the unexpected high rates of evapotranspiration on bright days at alpine habitats.

ACKNOWLEDGEMENTS

We would like to thank Dr J.A.P.V. de Moraes for his assistance in part of the field measurements. We are also indebted to Dr A. Polatschek and Dr S. Fröhner for their help in the identification of *Alchemilla* species and to R. Gapp for drawing the figures. Especially we would like to thank Dozent Dr M. Kuhn, Institute of Meteorology, Innsbruck, for his help in discussing effects of atmospheric pressure on diffusive conductance.

These investigations have been supported by the Austrian IBP and MaB programme.

REFERENCES

Au S.-F. (1969) Internal leaf surface and stomatal abundance in arctic and alpine populations of *Oxyria digyna*. *Ecology*, **50**, 131–134.

Berger-Landefeldt U. (1936) Der Wasserhaushalt der Alpenpflanzen. *Bibliotheca Botanica*, 115. Schweizerbart'sche Verlagsbuchhandlung, Stuttgart.

Cernusca A. (1976) Bestandesstruktur, Bioklima und Energiehaushalt von alpinen Zwergstrauchbeständen. *Oecologia Plantarum*, **11**, 71–102.

Dobesch H., Neuwirth F. & Weiss E. (1978) Klimatologie des Glocknergebietes, I. Arbeiten aus der Zentralanstalt für Meteorologie und Geodynamik, Wien, report 30.

Drake B.G. & Salisbury F.B. (1972) After effects of low and high temperature pretreatment on leaf resistance, transpiration, and leaf temperature in *Xanthium*. *Plant Physiology* **50**, 572–575.

Ehrendorfer F. (Ed.) (1973) *Liste der Gefäßpflanzen Mitteleuropas*. Gustav Fischer, Stuttgart.

Fliri F. (1975) Das Klima der Alpen im Raume von Tirol. *Monographien zur Landeskunde Tirols I* (Ed. by A. Leidlmair & F. Huter). Universitätsverlag Wagner, Innsbruck-München.

Gale I. (1973) Experimental evidence for the effect of barometric pressure on photosynthesis and transpiration. *Plant Response to Climatic Factors* (Ed. by R.O. Slatyer), pp. 289–294. Proceedings of the Uppsala Symposium 1970. UNESCO, Paris.

Hader F. (1954) Nordostalpine Seehöhenmittel der Niederschlagsmenge. *Archiv für Meteorologie, Geophysik und Bioklimatologie*, Serie **B5**, 331–343.

Koh S., Kumura A. & Murata Y. (1978) Studies on matter production in wheat plants. V, The mechanism involved in an after-effect of low night temperature. *Japanese Journal of Crop Science*, **47**, 75–81.

Körner Ch. (1976) Wasserhaushalt und Spaltenverhalten alpiner Zwergsträucher. *Verhandlungen der Gesellschaft für Ökologie* (Ed. by P. Müller), pp. 23–30. Proceedings of the Vienna Symposium 1975. Junk, The Hague.

Körner, Ch. (1977a) Evapotranspiration und Transpiration verschiedener Pflanzenbestände im Bereich des alpine Grasheidegürtels der Hohen Tauern. *Alpine Grasheide Hohe Tauern, Ergebnisse der Ökosystemstudie 1976* (Ed. by A. Cernusca), pp. 47–68. Veröffentlichungen des Österreichischen MaB-Hochgebirgsprogrammes Hohe Tauern, 1. Universitätsverlag Wagner, Innsbruck.

Körner Ch. (1977b) Blattdiffusionswiderstände verschiedener Pflanzen im alpinen Grasheidegürtel der Hohen Tauern. *Ibid.*, pp. 69–82.

Körner Ch. Mechanische Belastbarkeit alpiner Pflanzenbestände und deren ökophysiologische Ursachen. *Verhandlungen der Gesellschaft für Ökologie* (Ed. by P. Müller). Proceedings of the Freising Symposium 1979. Junk, The Hague. In press.

Körner Ch. & Cernusca A. (1976) A semi-automatic, recording diffusion porometer and its performance under alpine field conditions. *Photosynthetica*, **10**, 172–181.

Körner Ch., Jussel, U. & Schiffer K. (1978) Transpiration, Diffusionswiderstand und Wasserpotential in verschiedenen Schichten eines Grünerlenbestandes. *Ökologische Analysen von Almflächen im Gasteiner Tal* (Ed. by A. Cernusca), pp. 81–98. Veröffentlichungen des Österreichischen MaB-Hochgebirgsprogrammes Hohe Tauern, 2. Universitätsverlag Wagner, Innsbruck.

Körner Ch. & Schubert A. (1978) Spaltenverhalten verschiederner Pflanzen auf Almwiesen an der zentralalpinen Waldgrenze. *Ibid.*, pp. 99–112.

Körner Ch. & Moraes J.A.P.V. (1979) Water potential and diffusion resistance in alpine cushion plants on clear summerdays. *Oecologia Plantarum*, **14**, 109–120.

Körner Ch., Scheel J.A. & Bauer H. (1979) Maximum leaf diffusive conductance in vascular plants. *Photosynthetica* **13**, 45–82.

Larcher W. (1969) Die Bedeutung des Faktors 'Zeit' für die photosynthetische Stoffproduktion. *Berichte der Deutschen Botanischen Gesellschaft*, **82**, 71–80.

Larcher W. & Wagner J. (1976) Temperaturgrenzen der CO_2-Aufnahme und Temperaturresistenz der Blätter von Gebirgsflanzen im vegetationsaktiven Zustand. *Oecologia Plantarum*, **11**, 361–374.

Lauscher F. (1954) Durchschnittliche Häufigkeiten der Erdbodenzustände in verschiedenen Höhenlagen der Ostalpenländer. *Wetter und Leben*, **6**, 47–49.

Monteith J.L. (Ed.) (1973) *Principles of Environmental Physics*. Edward Arnold, London.

Moore W.J. (Ed.) (1972) *Physical Chemistry*, 4th Ed. Prentice Hall, Englewood Cliffs.

Nachuzrischwili G. (Ed.) (1971) Ekologiya vysokogornykh travyahistykh rastenii i fitozen-ozov zentralnogo Kavkaza. Vodnyi redzim. *Mezniereba, Tbilisi.*

Neilson R.E. & Jarvis P.G. (1975) Photosynthesis in Sitka spruce (*Picea sitchensis* (Bong.) Carr.). VI, Response of stomata to temperature. *Journal of Applied Ecology*, **12**, 879–891.

Scheel J.A. (1979) Variabilität des minimalen und maximalen Blattdiffusionswiderstandes. Theses, University of Innsbruck.

Schulze E.-D., Lange O.L., Evenary M., Kappen L. & Buschbom U. (1974) The role of air humidity and leaf temperature in controlling stomatal resistance of *Pruuus armeniaca* L. under desert conditions. I. A simulation of the daily course of stomatal resistance. *Oecologia*, **17**, 159–170.

Smithsonian Institution (1966) *Smithsonian Meteorological Tables*, 6th revised edition. Washington D.C.

Wagner A. (1892) Zur Kenntnis des Blattbaues der Alpenflanzen und dessen biologischer Bedeutung. *Sitzungsberichte der Kaiserlichen Akademie der Wissenschaften, Mathematisch- naturwissenschaftliche Klasse*, **100**, 487–547.

Walter H. (Ed.) (1960) *Einführung in die Phytologie*. III, Grundlagen der Pflanzenverbreitung. I, Standortslehre. Eugen Ulmer, Stuttgart.

Winkler E. & Moser W. (1967) Die Vegetationszeit in zentralalpinen Lagen Tirols in Abhängigkeit von den Temperatur- und Niederschlagsverhältnissen. *Veröffentlichungen des Museums Ferdinandeum, Innsbruck*, **47**, 121–147.

12. DIFFUSION RESISTANCES
IN BRYOPHYTES

M.C.F. PROCTOR

Department of Biological Sciences, University of Exeter

SUMMARY

Conocephalum conicum, *Marchantia polymorpha* and other liverworts with internal thallus differentiation and pores in the upper thallus surface show internal resistances in the same range as mesophytic angiosperm leaves (42–740 s m⁻¹), with fair agreement between measurements and values calculated from pore size and density. Some approximate estimates of cuticular resistances of liverwort thalli and moss leaves range from low values up to a few hundred s m⁻¹. Wind-tunnel measurements of water loss from moss cushions of various sizes and roughness are consistent with laminar flow over the surface of the cushion at low windspeeds, water loss increasing rapidly beyond a critical windspeed at which the surface irregularities of the cushion become comparable with boundary-layer thickness. The hair-points of the leaves projecting above the cushion surface of *Tortula intermedia* and *Grimmia pulvinata* reduce boundary-layer conductance by *c.* 20–35 %.

INTRODUCTION

Traditionally, the control of water loss in higher plants was seen simply in terms of stomata. Bryophyte gametophytes were consequently regarded (with a few notable exceptions) as lacking the means of regulating water loss. However, that ignores the importance of resistances other than the stomatal resistance in determining rates of water loss, and ignores too the extent to which these resistances may be adaptive, and potentially variable in evolution and ontogeny. The relationships of bryophyte structures and growth forms to environment (and by implication to evaporation) have been discussed by a number of authors (Goebel 1905; Buch 1945, 1947; Gimingham & Robertson 1950; Gimingham & Birse 1957; Mägdefrau 1969; Gimingham & Smith 1971; Dilks & Proctor 1979; Proctor 1979a). Buch (1945) and Gimingham & Smith (1971) present water loss curves for *Polytrichum* species and other mosses under more or less defined conditions, and a number of other measurements of evaporation rates from bryophytes are to be found in the literature (Wenzl 1933). However, apart from the measurements of water loss from *Reboulia hemispherica* by Hoffman & Gates (1970) and the estimates of CO_2 diffusion resistances in *Mnium ciliare* of Nobel

(1977) there seem to have been few attempts at detailed or quantitative analysis of relationships between bryophytes and their atmospheric environments. In this paper I shall consider data on three specific aspects of water loss in bryophytes, to show something of the range of problems in plant–atmosphere relations that these plants present.

METHODS

'Leaf resistances' of liverwort thalli were evaluated in essentially the manner described by Jarvis (1971). Replicas of sample portions of thallus were cut from Whatman No. 1 filter paper. The undersides of the thalli were vaselined, and the thalli and moistened replicas were placed on individual 25 × 75 mm glass slides. Water loss over a ten minute period in the arbitrary conditions of the laboratory atmosphere was determined by weighing. One weighing was made each minute, so that five thalli and their replicas could be dealt with as a single batch, and nine-tenths of the period for which a thallus was exposed were common with the corresponding period for its replica. Air temperature and relative humidity were measured with an Assmann psychrometer, and thallus and replica temperatures were measured with a thermistor probe, 0·75 mm diam., held parallel and in contact with the surface immediately after the second weighing.

It is not practicable to make an aerodynamically realistic replica of a moss shoot or cushion. However, an appropriate comparison can be made between an untreated plant and the same plant after moistening with water containing a small quantity of a wetting agent, which will remove the cuticular resistance with negligible effect on vapour pressure.

Measurements of water loss from moss cushions at different windspeeds were made using a simple open-circuit wind-tunnel. This provided a usable range of windspeeds from about 0·2 to 4·0 m s^{-1}. Windspeeds were checked against a recently calibrated Fuess hot-wire anemometer; routine measurements were made with a hot-thermistor air velocity meter (Prosser AVM 502). There was some free-stream turbulence, windspeed fluctuating within a few per cent of the mean value. The larger moss cushions were chosen to fit shallow 5 cm or 9 cm plastic Petri dish lids; the smaller cushions were placed on 25 × 75 mm glass slides, with a contoured base of 'Plasticine' to accommodate irregularities in the under surface of the cushion if necessary. Water loss was determined by weighing. The temperature of the surface of the moss cushion was measured with a thermistor probe either before or after the weighing period at each windspeed. A single temperature from the centre of the surface of the cushion was taken for calculation. This ranged with different material and different windspeeds from 0·5 to 3·0 °C above the wet-bulb temperature, but was closer to the latter near the leading edge.

'LEAF' RESISTANCES IN MARCHANTIALES

The nearest analogues of flowering-plant leaves amongst bryophytes are found in the large Marchantialean liverworts. The interior of the upper part of the thallus is occupied by a system of air chambers. These enclose a tissue

TABLE 12.1. 'Leaf' resistance, r_l, of liverworts to water-vapour loss (s m^{-1}).

	Measured		Calculated
	Range	Mean and 95% C.L.	
Conocephalum conicum			
Fingle Bridge, Drewsteignton, Devon, Aug. 1978	176–288	234 ± 51	—
Glasshouse weed, Exeter, Aug. 1978	447–740	528 ± 139	—
Glasshouse weed, Exeter, Mar. 1979	306–608	412 ± 143	313
Lunularia cruciata			
Fingle Bridge, Drewsteignton, Devon, Aug. 1978	42–130	79 ± 43	—
Holne Bridge, nr Ashburton, Devon, Feb. 1979	—	—	51
Marchantia paleacea			
Cambridge Botanic Garden, Mar. 1979 (turgid)	125–173	155 ± 25	178
Cambridge Botanic Garden, Mar. 1979 (wilted)	404–611	520 ± 104	—
Marchantia polymorpha			
Wicken Fen, Cambs., Mar. 1979	48–120	80 ± 12	—
Reboulia hemispherica			
Missouri (Hoffman & Gates 1970)		minimum 39	
Pellia epiphylla			
Lydford Gorge, Devon, Aug. 1978	67–147	103 ± 44	—
Stoke Woods, Exeter, Devon, Mar. 1979	8–52	26 ± 23	—
Stoke Woods, Exeter, Devon, Mar. 1979 (killed with liquid N$_2$)	0–39	23 ± 20	—

Calculated resistances are based on measured mean values of $n = 188$ pores cm^{-2} and pore diameter $d = 35·8$ μm for *Conocephalum conicum*, and $n = 1003$ pores cm^{-2}, $d = 40·6$ μm for *Lunlularia cruciata*. For *Marchantia paleacea* $n = 1251$ pores cm^{-2}; the depth of the outer pore has been taken as 80 μm, and mean diameter 70·3 μm. An approximate geometric mean diameter of 16 μm has been taken for the opening of the inner pore, which was commonly slit-like rather than cruciate. Pore-size was too variable in the Wicken Fen material of *M. polymorpha* for a satisfactory calculation, but suggests a resistance around 100 s m^{-1}.

of photosynthetic filaments functionally equivalent to a mesophyll, and each opens to the exterior by a pore in the upper epidermis. Schönherr & Ziegler (1975) have provided excellent illustrations of these pores and the water-repellent cuticular ridges around them.

Estimates of 'leaf resistance' for several species of Marchantiales are given in Table 12.1. The values are in the same range as for mesophytic flowering plants (Holmgren, Jarvis & Jarvis 1965; Rutter 1975); there are demonstrable differences between species and between populations.

In *Marchantia paleacea* the inner orifices of the compound pores close under water stress, as happens at least to some extent with the similar pores of *Preissia quadrata* (Walker & Pennington 1939). The resistance of the wilted thallus in Table 12.1 may give a fair estimate of the cuticular resistance. Wenzl (1933) found transpiration from the upper surface of *M. polymorpha* 2–3 times as great as from leaves of *Betula pendula* under the same conditions, which suggests a substantially lower resistance (though r_s for Wenzl's birch leaves may well have been higher than the published values for *B. pendula* if the stomata were partly closed). In the population of *M. polymorpha* from Wicken Fen, the pores remain permanently open, and so differ little functionally from the simple pores of most other Marchantiales. In these the cuticular resistance cannot be evaluated directly. However, the resistance of the pores can be calculated theoretically from measurements of their dimensions and frequency. In doing this for the examples in Table 12.1, I have accepted the argument of Monteith (1973) that the 'end correction' should be applied for one side only of the pore. The calculated resistances fall within the same range as the measured resistances for the same species. Where measurements and calculations have been made from the same material they imply that the cuticular resistance is substantially higher than the pore resistance.

CUTICULAR RESISTANCES OF BRYOPHYTES

The thalloid liverworts belonging to the order Metzgeriales lack the internal structural differentiation of the Marchantiales. Nevertheless, *Pellia epiphylla* also shows a measurable leaf resistance, apparently varying with site and season (Table 12.1). In a limited test, this resistance remained substantially unchanged in five thalli when the tissue was killed by chilling in liquid nitrogen (the same is true of *Mnium hornum*, considered below). Evidently we are dealing with a cuticular resistance, at first sight surprising in so delicate and aquatic-looking a plant, but less so when considered in relation to the comparable cuticular resistances of mesophyll cells in flowering plants (Jarvis & Slatyer 1970).

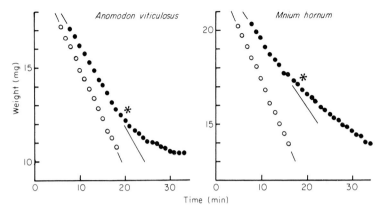

FIG. 12.1. Water loss from single shoots of *Anomodon viticulosus* and *Mnium hornum*; still air, 20 °C, relative humidity 50%. The asterisks indicate when leaves first became wilted. The left-hand curve in each case shows the rate of water loss from the same shoot completely wetted with water containing a wetting agent.

Most bryophytes have simple leaves one cell-layer thick, and evaporation can potentially take place over the whole surface. Buch (1945) considered that the 'endohydric' species with effective internal water conduction possess at least a thin cuticle, while the 'ectohydric' species directly dependent on precipitation for their water do not. Scanning electron microscopy shows well-developed epicuticular wax on the leaves of many endohydric species, and confirms Buch's supposition at least in broad terms (Proctor 1979b). The effect of this contrast in surface character on water loss is illustrated in Fig. 12.1. In *Anomodon viticulosus* (ectohydric) there is rather little difference between the rate of water loss of an untreated shoot and a shoot completely wetted with water containing a wetting agent. In *Mnium hornum* (endohydric) the contrast is strongly marked, and indicates a cuticular resistance in the region of 150 s m^{-1}. Measurements on an intact cushion of *Mnium hornum* showed a difference in resistance between the untreated and completely wetted cushion (on a 'cushion area' basis) of about 30 s m^{-1}. For a leaf area index of 18 (i.e. a total leaf surface 36 times the projected area of the cushion) this gives an average leaf resistance of about 1100 s m^{-1}. The implication is that evaporation is effectively concentrated over a fraction of the leaf area near the surface of the cushion, which would be consistent with the known pattern of water movement in Mniaceae (Zacherl 1956). *Anomodon viticulosus*, as an ectohydric species, has a requirement for the leaf surface to be wettable. Nevertheless, the measurements of Fig. 1 indicate a definite, though small, leaf resistance of about 40 s m^{-1}. This may reflect limited development of cuticle, perhaps especially on the tops of the papillae, for which there is some evidence in other species (Buch 1947; Proctor 1979b).

BOUNDARY LAYER RESISTANCES OF
MOSS CUSHIONS

In flowering plants it can often be assumed that the boundary layer resistance of individual leaves will be much higher than the resistance to transfer in the turbulent air between them. However, the scale of bryophytes is such that, for most species, molecular diffusion is likely to predominate in transfer processes in the spaces between the leaves. A compact moss cushion might therefore be expected to behave as a large simple object in an air stream. The surface area of such a cushion relates to the total surface area of the leaves within it in much the same way as the external surface area of a vascu-

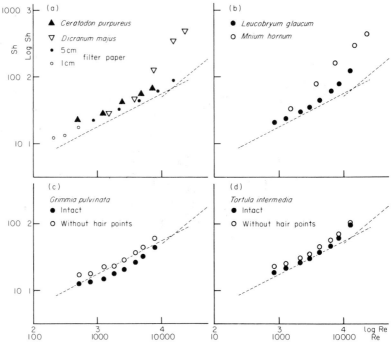

FIG. 12.2. Dimensionless plot of water loss measurements from wind-tunnel experiments. For a given object at constant temperature and pressure, Reynolds number (Re) is proportional to windspeed, and Sherwood number (Sh) to water vapour flux. (a) *Ceratodon purpureus*, $l = 2\cdot9$ cm, ▲; *Dicranum majus*, $l = 9$ cm, ▽; measurements from flat squares of wet filter paper, 1×1 cm, ○; 5×5 cm, ●. (b) *Leucobryum glaucum*, $l = 5$ cm, ●; *Mnium hornum* (completely wetted), $l = 9$ cm, ○. (c) *Grimmia pulvinata*, $l = 3$ cm; intact, ●; with hair points removed, ○. (d) *Tortula intermedia*, $l = 4\cdot8$ cm; intact, ●; with hair points removed, ○. The pecked lines are standard relations for flat plates in laminar flow (Sh $= 0\cdot58$ Re$^{0\cdot5}$) and turbulent flow (Sh$=0\cdot031$ Re$^{0\cdot8}$). See equations 2.3–2.5 for definitions of l, Re and Sh.

FIG. 12.3. (a) *Leucobryum glaucum*. Forms dense, whitish green cushions on rotten stumps or peaty soil; the surface humus under acid oak or beech woodland is a characteristic habitat. (b) *Dicranum majus*. Typically forms large gently domed cushions or patches on the ground in moist sheltered woodland on acid soils. (c) *Mnium hornum*. A common species of tree bases and banks in acid woodland, forming rather irregular cushions or extensive patches. (d) *Tortula intermedia*. A species of sunny dry calcareous rocks and walls, typically forming strongly domed cushions which are hoary with the projecting hair points of the leaves. *Grimmia pulvinata* forms similar but smaller cushions; it is a common pollution-tolerant moss of wall-tops. In *Ceratodon purpureus* the leaves are smaller and lack hair points, so that the small cushions and patches are relatively smooth. Scale x1 approx.

lar plant leaf relates to the internal mesophyll area—but without the convenient sharp discontinuity in the diffusion path provided by the stomata.

Measurements of rates of water loss from a number of moss cushions at different windspeeds are set out in Fig. 12.2; for the sake of generality they are plotted in terms of the dimensionless parameters, Reynolds number and Sherwood number (see Grace, this volume). The results show that for smooth moss cushions, or at low windspeeds, the rate of evaporation is close to that predicted from the standard relations for a smooth plane surface in laminar flow (Monteith 1973; Leyton 1975). The somewhat high values at the lowest windspeeds are perhaps partly a consequence of incipient mixed convection (Monteith 1973 shows how the Grashof number Gr can be used as a criterion for natural convection: in this case the ratio Gr/Re^2 is about 0·1 to 0·2). As

Leyton emphasises, roughness should have no effect on transfer rates as long as it remains submerged within a laminar boundary layer. However, the curves for all but the small and relatively smooth cushions of *Ceratodon purpureus* and *Grimmia pulvinata* break away from the standard laminar flow relation at high windspeeds to give progressively enhanced rates of water loss. This happens at approximately the windspeed at which the surface irregularities of the cushion become comparable in size with the boundary layer thickness—about 1·5 mm at a windspeed of about 1·0 m s^{-1} for the relatively smooth cushion of *Leucobryum glaucum* (Fig. 12.3a), just under 3 mm at 0·5 m s^{-1} for *Dicranum majus* (Fig. 12.3b), and about 4·5 mm at 0·2 m s^{-1} for the very rough cushion of *Mnium hornum* (Fig. 12.3c). Roughness will have the two effects of increasing the effective area for mass transfer, and promoting turbulence. The detailed interplay between these will depend on the size of the moss cushion as well as the windspeed. Under generally turbulent field conditions (the moss cushion forming part of an extended surface of comparable character), the thickness of the laminar sublayer may often be of more general relevance, but this would be of similar magnitude at the windspeeds in the examples quoted (0·9, 1·7 and 4·4 mm, respectively, from Leyton 1975, eqn 4·8). A further possible effect is that small-scale turbulence may penetrate down into the spaces between leaves, so enhancing diffusion in the outer layers of the moss cushion. A mechanism of this kind was postulated by Cannon *et al.* (1979) to account for their results in experiments on water vapour transfer into the airflow over perforated plates.

A number of mosses have prominent white hair points of elongated dead cells at their leaf apices, giving the plant a greyish appearance. Fig. 12.2 (c) and (d) show the results of experiments on two such species, *Grimmia pulvinata* and *Tortula intermedia* (Fig. 12.3d), in which rates of water loss from intact cushions are compared with those from the same cushion after the projecting hair points have been removed with fine scissors. The difference is striking; the hairs reduce water loss by about 35% in *G. pulvinata* and by about 20% in *T. intermedia*. These figures are likely to be conservative, because it is virtually impossible to remove the hair points completely. The result is in line with the findings of Woolley (1964) that dry hairs on soybean leaves reduced transpiration by about 20% compared with leaves that had been shaved with an electric razor. As Johnson (1975) pointed out, pubescence does not always reduce the transpiration of angiosperm leaves, and indeed Woolley observed that when the leaf hairs of soybean were liquid-filled they contributed slightly to cuticular transpiration. However, in the bryophytes under consideration the hair points appear to form no direct part of the pathway of evaporation. Some evidence on the distribution of evaporation was obtained using dilute solutions of acid fuchsin (Buch 1945, 1947) or lead EDTA (used by Crowdy & Tanton (1970) as a water-tracing solute in

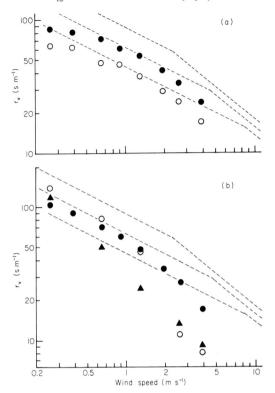

FIG. 12.4. Boundary layer resistances plotted against windspeed from wind-tunnel measurements. (a) *Grimmia pulvinata*, $l = 3$ cm; intact, ●; hair-points removed, ○. (b) *Leucobryum glaucum*, $l = 5$ cm, ●; *Dicranum maius*, $l = 9$ cm, ○; *Mnium hornum*, $l = 9$ cm, ▲. The pecked lines show theoretical forced convection relationships for flat plates at 3 different characteristic dimensions (10 cm (top), 5 cm, 2.5 cm). See equation 2.3 for definition of l.

wheat leaves). Both solutes accumulated on the papillose surfaces of the upper parts of the leaf laminae, with no observable movement into the hair points. The behaviour of other mosses with hair points remains to be studied.

In Fig. 12.4, boundary layer resistance is plotted against windspeed for a selection of the material studied. At high windspeeds in exposed situations a growth-form which will maintain laminar flow over the outline contour of the cushion is likely to be advantageous, as are the hair points of *Grimmia pulvinata* or *Tortula intermedia*. *Mnium hornum*, with its coarse leaves, loses the advantage of a relatively high boundary layer resistance, but recoups it by adding a cuticular resistance. The preference of *Dicranum majus* for moist, sheltered woods is readily intelligible from the diagram. Presumably the large woodland species that form rougher and more open cushions and mats than those considered here have potentially even higher evaporation rates.

They would probably be most readily analysed in terms of individual shoots, and may approach in behaviour the coniferous foliage studied by Tibbals *et al.* (1964), Gates, Tibbals & Kreith (1965), Landsberg & Thom (1971) and Schuepp (1972).

ACKNOWLEDGEMENTS

My thanks are due to Dr P.J. Grubb and the Director of the University Botanic Garden, Cambridge, and to Dr H.L.K. Whitehouse, for material of the *Marchantia* species, to my colleague Dr R.J. Wootton who devised the wind-tunnel used in this work, originally for studies on insect flight, and to Dr M.A. Patrick of the Department of Chemical Engineering, University of Exeter, for much helpful discussion.

REFERENCES

Buch H. (1945, 1947) Über die Wasser- und Mineralstoffversorgung der Moose. I. *Commentationes Biologicae*, 9(16), 1–44; II. *ibid.*, 9(20), 1–61.
Cannon J.N., Krantz W.B., Kreith F. & Naot D. (1979) A study of transpiration from porous flat plates simulating plant leaves. *International Journal of Heat and Mass Transfer*, 22, 469–483.
Crowdy S.H. & Tanton T.W. (1970) Water pathways in higher plants. I. Free space in wheat leaves. *Journal of Experimental Botany*, 21, 102–111.
Dilks T.J.K. & Proctor M.C.F. (1979). Photosynthesis, respiration and water content in bryophytes. *New Phytologist*, 82, 97–114.
Gates D.M., Tibbals E.C. & Kreith F. (1965) Radiation and convection for Ponderosa pine. *American Journal of Botany*, 52, 66–71.
Gimingham C.H. & Birse E.M. (1957) Ecological studies on growth-form in bryophytes. I. Correlations between growth-form and habitat. *Journal of Ecology*, 45, 533–545.
Gimingham C.H. & Robertson E.T. (1950) Preliminary investigations on the structure of bryophyte communities. *Transactions of the British Bryological Society*, 1, 330–344.
Gimingham C.H. & Smith R.I.L. (1971) Growth form and water relations of mosses in the maritime Antarctic. *British Antarctic Survey Bulletin*, 25, 1–21.
Goebel K. (1905) *Organography of Plants.* Part II. Special organography (Trans. by I. Bayley Balfour). Clarendon Press, Oxford.
Hoffman G.R. & Gates D.M. (1970) An energy budget approach to the study of water loss in cryptograms. *Bulletin of the Torrey Botanical Club*, 97, 361–366.
Holmgren P., Jarvis P.G. & Jarvis M.S. (1965) Resistances to carbon dioxide and water vapour transfer in leaves of different plant species. *Physiologia Plantarum*, 18, 557–573.
Jarvis P.G. (1971) The estimation of resistances to carbon dioxide transfer. *Plant Photosynthetic Production: Manual of Methods* (Ed. by Z. Šesták, J. Čatský & P.G. Jarvis), pp. 566–631. W. Junk, The Hague.
Jarvis P.G. & Slatyer R.O. (1970) The role of the mesophyll cell wall in leaf transpiration. *Planta*, 90, 303–322.
Johnson H.B. (1975) Plant pubescence: an ecological perspective. *Botanical Review*, 41, 233–258.

Landsberg J.J. & Thom A.S. (1971) Aerodynamic properties of a plant of complex structure. *Quarterly Journal of the Royal Meteorological Society*, **97**, 565–570.

Leyton L. (1975) *Fluid Behaviour in Biological Systems*. Clarendon Press, Oxford.

Mägdefrau K. (1969) Die Lebensformen der Laubmoose. *Vegetatio*, **16**, 285–297.

Monteith J.L. (1973) *Principles of Environmental Physics*. Edward Arnold, London.

Nobel P.S. (1977) Internal leaf area and cellular CO_2 resistance: photosynthetic implications of variations with growth conditions and plant species. *Physiologia Plantarum*, **40**, 137–144.

Proctor M.C.F. (1979a) Structure and eco-physiological adaptation in bryophytes. *Bryophyte Systematics* (Ed. by G.C.S. Clarke & J. G. Duckett), pp. 479–509. Academic Press, London.

Proctor M.C.F. (1979b) Surface wax on the leaves of some mosses. *Journal of Bryology*, **10**, 531–538.

Rutter A.J. (1975) The hydrological cycle in vegetation. *Vegetation and the Atmosphere* (Ed. by J.L. Monteith), pp. 111–154. Academic Press, London.

Schönherr J. & Ziegler H. (1975) Hydrophobic cuticular wedges prevent water entering the air pores of liverwort thalli. *Planta*, **124**, 51–60.

Schuepp P.H. (1972) Studies of forced-convection heat and mass transfer of fluttering realistic leaf models. *Boundary-Layer Meteorology*, **2**, 263–274.

Tibbals E.C., Carr E.K., Gates D.M. & Kreith F. (1964) Radiation and convection in conifers. *American Journal of Botany*, **51**, 529–538.

Walker R. & Pennington W. (1939) The movement of the air pores of *Preissia quadrata* (Scop.). *New Phytologist*, **38**, 62–68.

Wenzl H. (1933) Untersuchungen über den Wasserhaushalt von *Marchantia polymorpha*. *Jahrbuch für wissenschaftliche Botanik*, **79**, 311–352.

Woolley J.T. (1964) Water relations of soybean leaf hairs. *Agronomy Journal*, **59**, 427–432.

Zacherl H. (1956) Physiologische und ökologische Untersuchungen über die innere Wasserleitung bei Laubmoosen. *Zeitschrift für Botanik*, **44**, 409–436.

13. THE ROLE OF INTERNAL FACTORS IN CONTROLLING EVAPORATION FROM INTERTIDAL ALGAE

HAMLYN G. JONES[1] AND T. A. NORTON[2]

[1]*East Malling Research Station, Maidstone, Kent, ME19 6BJ*
and
[2]*Department of Botany, University of Glasgow, Glasgow G12 8QQ*

SUMMARY

The resistances to water loss from the fronds of three intertidal algae were studied as they dried. In all species the turgid fronds had a very low resistance to water loss, but as the tissue dried, resistance to water loss increased. *Laminaria digitata* tended to have a lower resistance at any water content than either *Fucus serratus* or *Fucus spiralis*. With rapid desiccation, the resistance to water loss at each water content was much greater than when tissue was allowed to dry slowly allowing water to equilibrate through the tissue before each measurement of resistance. For example, in *L. digitata* the resistance at 50% relative water content was 0.16 s m^{-1} when drying at 0.7 g m^{-2} s^{-1} but the equilibrium resistance was only 0.02 s m^{-1}.

INTRODUCTION

Intertidal algae usually inhabit well-defined zones on the sea shore (Zaneveld 1937). Although the reasons are complex, it is likely that the desiccation at low tide is an important factor determining the distribution of different species (Lewis 1964).

Species differences in the control of water loss may involve several mechanisms. These could include differences in (a) the degree of self-protection offered by multilayer canopies (Isaac 1933; Schonbeck & Norton, in press), (b) microsite preferences, which determine degree of exposure, (c) surface to volume ratio (Schonbeck 1976), (d) water content per unit dry matter (Jones & Norton 1979) and, (e) the degree of surface area reduction as water is lost (Jones & Norton 1979).

Although littoral algae have neither the thick cuticle nor the stomata of higher plants for controlling water loss, there remains the further possibility that under certain circumstances there may be an additional 'surface' or 'wall' resistance to water loss from the thallus, resulting from the presence of a significant cuticle or the withdrawal of water menisci into the wall (Jones & Norton 1979).

In this paper we report on the magnitude of this surface resistance in three brown algae from different levels on the shore, and the dependence of this resistance on desiccation rate and tissue water content.

MATERIALS AND METHODS

The three species used were *Fucus spiralis* L., *Fucus serratus* L. and *Laminaria digitata*, of which *F. spiralis* occurs highest on the shore and *L. digitata* the lowest, being rarely uncovered. Plants were collected at Millport, Strathclyde, Scotland and stored damp at *c*. 5 °C for up to five days. Measurements of evaporation and diffusive resistance were made in a one cm diameter, stirred chamber of a continuous flow diffusion porometer (Day 1977). The surface resistance (r_s) was obtained from

$$r_s = (1/h - 1) A/f - r_a$$

where h is the water vapour concentration in the porometer outlet as a fraction of the saturation concentration, A is the frond area exposed in the chamber (m^2), f is the air flow rate through the chamber ($m^3 s^{-1}$) and r_a is the calculated resistance when moist filter paper is used in place of the leaf. Filtered sea water (total salinity $32°/_{oo}$) was used for saturating algae and filter paper. No correction was made for the effect of salt on saturation water vapour pressure. The equilibrium relationship between r_s and relative water content, R, defined as the water content as a percentage of the turgid water content, was distinguished from the rather larger increases in r_s observed with rapid desiccation. An approximation to the equilibrium relationship was obtained by drying tissue on the laboratory bench at less than $0.07 \text{ g m}^{-2} \text{ s}^{-1}$, then covering for 5 min before each measurement with thin PVC 'cling-film' (Borden UK Ltd., Bridgwater, Somerset) to permit equilibration. More rapid desiccation, reaching $0.7 \text{ g m}^{-2} \text{ s}^{-1}$, was achieved by keeping the tissue continuously in the porometer. Further details of the plant material and experimental techniques are given elsewhere (Jones & Norton 1979).

RESULTS AND DISCUSSION

The total resistance to water loss from turgid frond tissue, average of from 12–24 samples, was $13.7 \pm 0.35 \text{ s m}^{-1}$ for *L. digitata*, $14.7 \pm 0.63 \text{ s m}^{-1}$ for *F. serratus* and $16.5 \pm 0.52 \text{ s m}^{-1}$ for *F. spiralis*. These values were all significantly ($P < 0.05$) larger than the boundary layer resistance of

12.6 ± 0.18 s m^{-1} obtained with filter paper. The differences between the resistances for tissue and filter paper have been ascribed to the 'surface' resistance, r_s, since similar results were obtained with a range of filter papers with differing surface topography, including Whatman 1, Whatman 3 and Millipore HA. In turgid tissue the observed values of r_s, which ranged from 1.1 to 3.9 s m^{-1}, would exert negligible control over water loss, since it is likely that boundary layer resistances in the field would be significantly higher than these values of r_s even on the windiest days (Jenik & Lawson 1967; Monteith 1973). These values of r_s are much less than the typical minimum values for the gas phase resistance, mainly stomatal, of 50–200 s m^{-1} in mesophytes (Cowan & Milthorpe 1968).

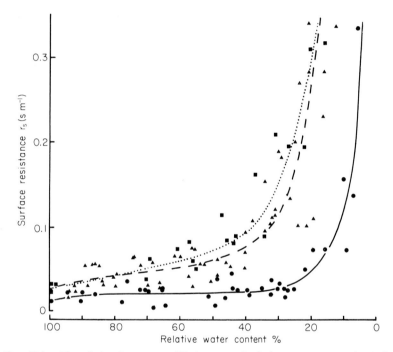

FIG. 13.1.   The relation between equilibrium r_s and relative water content for 8 discs of *L. digitata* (———●———), 11 discs of *F. serratus* (– – ▲ – –) and 7 discs of *F. spiralis* (· · · · ■ · · · ·). Curves were fitted by eye.

As the tissue dried out r_s increased dramatically for each of the three species. Fig. 13.1 shows the equilibrium relationships between r_s and relative water content. Each measurement shown in Fig. 13.1 was made immediately following a period of equilibration with evaporation stopped by covering the tissue with 'cling-film'. The surface resistance at any R was highest in *F. spiralis* followed by *F. serratus* and then *L. digitata* (though the difference between the first two species was not significant). This ranking corresponds to the natural positions of these species on the seashore. The development of higher resistances in *F. spiralis* may be a factor allowing it to exist higher on the shore than the other species. However, species usually found lower on the shore have higher water contents when first exposed on the shore, e.g. twice as high in *L. digitata* as in *F. spiralis*, and consequently, the lower shore species take longer to dry to any given water content. When r_s is calculated at equal water content per unit dry weight or per unit initial area, the species differences become small and non-significant (Jones & Norton 1979).

As the evaporation rate from the fronds was increased, the r_s at each relative water content also increased. Fig. 13.2 shows the effect of the two desiccation rates on the relationship between r_s and R for *L. digitata*. At the high evaporation rate, which was similar to that expected on a hot dry day on the shore, r_s increased significantly before R had decreased by 5%. Similar results were obtained for the other species, though *L. digitata* tended to be least sensitive to evaporation rate. In this case, species from higher up the shore would have had the most efficient control of water loss on hot dry days.

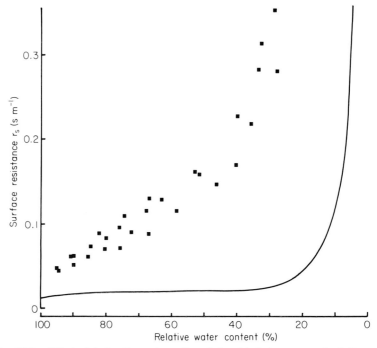

FIG. 13.2. Effect of desiccation rate on the relationship between r_s and relative water content for *L. digitata*. The solid line is the 'equilibrium curve' from Fig. 13.1, while the squares represent similar material dried continuously in the porometer (initially at about $0.7 \text{ g m}^{-2} \text{ s}^{-1}$).

The increase in calculated r_s as the algal tissue dried may be the result of one or both of the following mechanisms (Jarvis & Slatyer 1970; Jones & Norton 1979); (a) Increased matric and osmotic forces which would lower the water potential and thus lower the water vapour pressure at the evaporating surfaces. (b) The development of an extra water transfer resistance either in the liquid phase up to the evaporating surface or in the gas phase within the drying cell walls. The sensitivity of r_s to evaporation rate indicates that the latter mechanism may be involved, at least in retarding evaporation when there is a high evaporative demand. Lowered surface water vapour pressure must also have been involved in the increase of equilibrium r_s as the water potentials of the fronds reached very low values (Jones & Norton 1979; Jones & Higgs 1980).

Internal factors have been shown to affect the resistance to water loss from littoral algae, and significant differences between the species do occur. However, factors such as the degree of frond overlap (reducing evaporation from the lower fronds) or the tolerance of desiccation shown by the different species, are likely to be more important in determining species distribution on the sea shore (Jones & Norton 1979).

REFERENCES

Cowan I.R. & Milthorpe F.L. (1968) Plant factors influencing the water status of plant tissues. *Water Deficits and Plant Growth*, Vol. 1 (Ed. by T.T. Kozlowski), pp. 137–193. Academic Press, New York.

Day W. (1977) A direct reading continuous flow porometer. *Agricultural Meteorology*, **18**, 81–89.

Isaac W.E. (1933) Some observations and experiments on the drought resistance of *Pelvetia canaliculata*. *Annals of Botany*, **47**, 343–348.

Jarvis P.G. & Slatyer R.O. (1970) The role of the mesophyll cell wall in leaf transpiration. *Planta*, **90**, 303–322.

Jenik J. & Lawson G.W. (1967) Observations on water loss of seaweeds in relation to microclimate on a tropical shore (Ghana). *Journal of Phycology*, **3**, 113–116.

Jones H.G. & Higgs K.H. (1980) Resistance to water loss from the mesophyll cell surface in plant leaves. *Journal of Experimental Botany*, **31**, 545–553.

Jones H.G. & Norton T.A. (1979) Internal factors controlling the rate of evaporation from fronds of some intertidal algae. *New Phytologist*, **83**, 771–781.

Lewis J.R. (1964) *Ecology of Rocky Shores*. English University Press, London.

Monteith J.L. (1973) *Principles of Environmental Physics*. Edward Arnold, London.

Schonbeck M.W. (1976) A study of the environmental factors governing the vertical distribution of intertidal fucoids. Ph.D. Dissertation, University of Glasgow.

Schonbeck M.W. & Norton T.A. An investigation of drought avoidance in intertidal fucoid algae. *Botanica Marina*. In press.

Zaneveld J.S. (1937) The littoral zonation of some Fucaceae in relation to desiccation. *Journal of Ecology*, **25**, 431–468.

14. WATER STRESS AS AN ECOLOGICAL FACTOR

M.T. TYREE[1] AND A.J. KARAMANOS[2]

[1]*Department of Botany, University of Toronto, Toronto, Canada M5S 1A1*
and [2]*Agricultural College of Athens, Department of Plant Physiology, Votanikos, Athens 301, Greece*

SUMMARY

There is still disagreement in the current literature on how best to define and measure the components of water potential (ψ) in plant tissue. A consensus on these matters is desirable before attempts are made to investigate the ecological significance of the components. In this paper arguments are put forward relating to the identity of the components and the following conclusions are reached:

1. The matric potential, τ, is best defined as the energy of interaction between the water dipole and the electric field in the vicinity of charged surfaces; even in cell walls, where τ might be the largest component potential, most of the water is held by negative pressure, P, arising from the surface energy of air-water interfaces (i.e. the so-called capillary effect).

2. It is wrong to identify τ with the capillary effect or with the osmotic potentials near charged surfaces.

3. The bound water concept should be abandoned in the analysis of τ and tissue-water-potential isotherms.

4. The most reasonable approximation to the components of ψ can be derived from the analysis of isotherms entirely on living tissue; the estimation of turgor pressure, P_t, and osmotic potential, π, from measurements of ψ on dead tissue is likely to cause substantial errors.

5. The water content of cell walls can *not* act as a reservoir of water for drought avoidance; it is necessary to invoke τ in order to explain how water in one part of a tissue can be in equilibrium with that in another part, but τ does not play much of a role in the analysis of tissue-water-potential isotherms and can be safely ignored as an ecological parameter.

6. π could be an important ecological parameter entering into the drought resistance of plants; osmoregulation is probably most important for the maintenance of high turgor pressure for growth.

7. It is too early to tell if differences between plants in the bulk elastic modulus of cell walls are of any ecological significance, but the role of elastic cell walls in bringing protoplasmic water into equilibrium with apoplastic and soil water is not disputed; i.e. with cell walls plants can develop high turgor pressures whereas without cell walls plant cells would have to continually pump water out or suffer excessive osmotic dilution.

8. Leaf structure, root structure and stomatal physiology may turn out to be more important ecologically than the components of ψ in understanding the way plants cope with water stress.

INTRODUCTION

Few people would dispute the conclusion that water availability is one of the two or three major factors determining the distribution of plants throughout the world. Water is available to a plant if the water potential in the soil or other source, ψ_{source}, is less negative than it is in the plant ψ_{plant}. When the water reserves decline, i.e. when ψ_{source} becomes too negative, then many plants die. It seems that some plant species can survive and grow under drier conditions than others. This is often, but not always, because the drought resistant species is capable of maintaining a lower ψ_{plant} than the prevailing ψ_{source} and a lower ψ_{plant} than the less drought resistant plant.

Water potential is connected thermodynamically to the concept of free energy, but unlike free energy it is an *intensive quantity*, i.e., it measures the magnitude or intensity of energy of water *at a point* in a system relative to pure water at atmospheric pressure and at the same temperature as the system in question. Water potential can be measured in units of pressure (the megapascal, MPa) which are dimensionally equivalent to energy per unit volume. Water potential is usually negative in the soil–plant atmosphere continuum.

Water potential is commonly split into several components although the thermodynamic basis of the splitting is questionable. Neglecting the gravitational component, these are:

$$\psi = P - \pi - \tau$$

where P is the pressure (zero being taken as atmospheric pressure), π is the osmotic pressure and τ is the so-called 'matric' potential. We define π as $-RT (\ln a_w)/V_w$ where R is the gas constant, T is absolute temperature, V_w is approximately the volume of a mole of water, viz. 18 cm^3 mol^{-1}, and a_w is the activity of water. Please note that our definition of π for osmotic pressure is opposite in sign to that used by some other people for osmotic potential ($= \psi_\pi$). Sign conventions used in the literature are arbitrary but our usage is in conformity with many recent fundamental contributions, e.g. Slatyer (1967, equations 1·67 and 3·15). The activity of water a_w is a thermodynamic quantity of no dimension. It equals the mole fraction of water in a solution times an activity coefficient which corrects for nonideality. The mole fraction x_w is $n_w/(n_w + n_s)$ where $n_w =$ the number of moles of water in a solution and n_s is the number of moles of dissolved solutes in the solution.

An osmole is defined here as the sum of vM for all solute species where $M =$ the number of gram formula weights of each species in solution and $v =$ the number of dissociated particles produced by each species in water, e.g. for sucrose $v = 1$, for KCl $v = 2$. If the system is ideal then π can be approximated by RTC where C is the osmolal solute concentration in osmoles per kg *of water*.

Plant ecologists and physiologists have been focusing their attention in recent years on these components of water potential to see if one or more of the components correlate with drought resistance. The measurement and interpretation of P, π, and τ in plant tissue is often closely tied to the analysis of 'water-potential isotherms', i.e., how ψ, P, π and τ depend on water content.

There has been considerable disagreement in the literature about how best to analyse these isotherms and to measure the components (Warren Wilson 1967a, b, c; Noy-Meir & Ginzburg 1967; Weatherley 1970; Tyree & Hammel 1972; Spanner 1973; Acock 1975; Tyree 1976a). There is also a great deal of confusion over what the matric potential really is. Since we cannot proceed with a discussion of the ecological importance of water stress without understanding these terms, we must first define our terms and critically review the literature quoted above.

MATRIC POTENTIAL

We begin with matric potential because it is the most talked-about but least understood concept in the field of plant–water relationships. Matric potential has been variously described as an 'adsorption force', 'imbition force', or 'hydration' energy. All of these words falsely instil in us cognitive confidence but carry no real meaning. In order to understand τ we must apply physical theories to the kinds of systems in which 'matric forces' are thought to arise.

'Matric forces' are interfacial forces, they occur between liquid and solid phases. They are prominent in systems with large surface to volume ratios, e.g. soils (especially clay), plant cell walls, gels and colloids. Systems with large surface to volume ratios also tend to have substantial surface charge densities. Since we are dealing here with plant tissues we will critically examine the situation in plant cell walls.

Plant cell walls have a high charge density associated with dissociated polyuronic acids. The uronic acid polymers are in ionic equilibrium with a large concentration (~ 0.5 M) of cations (Dainty & Hope 1959; Dainty, Hope & Denby 1960; Clymo 1963; Collins 1969; Tyree 1972). This high ion exchange capacity is currently associated with the concept of apparent free space or Donnan free space (DFS) (Briggs 1957; Pitman 1965). Dainty & Hope (1961) experimentally determined the magnitude of the surface charge density σ in the micro-capillary wall of *Chara* cell walls. The charge density is $-4 \times 10^{-5} \times C$ cm^{-2} and the authors conclude that the charged surfaces are so widely separated in *Chara* that the DFS is not continuous and is most accurately treated as a Gouy-Chapman double layer. The exact structural relationship between the polyuronic acids in cell walls and the idealized charged surfaces in the Gouy-Chapman double layer is vague in the literature. The charges are envisaged to coat the surfaces of micro-capillaries inside cell walls. The true structure is probably more complex, but the justification for treating it as charges confined to many thin surfaces in the cell wall is that the Gouy-Chapman model for such a structure does explain the observed ionic distribution between cell walls and bathing media.

A surface charge will produce an electrical field in the vicinity of the solid–liquid interface. The narrow region at an interface where electric potential charges decline rapidly with distance from a charged surface is called a double layer. The depth to which the electric field extends into the solution depends on the concentration and valence of the ions beyond the range of the field. (See the Appendix to this chapter for details of the theory and computations leading to Figs. 14.1 and 14.2.) In Fig. 14.1a we have plotted the profile of the electrical potential, E in mV, versus distance from a charged surface when the solution contains 10 mM or 250 mM salts in which the cations and anions are all univalent. The 10 mM line might be the usual case for cell walls in ionic equilibrium with the solution in the xylem. The 250 mM case might represent the profile in dead tissue (e.g. killed by freezing) in which the cell wall is in ionic equilibrium with the cell sap released by killing.

Within the double layer, where the voltage is nonzero, the ionic concentrations will not be the same as outside the double layer; although the concentration of uncharged species will be the same. The negative double layer will attract cations and repel anions. The ionic concentrations within the double layer will be governed by the Boltzman equation, i.e., the cationic concentration will be increased by a factor equal to exp $(z\mathscr{F}E/RT)$ times the concentration outside the double layer (C_o) and the anionic concentration will be decreased by a factor of exp $(-z\mathscr{F}E/RT)$ times C_o, where $E =$ the voltage, $z =$ the valence *without* sign, and $\mathscr{F} =$ the Faraday constant. The total osmolal concentration inside the field is given by

$$C = C_o \left(\exp \frac{z\mathscr{F}E}{RT} + \exp - \frac{z\mathscr{F}E}{RT} \right) + C_o{}^* \qquad (14.1)$$

where $C_o{}^*$ is the concentration of nonionic solutes. An estimate of the osmotic pressure inside the field can be obtained by multiplying the above equation by RT. This has been done and is plotted in Fig. 14.1b. It can be seen that most of the osmotic effects of the double layer are within a distance of 3 nm of the charged surface whereas the 'pores' in cell walls are likely to be 10 to 30 nm in diameter; therefore a large portion of the water of cell walls is likely to be outside the influence of the osmotic effects in the double layer.

One might be inclined to define τ partly in terms of the osmotic effects

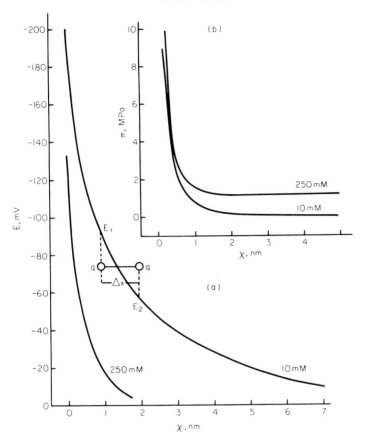

FIG. 14.1.   (a) The voltage profile near a charged surface according to the Gouy-Chapman double layer theory calculated from equation A1 with a surface charge of -4×10^{-5} C m^{-2} and univalent ion concentrations of 10 mM and 250 mM; (b) The ideal osmotic pressure in the double layer as given by the Boltzman distribution (eqn. 14.1). A dipole is illustrated in the graph. The energy of interaction of this dipole with the field equals $(E_1 q - E_2 q)$.

inside the double layer. Warren Wilson (1967a) and Acock (1975) implicitly include π within the double layer in τ, but we feel this is inelegant, misleading and perhaps wrong. If we have already decided to split ψ into P, π and τ it would be inconsistent to say that π contributes to τ inside the double layer but not outside it (e.g. in the xylem lumen or vacuole).

What then is τ? Surely it must be something physically separable from π and P if it is to have any real meaning. Fortunately, there is a quantity that is physically separable from π and P. This is the energy of interaction of the dipole of water with the electrical field, $F = dE/dx$, where F is the gradient of voltage. A dipole is two charges of equal magnitude, q, but opposite sign separated by a distance Δx. The magnitude of a dipole, P_o, is defined as $q \times \Delta x$. The positive charge q^+ is nearest the negative surface; the negative charge q^- is farthest away (because opposite charges attract and like charges repel). For a single dipole oriented perpendicular to the charged surface as in Fig. 14.1a the energy of interaction can be calculated by the voltage, E, at each charge q^+ and q^- because voltage by definition is the energy possessed by a unit charge by virtue of being at a point in space (dimensionally a volt = joules of energy per coulomb of charge). The energy of interaction, ε, is thus $\varepsilon = qE_1 - qE_2 = q(E_1 - E_2) = -q \, \Delta x \, (dE/dx) = -P_o F$. The energy ε is negative and it is the lowest energy that the dipole will possess. If the dipole is oriented in any other direction the ε is larger (i.e. less negative).

We propose that τ should be equated primarily with this energy of interaction of the dipole with the field. Since the thermal energy of water molecules will tend to disalign their dipoles with respect to the field and thus raise their energy (see Appendix), then the expression for τ (in MPa) becomes

$$\tau = \left(\frac{N_o P_o{}^2}{3 V_w kT}\right) F^2 \tag{14.2}$$

where N_o is Avagadro's number and k = the Boltzman constant ($= R/N_o$). Equation 14.2 has been evaluated and plotted together with π in Fig. 14.2. In this figure we have plotted the variables versus the log of the distance in order to expand that portion of the scale nearest the surface charge. The value of $-\tau$ will be lower (more negative) than given by equation 14.2 at the first molecular layer near the charged surface because we have neglected the impact of hydrogen bonds between the water and solids. If the energy of the hydrogen bond between the water and solid, H_{ws}, is larger than that between water and water H_{ww}, then the water potential of the first layer of water will be lowered by $(H_{ww} - H_{ws})$. The value of H_{ws} does not need to differ much from H_{ww} to have a considerable impact because H_{ww} is of the order 10^2 MPa. So if H_{ws} is just 10% larger than H_{ww} it will decrease τ by ~ 10 MPa. A surface in which $H_{ws} > H_{ww}$ is hydrophilic and one in which $H_{ws} < H_{ww}$ is hydrophobic.

If the water near a charged surface is to be in equilibrium with water elsewhere (e.g. the xylem lumen, or cell sap) then the hydrostatic pressure, P, must compensate for the changing values of π and τ. This is illustrated in Fig. 14.2. If we suppose that ψ is -1.5 MPa everywhere, then P must equal $\psi + \pi + \tau$ everywhere in the profile. Near the charged surface P becomes very large and positive whereas beyond the field P becomes negative and approaches ψ. The large positive pressures that develop near charged surfaces might be identified as the swelling pressure of gels. The pressure arises because of the force with which water molecules are drawn towards the charged surfaces. The molecules piling on top of each other produce the

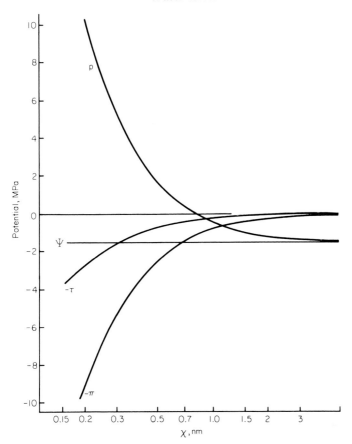

FIG. 14.2. The components of water potential, ψ, near a charged surface $(-4 \times 10^{-5}\,\mathrm{C\,cm^{-2}})$. The concentration outside the double layer is taken as 10 mM. The osmotic pressure, π, is calculated in the following way. The electrical potential, E, at a distance x from the surface is computed from equation A2. This value of E is then used in equation 14.1 to compute the concentrations C, at distance x. The value of π is RTC. The value of τ can also be calculated at x using the same E. The electric field, F, is calculated from equation A3 and this value of F is used in equation 14.2 to calculate τ. $P =$ the hydrostatic pressure $= \psi + \pi + \tau$.

pressure in much the same way that positive pressures develop at the bottom of lakes because of the gravitational forces pushing water down from above.

What then produces the negative pressure beyond the electric field? This must be surface tension transmitted from air–water interfaces. Warren Wilson (1967a) and others suggest that surface tension effects ought to be included in τ, but we feel this is no more justified than including π in τ. Why should negative pressure be considered part of τ when they occur in cell walls but not part of τ when they occur in the xylem lumen? For purposes of consistency

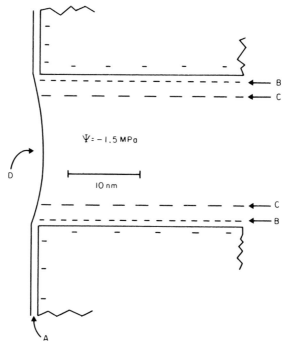

FIG. 14.3.   A drawing, approximately to scale, of the air-water interface near a water-filled, charged pore. The pore is 20 nm in diameter and is filled with water at $\psi = -1.5$ MPa. The other parameters are as in Fig. 14.2. A, a thin layer of water on the charged surface held by π and τ; B, the distance from the charged surface at which $P = 0$; C, the distance at which ψ is more than 90% due to negative P; D, the air-water interface with a radius of curvature of 96 nm.

and clarity we feel it is necessary to include surface tension effects in P. The surface tension contribution to P can be easily calculated from the radius of curvature of the interface, r, and the surface energy, T, i.e., $P = -2T/r$. In Fig. 14.3 we illustrate what the air–water interface might look like near a charged pore in hydrophilic walls. It can be seen that most of the water in a 20 nm diameter pore will be held by surface tension effects (= negative pressure).

Our approach to the meaning of τ near charged surfaces is essentially like that taken by Bolt & Miller (1958); Warren Wilson (1967a) quoted this paper but did not rationalize these ideas to his own.

WATER-POTENTIAL ISOTHERMS

There are basically two approaches to the analysis of water-potential isotherms of plant tissue. The ultimate aim in all treatments is to partition ψ into its components (P, π and τ) and to determine how the components change with relative water content ($R_w = w/w_o$, where $w_o =$ the total water content of a tissue sample at full turgidity and $w =$ the water content at lower turgidity). The first approach is to deduce the components (π, P, and τ) from a paired analysis of the change in ψ between living and dead tissue at similar R_w (Warren Wilson 1967a, b, c; Noy-Meir & Ginzburg 1967; Acock 1975); we will call this model A. The more recent approach (Tyree & Hammel 1972; Hellkvist, Richards & Jarvis 1974; Tyree 1976a) is to deduce the components (π, P, and τ) from the analysis of the dependence of ψ on R_w *on living tissue*; we will call this model B. In addition Spanner (1973) has proposed a new way of splitting ψ into components. This will also be discussed. Models A and B are both fraught with difficulties arising from the basic heterogeneity of plant tissue and all authors have recognized the problem. The problem is that although ψ might be the same everywhere in a tissue the components contributing to ψ will differ between cellular phases, i.e. vacuole, cytoplasm, cell wall and xylem vessel lumen. In addition the π, P, and τ components might differ from cell to cell within the vacuolar and cytoplasmic phases. Models A and B both attempt to handle the problem of heterogeneity by making some simplifying assumptions and by attempting to 'volume average' the components. The validity of the various models stands or falls on the reasonableness of the assumptions made. We will therefore examine these assumptions.

Model A

The assumptions of Acock and Warren Wilson

Both Warren Wilson (1967a) and Acock (1975) made a central assumption around which all of their equations revolve. Warren Wilson (1967a) writes: 'It is assumed . . . that (leaf tissue) consists of two phases. One phase contains matrix and water, *but no solutes*; the other contains solute and water, *but no matrix*'. He then proceeds to identify the water in the matrix as 'bound water' of volume fraction B and the osmotic water as $R_W - B$. He also states that the water potential of dead tissue, ψ_d, equals $-\pi - \tau$. His analysis then proceeds as follows. If ψ_d were entirely equal to $-\pi$ then $\psi_d \times R_w$ would be constant for all values of R_w. Experimentally $\psi_d \times R_w$ is not a constant, so Warren Wilson attributes all deviation from constancy to the presence of bound water, i.e., he finds a value of B such that $\psi_d(R_w - B)$ is constant for the experimental data. Acock (1975) does basically the same thing but a bit more elaborately, i.e., he allows the apoplastic and symplasmic phases to have separate bound water fractions.

The basic weakness of this analysis is the assumption that the matrix water contains no solutes. Warren Wilson states that this is only a first approximation, but it is much worse than that! It is diametrically opposite to the real situation. By reference to Figs 14.1 and 14.2 it can be seen that in the matric phase the ionic concentrations are uniform outside the field and even higher inside the field. Also non-electrolytes would be free to diffuse into the field. It therefore follows that systems of equations based on this false assumption are wrong. This assumption would make an error of typically 20% in the value of π between models A and B and will lead to negative values of P which may not really exist.

The bound water method of analysis used by Warren Wilson is similar to that used by Kramer (1955) and has been used by many people to explain the deviation of π from ideality (in an ideal solution $\pi = -RT \ln x_w/V_w$ where x_w is the mole fraction of water). However this is not the accepted approach. The usual method is to assign deviations to the activity of water, a_w, or to an osmotic coefficient (Robinson & Stokes 1959). Such a method is thermodynamically more sound and is to be preferred.

The thermodynamic approach of Noy-Meir and Ginzburg

Within the biological literature, Noy-Meir & Ginzburg (1967) have published the most rigorous thermodynamic treatment of the meaning of the components of water potential, π, P, and τ. One of the most important points made in their paper is that there are thermodynamically six different ways of defining π, P, and τ, but that only a couple of the definitions have any relation to the experimental operations that might be carried out in a laboratory. In their paper π, P, and τ are defined explicitly by integrals with well defined limits (cf. their equations 8 and 10). However the integrals cannot be integrated analytically for all of the terms in any of the two meaningful definitions of π, P, and τ. The authors do propose, however, that the values of the integrals can be assessed experimentally, so that properly chosen thermodynamic definitions of π, P, and τ may have corresponding experimental meaning. But the converse is not true: not every experimental definition of the components is thermodynamically meaningful. For example, Boyer's pressure bomb definition of τ (Boyer 1967) has no thermodynamic meaning.

Noy-Meir & Ginzburg (1967) split their analysis of isotherms up into two sections. In the first section they treated hypothetical homogeneous systems in which the components π, P, and τ were defined. The homogeneous system consisted of water, solutes and matrix (i.e. insoluble substances). The system was assumed homogeneous with regard to pressure and concentration of all the substances, i.e., P and C were treated as uniform throughout the phase. The authors realized that real plant tissues are heterogeneous between vacuole, cytoplasm and cell wall. They treated the cell wall as a homogeneous phase, but considering that the Gouy-Chapman theory predicts quite large localized differences in P, π, and τ, it seems that it is incorrect to treat the cell wall as thermodynamically homogeneous. Indeed it is difficult to imagine many real systems containing an insoluble matrix that is even approximately homogeneous. Thus it is rather difficult to get a proper intuitive grasp of what π, P, and τ really are as defined by Noy-Meir & Ginzburg.

These authors have considerable difficulty in treating theoretically the heterogeneity between cellular compartments. The reason for this is that they are obliged to average values of π, P, and τ over the heterogeneous space. In order to get an *experimental* evaluation of the integrals defining π and τ they must first eliminate P experimentally by killing the tissue, i.e., by rendering the membranes permeable to all solutes. Tissue killing has the effect of allowing vacuolar and cytoplasmic solutes to intermingle with apoplastic and symplasmic water. This forces us to 'make do' with volume averaged values of P, π, and τ and also obliges us to accept a certain amount of error as a result of interaction between solutes and matrix when π and τ are evaluated on dead tissue. The magnitude of these errors is unknown, but it can be seen from Fig. 14.1 that part of the effect of killing tissue is that the width of the electric field is contracted nearer to the charged surfaces in the cell walls. This will in turn change all the profiles of π, P, and τ in Fig. 14.2. Within the framework of the Noy-Meir & Ginzburg analysis this will change the volume averaged values of π and τ by an unknown amount.

This need to average π, P, and τ over the entire tissue water-volume is unfortunate, but it is a circumstance to which all proponents of model A are reduced (Noy-Meir & Ginzburg 1967; Warren Wilson 1967a, b, c; Acock 1975).

Surely as plant biologists we are really more interested in the values of P and π within the vacuole and cytoplasm rather than values averaged over the entire water volume. The averaging error could be quite large in drought-adapted plants with a large cell wall volume.

We are left wondering if there could be any way out of this dilemma of partitioning ψ into components and/or the necessity of averaging. Model B affords some relief to this 'law of averages', but let us first see if there is a viable alternative to partitioning ψ into π, P, and τ.

Spanner's model

Spanner (1973) suggested an alternative of splitting ψ into energetic and entropic components:

$$\psi = \psi_H + \psi_S$$

Spanner identified the entropic component (ψ_S) primarily with osmotic pressure and also suggested that the energetic component (ψ_H) might be split into pressure and internal energy terms. Spanner suggested that ψ_S could be measured by the change of ψ with temperature, i.e.,

$$\psi_S = T\left(\frac{\delta\psi}{\delta T}\right)_P \cong \frac{T_1 + T_2}{2}\frac{\psi_1 - \psi_2}{T_1 - T_2}$$

In terms of our earlier partitioning of ψ, this definition of ψ_S would be exactly equal to $-\pi$ if P and τ were temperature independent, because then

$$\frac{\mathrm{d}\psi}{\mathrm{d}T} = -\frac{\mathrm{d}\pi}{\mathrm{d}T} = -RC \quad \text{so} \quad \psi_S = -RTC = -\pi$$

Because of this possible correspondence between ψ_S and $-\pi$ Spanner's suggestion might be quite acceptable to plant biologists, particularly if the assessment of ψ_S could be made on living tissue.

Tyree, Dainty & Hunter (1974) have studied the temperature dependence of ψ with the pressure bomb. Examination of these data reveals that ψ_S is >0 for samples of high turgidity and <0 for samples at low turgidity: this corresponds to positive and negative concentrations, respectively. In terms of the earlier components, the reason for this odd behaviour is an unexpectedly large temperature dependence of P.

The only way to make ψ_S correspond more closely to $-\pi$ would be to measure ψ_S on dead tissue, but this brings us back to the problem of volume averaging.

Model B

The detailed theory for model B was originally derived in the context of the pressure bomb technique (Tyree & Hammel 1972; Hellkvist et al. 1974) to explain how the water potential of living plant tissue changes with water loss. It has subsequently been argued that the same theory applies to any technique of measuring ψ versus relative water content (Talbot, Tyree & Dainty 1975; Tyree 1976a; Richter 1978).

Model B has the advantage that it purports to yield values of π and P for the symplasm. The details of the analysis makes it obligatory to average these values over the total volume of water in the symplasm, but this is a decided improvement over averaging the same values over the water contents of both the apoplast and symplasm.

The fundamental procedural difference between models A and B is that in model A the turgor pressure is computed from the difference between the water potential of living tissue and the same (or similar) sample that is dead whereas in model B the turgor pressure is deduced from the way in which water potential changes with water loss from living tissue. The fact that π and P are computed from data derived entirely from living tissue should give us confidence because we do not have to be concerned with errors and interferences that might arise from the killing procedure. The errors that might arise from killing are (1) incomplete killing of all the cells; (2) dilution of symplasmic contents (which requires volume averaging of the symplasmic and apoplastic water contents); (3) incomplete mixing of macromolecules between the apoplast and symplasm because the cell wall will act like dialysis tubing; and (4) interference of the symplasmic ions with the charged surfaces.

The theoretical basis of model B rests on four central assumptions: (1) the osmotic pressure of the symplasm varies inversely with the symplasmic water volume (or mass); (2) a matric potential does not contribute significantly to ψ in the symplasm; (3) the apoplastic solute concentration beyond the double layer is small compared to the symplasmic concentration; and (4) most of the water lost from living tissue comes from the symplasm, i.e., very little water is lost from the apoplast. All of these assumptions apply within the experimental range of the measured water potential isotherm (typically from 0 to -3 MPa in mesophytes and down to -6 or -10 MPa in some xerophytes).

Solutions of low molecular weight electrolytes and non-electrolytes follow ideal osmotic behaviour within an error of 10% or less for values of π between 0 and -3 MPa (Robinson & Stokes 1959). Thus we might have some confidence in expecting that the first assumption is met.

With regard to the second assumption we can safely say that τ will be less important in the vacuole than in the cytoplasm because the vacuole does not contain a matrix. Within the cytoplasm we would expect τ to be important only in the matric double layers. But the total amount of matrix in the cytoplasm is probably less than in the cell wall. Since we have already concluded that τ influences the state of only a small fraction of the total cell wall water the overall affect in the cytoplasm is probably even smaller. Therefore we can ignore the impact of τ in the symplasm.

Regarding the third assumption, the concentration of the solution in the apoplast can be measured experimentally. In mesophytes we have found it to be equivalent to an osmotic pressure of less than 0·050 MPa in most cases.

It is much more difficult to assess experimentally the amount of water lost from the apoplast. If a vapour phase equilibrium method is used to measure ψ on detached leaves or leaf subsamples, then much of the lost water is probably lost from the xylem vessels. The water in the vessel lumens would be sucked into the dehydrating cells. But this process would probably be completed before ψ falls below -0.2 MPa (except in vessels and tracheids isolated on all sides by imperforate walls at the pits and vessel ends in which case the lumens might not drain at all). In the pressure bomb technique, however, the vessels probably would not drain. What then is the possibility of substantial water loss from the cell walls? In *Eucalyptus* leaves the apoplastic water content has been reported to be 40% of the total leaf water content (Gaff & Carr 1961), and it has been argued (Cutler, Rains & Loomis 1977) that a high apoplastic water content is a common feature of drought adapted plants. The implication would seem to be that this apoplastic water could act as a reservoir of water making up water loss from the symplasm. But is there evidence that the volume of water in cell walls can change with ψ? Acock (1975) has predicted a large change in *cell wall water volume* on the basis of his analysis of water potential isotherms. However, these conclusions are based entirely upon his rather unreasonable concepts of bound water and matric potential and thus must be discounted. Weatherley (1970, p. 194) has very effectively argued that the rate of net water loss from leaves accounts for less than 5% of the evapotranspiration rate of plants during daylight hours; from this it follows that no part of the leaf water content could ever act as an effective reservoir of water. Thus even the circumstantial evidence of a reservoir effect has no basis. Finally Boyer & Potter (1973) have compared pressure bomb estimates of π (by model B) to psychrometer determinations of π on dead tissue (model A) in sunflower leaves. Comparisons of π were made in the range of $-1\cdot0$ to $-1\cdot8$ MPa. Pressure bomb estimates of π were always larger than psychrometer estimates of π. The difference between the results were accounted for by a constant apoplastic water volume, i.e., there was no evidence that apoplastic water volume changed with leaf water potential.

One other piece of evidence can be cited in favour of model B. The four assumptions of model B lead us to predict that a plot of $1/\psi$ versus water loss or versus R_w will yield a straight line when P falls to zero. The model also predicts that if one or more of these assumptions is wrong then a straight line would not result. But since we always do get a straight line (consult Tyree, 1976a, for examples), it logically implies that all of the assumptions are approximately correct. This puts model B on much firmer logical grounds than model A. The assumptions of model A do not lead to any definite predictions of how ψ will change with water content, therefore we cannot check the correctness of the assumptions by the internal consistency of the data.

Conclusions regarding isotherms

There has been a great deal of ambiguity and confusion in the literature over the meaning and importance of the matric potential, τ. We feel that much of this ambiguity can be overcome by viewing τ in the context of the Gouy-Chapman electric double layer theory. In this context τ is identified as the energy of interaction of the water dipole with the electric field in the double layer (see also Bolt & Miller 1958). It is argued that τ has less of an impact on ψ in cell walls than π and P; indeed, most of the cell wall water appears to be under the influence of negative pressure arising from the energy of the air–water interface (= the so-called capillary effect). We argue that it is not appropriate to equate this 'capillary effect' to τ. For similar reasons τ would have little impact on the water potential of cytoplasm.

It is also argued that the water content of plant cell walls does not change much with ψ. This allows us to ignore the contribution of cell wall water to water potential isotherms of living tissue. Thus we can explain water potential isotherms primarily in terms of changes in P and π in the symplasm. The contribution of τ will be unimportant in the symplasm.

Finally, it is argued that the analysis of isotherms embodied in the pressure bomb theory (model B) will give the most biologically meaningful estimates of P and π in the symplasm.

MECHANISMS OF DROUGHT TOLERANCE

The mechanisms of drought tolerance act whenever plants have little or no means to avoid low water potentials. Their primary aim is to maintain satisfactory hydration levels in the tissue despite the overall adverse environmental conditions. It is important to distinguish between dehydration avoidance and dehydration tolerance mechanisms (Levitt 1972). The latter is associated with the ability of various physiological processes to proceed relatively unimpaired in already dehydrated tissues while the former prevents dehydration.

It has been argued that one way to avoid dehydration, i.e. one way to maintain high turgor pressure and hence growth, is by maintaining a high osmotic pressure π (= RTC) (Levitt 1972). It is generally recognized that plants growing in drier habitats have a higher π (Maximov 1929; Iljin 1957; Levitt 1972). Iljin (1957) showed that the sugar content on a dry weight basis for a number of species from different habitats was minimum for succulents (0·72%) and maximum for xerophytic trees and shrubs (6·89%), while mesophytes and herbaceous xerophytes had intermediate values (1·25 and 2·64% respectively). Eaton & Ergle (1948) found a higher sugar content on a dry weight basis in unirrigated cotton plants when compared with irrigated ones. The sugar content was modified whenever a species grew in a habitat different from its native habitat (Iljin 1957; Slatyer 1963). It has also been suggested that plants with high values of π avoid dehydration by virtue of being able to extract a greater fraction of the total soil water content. How advantageous this is, however, is open to question because extraction of soil water depends on the resistance to water flow through soils and plant and on the water-potential isotherms of the soil in question, e.g. a high π is of little drought avoidance advantage in sandy soils (Tyree 1976b).

The osmotic pressure, π

Plants do experience changes in their osmotic pressure. Since π might be involved in drought avoidance it is appropriate to discuss the factors that bring about changes in π.

All plants are subjected to passive changes in π, i.e., changes in π that are purely the consequence of the much greater volatility of water than the solutes commonly found in plants. As water evaporates from protoplasts the solution concentration and π increase, turgor pressure decreases, and ψ becomes more negative. The magnitude of passive changes in π depends on many factors, but it is well known that protoplasts surrounded by cell walls experience much smaller changes in π for a given change of ψ than protoplasts without cell walls, and protoplasts surrounded by rigid cell walls undergo smaller changes in π than cells with less rigid walls. It can be argued that plants have evolved cell walls in order to avoid the large scale dilution that would occur in freshwater in the absence of any other method to counteract osmosis. So it can be argued that too much dilution of cell sap is a disadvantage. Because of cell walls, plant protoplasts typically experience small volume changes, and thus small passive changes in π, of only 10 to 30%. One is tempted to argue that passive changes in π are of no advantage as a species selective force because most plants experience comparable and relatively small passive changes in π.

It is very important to know whether or not plants can play an active role in regulating π or turgor pressure, P_t. A drought induced reduction in cell volume is accompanied by a decrease in P_t which tends to reduce growth. If plants can actively regulate π then under drought conditions an increase in π can maintain water supply by reducing ψ_{plant} with little or no volume change, in other words without a lowering of P_t. The ability to change π or to maintain constant P_t in spite of volume changes can be loosely termed osmoregulation. Osmoregulation is known to occur in fungi and algae growing under conditions of high salinity and drought (Hellebust 1976). Among higher plants, both glycophytes and halophytes possess osmoregulatory mechanisms under salt stress associated either with the accumulation of solutes or with the control of internal salt concentration (Bernstein & Hayward 1958). Osmoregulation in higher plants may also be a fairly widespread phenomenon. It was first proposed as a response to water stress about thirty years ago (Crafts, Currier & Stocking 1949; Iljin 1957). It is now clear that the ability to osmoregulate may be an inheritable property; intra-specific difference for drought-induced osmoregulation has been reported in wheat (Morgan 1977) and sorghum (Stout & Simpson 1978) but not in soybean (Turner *et al.* 1978).

Osmoregulation may take on three forms. It may be ontogenetically induced, drought or stress induced, or diurnal. Within the context of the present literature it is often difficult to decide if an observed change in π is a passive change or one of the three types of osmoregulation. A way to separate passive changes in π from osmoregulation in higher plants is to examine the patterns of π at standard reference levels of tissue hydration, as at maximum turgor (π_o) or when the turgor pressure first reaches zero (π_p).

It is difficult to separate ontogenetic and diurnal osmoregulation from drought-induced osmoregulation. This is partly because it is often difficult to grow plants over day-long or season-long periods without variations in water stress. Some reasonably clear cases of ontogenetic changes in π have been reported by Knipling (1967), Millar, Duysen & Milkinson (1968) and Tyree *et al.* (1978).

It is usually easier to demonstrate drought-induced osmoregulation by comparing droughted plants to well-watered controls. Drought-induced changes in π_o or π_p have been demonstrated in field beans (Karamanos 1978), sorghum (Hsiao *et al.* 1976; Jones & Turner 1978), and cotton (Cutler & Rains 1978). Not all plants examined show this phenomenon. Acropetal increases in π_o have also been demonstrated in a number of plants (Kassam & Elston 1976; Hellkvist *et al.* 1974; Hsiao *et al.* 1976); this may also reflect drought-induced osmoregulation.

Many attempts to demonstrate diurnal osmoregulation have failed. Usually diurnal changes in π are passive (Goode & Higgs 1973; Cline & Campbell 1976; Campbell & Harris 1977). Some clear cut cases have been demonstrated however in maize and soybeans (Hsiao *et al.* 1976; Wenkert, Lemon & Sinclair 1978).

Diurnal osmoregulation has been reported to be more pronounced in water-stressed cotton than in irrigated cotton plants (Cutler, Rains & Loomis 1977). It is possible that diurnal osmoregulation is only a fast form of drought-induced osmoregulation. But in maize it has been shown that diurnal osmoregulation is linked to light (Wenkert, personal communication).

The turgor pressure, P_t

The contribution of P_t to cell and tissue water relations is the direct result of the changes in cell volume. The dehydration of a cell results in a volume decrease with a concomitant decrease in the turgor pressure. The extent to which P_t is reduced for a given amount of water lost is determined by the elastic properties of the cell walls. The elastic properties of the cell walls determine the shape of the water potential isotherms more than does π. The fluctuations in P_t are extremely important because of the decisive impact of P_t on many physiological processes (Hsiao 1973).

In most cases, the parameter used for the description and interpretation of the behaviour of P_t is the bulk elastic modulus of the cell walls (ε). The value of ε determines the slope of the pressure–volume curve of a plant cell and can be defined by the following equation:

$$dP_t = \varepsilon \frac{dV}{V}$$

where dP_t is the change in turgor pressure for an infinitesimal change in volume dV relative to the initial volume V. The bulk modulus ε is *not* related in any simple way to Young's modulus for the cell wall. For the simplest case, a spherical cell of radius r and an *isotropic* wall of thickness δ, ε equals $2\delta\varepsilon^*/3r$ where ε^* is Young's modulus. In reality cell walls are not spherical shells, they are very anisotropic (Vinters, Dainty & Tyree 1977), and they are subject to elastic interactions with neighbouring cells. Consequently the bulk elastic modulus for a complex tissue is not the same as that of the individual cells even if each cell has the same value of ε (Cheung, Tyree & Dainty 1976).

The value of ε can be estimated from changes of P_t versus volume in plant tissues, namely from plots relating P_t to the corresponding changes in symplasmic water volume. Usually P_t is derived by adding π to ψ, but some direct methods of measuring and manipulating cell hydrostatic pressure have also been suggested (for a review see Zimmermann 1978). The relative water content (R_w) is also widely used to account for changes in water volume. It is generally recognized that the P_t versus volume relationship is non-linear (Haines 1950; Gardner & Ehlig 1965; Noy-Meir & Ginzburg 1969; Tyree & Hammell 1972; Hellkvist *et al.* 1974; Cheung *et al.* 1976), and this departure from linearity is attributed to the dependence of ε on both P_t and cell volume (Zimmermann 1978). The value of ε increases in an approximately linear fashion with increasing P_t and approaches a constant value at high pressures (Hellkvist *et al.* 1974; Cheung *et al.* 1976; Steudle, Zimmermann & Lüttge 1977). In addition ε also increases with cell volume, its dependence being more pronounced at higher pressures (Steudle *et al.* 1977; Zimmermann 1978). The pressure dependence of ε makes the use of ε difficult for comparative purposes and for the interpretation of its ecological significance. It is possible to get around this problem by considering only high values of P_t where ε is generally constant (Cheung *et al.* 1976; Tyree *et al.* 1978). On several occasions a linear relationship between P_t and R_w has also been assumed, perhaps without justification (Broyer 1952; Philip 1958; Kassam & Elston 1974; Elston *et al.* 1976). Such an assumption appears to hold satisfactorily for only a narrow range of values of P_t (Philip 1958; Gardner & Ehlig 1965; Noy-Meir & Ginzburg 1969; Jones & Turner 1978; Karamanos 1978).

Since drier conditions of growth are associated with more rigid cell walls

(Slatyer 1967), one should normally expect ε to increase with drought. Warren Wilson (1967b) reported that the values of ε were much higher in xerophytic species than those of some mesophytes. Sanchez-Diaz & Kramer (1971, 1973) found that cell wall 'elasticity' was lower (i.e. ε was higher) in sorghum, a drought-resistant species, in comparison with the less resistant maize. Furthermore, preconditioning to water stress resulted in an increase in ε in cotton (Brown, Jordan & Thomas 1976) and sorghum (Jones & Turner 1978). Such an increase in ε may be associated with an acceleration of leaf ageing which is known to occur under water stress (Brady, Scott & Munns 1974). The secondary thickening imparts rigidity to the walls and thus reduces their elasticity. This assumption is supported by the fact that ε increases systematically with leaf age (Kassam & Elston 1976; Cutler & Rains, 1977; Tyree *et al.* 1978). In contrast with the results reported above, a consistent lowering of ε was observed on a seasonal scale in unirrigated field beans in comparison with well-irrigated controls (Elston *et al.* 1976). This lowering of ε was more pronounced in a wet than in a dry season (Karamanos 1978). It appears that interspecific differences may play a role in the response of cell wall elasticity to water stress.

There are different approaches to the ecological importance of the fluctuations in ε. For example, a cell with a high ε will drop its P_t for a given change in its water content more quickly than one with a low ε. Thus, for a certain level of dehydration, the changes in ψ above incipient plasmolysis are mainly accomplished by the changes in P_t and, consequently, π is maintained rather steady (Zimmermann 1978). Cheung, Tyree & Dainty (1975) characterized this behaviour of the cell walls as a kind of osmoregulation.

On the other hand, a lower ε allows longer times for water exchange. According to Philip (1958):

$$t_{1/2} = \frac{0.693\,V}{AL_p(\varepsilon + \pi)} \tag{14.3}$$

where $t_{1/2}$ is the half-time for water exchange in a cell, V and A are the cell volume and area respectively, π is the internal osmotic pressure, and L_p is the hydraulic conductivity for water flow into or out of the cell. A long $t_{1/2}$ may be advantageous because it implies that the cell is physically protected against short-term water stress (Zimmermann 1978). A lower ε also means that more water must be lost before its turgor pressure falls to zero. But the advantages of a low ε in terms of increased half-time may well be minimal. For example, we can choose a set of values designed to maximize $t_{1/2}$ in equation 14.3 for a vascular plant cell. For a long half-time we want a large V/A ratio and small values of L_p, ε and π. If we choose a large $V/A = 6.7 \times 10^{-4}$ cm (for a spherical cell 40 μm in diameter) and the minimum likely value of L_p ($= 10^{-6}$ cm s^{-1} MPa^{-1}) then $t_{1/2}$ is 6 min if $\varepsilon + \pi =$

4 MPa (a small value) or 0·6 min if $\varepsilon + \pi = 40$ MPa (a large value). For smaller more permeable cells $t_{1/2}$ is likely to be even smaller. On the basis of this superficial examination it would seem that the buffering effect is of advantage only for rather short period fluctuations in external water potential.

Equation 14.3 has been invoked by many people to discuss the possible significance of the response time of higher plants to change in evaporative demand (Dainty 1976; Zimmermann 1978). However, conclusions should not be drawn too quickly from such calculations. Strictly speaking, equation 14.3 describes only the response time of a single cell in an aqueous medium (e.g. an algal cell subjected to rapid changes of external osmotic pressure); in this situation the surface area of exchange, the cell volume, the driving force, and the L_p are well defined. For water loss by evaporation from a leaf cell, the situation is not so clearly defined. The site of evaporation is restricted and the resulting pathway of liquid water flow is not known; this may make the area of water loss less than the cell surface area, A. The water loss comes from both the cell and the soil, so the effective volume of the system may not equal the cell volume, V. Equation 14.3 was derived in terms of driving forces in water potential units, but in the vapour phase water potential is not clearly defined when there is a temperature difference between the liquid and the vapour phase. Also L_p would be dominated by the vapour phase conductivity, but the conductivity expressed in water potential units is strongly a function of water potential in the vapour phase whereas equation 14.3 is derived assuming a constant L_p. In short, the derivation of equation 14.3 needs to be rethought before it can be applied with certainty.

In conclusion, we tend to think that the views cited above are merely speculative and that the ecological importance of ε in plant water relations has not been elucidated yet. Whether a low or a high value of ε is more preferable depends on the physiological importance of the displacements involved (Weatherley 1970) which may be quite different among plant species.

We have confined our discussion only to two possible mechanisms of drought tolerance but do not wish to imply that these are the most important means by which plants survive in dry habitats. The reader is referred to Levitt (1972) for a more detailed discussion of other important mechanisms of drought avoidance and drought tolerance.

APPENDIX

The theory regarding the behaviour of electrolytes near charged surfaces has been discussed at length elsewhere (Eriksson 1952; Bolt 1955; Overbeek 1956). The equations from this theory that have been used to calculate the curves in Figs. 14.1 and 14.2 are as follows:

The surface voltage at the solid–liquid interface was obtained implicitly from

$$\sigma = (\omega RTC_0/2\pi)^{1/2} (\exp (E_0\mathscr{F}/2RT) - \exp (-E_0\mathscr{F}/2RT)) \qquad (A1)$$

where σ = the surface charge density in C cm^{-2}.

ω = the permittivity of the medium $(80/(9 \times 10^{11})$ C V^{-1} cm^{-1} for water).

R = the gas constant (8·31 J mol^{-1} K^{-1}).

T = the temperature in (K).

E_0 = the electric potential of the charged surface, relative to the bulk phase (V).

C_0 = the ionic concentration in the bulk phase (equiv. cm^{-3}).

$\mathscr{F}$ = 96 500 C equiv.$^{-1}$ and π here is 3·142.

Once E_0 is determined then the voltage, E, as a function of the distance from the charged surface x (in cm), is computed from:

$$x = -(\omega RT/8\pi\mathscr{F}^2C_0)^{1/2} \ln ((\tanh (E\mathscr{F}/4RT))/(\tanh (E_0\mathscr{F}/4RT))) \quad (A2)$$

The above equations apply to salt solution containing only univalent salts but analogous equations for divalent salts can be derived. In these equations a number of simplifying assumptions are made, and it is well known that the theory should be modified to account for ionic interactions, polarization of the ions, dielectric saturation, and finite ion size which determines the distance of closest approach. This has been done in a more elaborate theory but the differences between the corrected and uncorrected theory cancel and the latter gives good agreement if the charge density does not exceed about 3×10^{-5} C cm^{-2}.

Equation A2 can be differentiated with respect to E to obtain an expression for the field, $F = dE/dx$. The answer is

$$F = (8\pi RTC_0/\omega)^{1/2} (\exp (E\mathscr{F}/2RT) + \exp (-E\mathscr{F}/2RT)) \qquad (A3)$$

when the electric field is small the average dipole moment of the water is small because of random thermal rotations of the water molecules. Within the electric field, however, it tends to align itself with the field so that the average component of the dipole parallel to the field equals $\bar{p}$. King (1976) states that if the system is in thermal equilibrium, then $\bar{p}$ is given by

$$\bar{p} = p_0 (\coth u - 1/u) \qquad (A4)$$

where $u = FP_0/KT$, P_0 = the dipole moment of water ($6·14 \times 10^{-28}$ C cm) and k = the Boltzman constant ($1·38 \times 10^{-23}$ J °K^{-1}). When $u < 1$ then equation A4 reduces to

$$\bar{p} \simeq P_0u/3 = FP_0^2/3kT \qquad (A5)$$

In Fig. 14.1 F ranges from $1·5 \times 10^6$ to 3×10^4 V cm^{-1} which gives u a range of from 0·224 to 0·0048. In this range equation A5 is accurate within a few per cent. The energy of interaction of the average dipole with the field is

$$F_{\bar{p}} = \frac{F^2 P_o{}^2}{3kT} \quad \text{J per molecule.} \tag{A6}$$

If we multiply by Avogadros number, N_o, and divide by the partial molar volume of water, V_w, we obtain units of J cm^{-3} which is equivalent to MPa. This quantity is identified with τ in equation 14.2

REFERENCES

Acock B. (1975) An equilibrium model of leaf water potentials which separates intra- and extracellular potentials. *Australian Journal of Plant Physiology*, **2**, 253–263.

Bernstein L. & Hayward H.E. (1958) Physiology of salt tolerance. *Annual Review of Plant Physiology*, **9**, 25–46.

Bolt G.H. (1955) Ion adsorption by clays. *Soil Science*, **79**, 267–276.

Bolt G.H. & Miller R.D. (1958) Calculation of total and component potentials of water in soil. *Transactions of the American Geophysical Union*, **39**, 917–928.

Boyer J.S. (1967) Matric potentials in leaves. *Plant Physiology*, **42**, 213–217.

Boyer J.S. & Potter J.R. (1973) Chloroplast response to low leaf water potentials. I. Role of turgor. *Plant Physiology*, **51**, 989–992.

Brady C.J., Scott N.S. & Munns R. (1974) The interaction of water stress with the senescence pattern of leaves. *Mechanisms of Regulation of Plant Growth* (Eds R.L. Bieleski *et al.*), pp. 403–409. The Royal Society of New Zealand, Wellington.

Briggs G.E. (1957) Some aspects of free space in plant tissue. *The New Phytologist*, **56**, 305–324.

Brown K.W., Jordan W.R. & Thomas J.C. (1976) Water stress induced alterations of the stomatal response to decreases in leaf water potential. *Physiologia Plantarum*, **37**, 1–5.

Broyer T.C. (1952) On volume enlargement and work expenditure by an osmotic system in plants. *Physiologia Plantarum*, **5**, 459–469.

Campbell G.S., Harris G.A. (1977) Water relations and water use patterns for *Artemisia tridentata* Nutl. in wet and dry years. *Ecology*, **58**, 652–659.

Cheung Y.N.S., Tyree M.T. & Dainty J. (1975) Water relations parameters on single leaves obtained in a pressure bomb and some ecological interpretations. *Canadian Journal of Botany*, **53**, 1342–1346.

Cheung Y.N.S., Tyree M.T. & Dainty J. (1976) Some possible sources of error in determining bulk elastic moduli and other parameters from pressure-volume curves of shoots and leaves. *Canadian Journal of Botany*, **54**, 758–765.

Cline R.G. & Campbell G.S. (1976) Seasonal and diurnal water relations of selected forest species. *Ecology*, **57**, 367–373.

Clymo J.C. (1963) Ion exchange in *Sphagnum* and its relation to bog ecology. *Annals of Botany*, **27**, 309–324.

Collins J.C. (1969) The ion exchange properties of plant cell walls. Ph.D. thesis., University of East Anglia, Norwich, England.

Crafts A.S., Currier H.B. & Stocking C.R. (1949) *Water in the Physiology of Plants*. The Ronald Press Co., New York.

Cutler J.M. & Rains D.W. (1977) Effects of irrigation history on responses of cotton to subsequent water stress. *Crop Science*, **17**, 329–335.

Cutler J.M. & Rains D.W. (1978) Effects of water stress and hardening on the internal water relations and osmotic constituents of cotton leaves. *Physiologia Plantarum*, **42**, 261–268.

Cutler J.M., Rains D.W. & Loomis R.S. (1977) Role of changes in solute concentration in maintaining favourable water balance in field-grown cotton. *Agronomy Journal*, **69**, 773–779.

Dainty J. (1976) Water relations of plant cells. *Encyclopedia of Plant Physiology*. Vol. 2 (Eds V. Lüttge and M.G. Pitman), pp. 12–35. Springer-Verlag, New York.

Dainty J. & Hope A.B. (1959) Ionic relations of cells of *Chara australis*, I. Ion exchange in the cell wall. *Australian Journal of Biological Sciences*, **12**, 395–411.

Dainty J. & Hope A.B. (1961) The electric double layer and the Donnan equilibrium in relation to plant cell walls. *Australian Journal of Biological Sciences*, **14**, 541–551.

Dainty J., Hope A.B. & Denby C. (1960) Ionic relations of cells of *Chara australis*, II. The indiffusible anions of the cell wall. *Australian Journal of Biological Sciences*, **13**, 267–276.

Eaton F.M. & Ergle D.R. (1948) Carbohydrate accumulation in the cotton plant at low moisture levels. *Plant Physiology*, **23**, 169–187.

Elston J., Karamanos A.J., Kassam A.H. & Wadsworth R.M. (1976) The water relations of the field bean crop. *Philosophical Transactions of the Royal Society of London*, **B273**, 581–591.

Eriksson E. (1952) Cation-exchange equilibria on clay minerals. *Soil Science*, **74**, 103–113.

Gaff D.F. & Carr D.J. (1961) The quantity of water in the cell walls and its significance. *Australian Journal of Biological Sciences*, **14**, 299–311.

Gardiner W.R. & Ehlig C.F. (1965) Physical aspects of the internal water relations of plant leaves. *Plant Physiology*, **40**, 705–710.

Goode J.E. & Higgs K.H. (1973) Water, osmotic and pressure potential relationships in apple leaves. *Journal of Horticultural Science*, **48**, 203–215.

Haines F.M. (1950) The relation between cell dimensions, osmotic pressure and turgor pressure. *Annals of Botany*, **14**, 385–394.

Hellebust J.A. (1976) Osmoregulation. *Annual Review of Plant Physiology*, **27**, 485–505.

Hellkvist J., Richards G.P. & Jarvis P.G. (1974) Vertical gradients of water potential and tissue water relations in Sitka spruce trees measured with the pressure chamber. *Journal of Applied Ecology*, **11**, 637–667.

Hsiao T.C. (1973) Plant responses to water stress. *Annual Review of Plant Physiology*, **24**, 519–570.

Hsiao T.C., Acevedo E., Fereres E. & Henderson D.W. (1976) Water stress, growth and osmotic adjustment. *Philosophical Transactions of the Royal Society of London*, **B273**, 470–500.

Iljin W.S. (1957) Drought resistance in plants and physiological processes. *Annual Review of Plant Physiology*, **9**, 257–274.

Jones M.M. & Turner N.C. (1978) Osmotic adjustment in leaves of sorghum in response to water deficits. *Plant Physiology*, **61**, 122–126.

Karamanos A.J. (1978) Understanding the origin of the responses of plants to water stress by means of an equilibrium model. *Praktika of the Academy of Athens*, **53**, 308–341.

Kassam A.H. & Elston J. (1974) Seasonal changes in the status of water and tissue characteristics of leaves of *Vicia faba* L. *Annals of Botany*, **38**, 419–429.

Kassam A.H. & Elston J. (1976) Changes with age in the status of water and tissue characteristics in individual leaves of *Vicia faba* L. *Annals of Botany*, **40**, 669–679.

King A.L. (1976) Values for the Langevin function. *Handbook of Chemistry and Physics*, 57th Edition, p. E115. CRC Press, Cleveland, Ohio.

Knipling E.B. (1967) Effects of leaf ageing on water deficit–water potential relationships of dogwood leaves growing in two environments. *Physiologia Plantarum*, **20**, 65–72.

Kramer P. (1955) Bound water. *Handbuch der Pflanzenphysiologie*. Vol. 1, pp. 223–242. Springer-Verlag, Berlin.

Levitt J. (1972) *Responses of Plants to Environmental Stresses*. Academic Press, New York.

Maximov N.A. (1929) *The Plant in Relation to Water*. Allen & Unwin, London.

Millar A.A., Duysen M.E. & Wilkinson G.E. (1968) Internal water balance of barley under soil moisture stress. *Plant Physiology*, **43**, 968–972.

Morgan J.M. (1977) Differences in osmoregulation between wheat genotypes. *Nature*, **270**, 234–235.

Noy-Meir I. & Ginzburg B.Z. (1967) An analysis of the water potential isotherm in plant tissue. I. The theory. *Australian Journal of Biological Sciences*, **20**, 695–721.

Noy-Meir I. & Ginzburg B.Z. (1969) An analysis of the water potential isotherms in plant tissue. II. Comparative studies on leaves of different types. *Australian Journal of Biological Sciences*, **22**, 35–52.

Overbeek J.Th.G. (1956) The Donnan equilibrium. *Progress in Biophysics and Biophysical Chemistry*, **6**, 57–84.

Philip J.R. (1958) The osmotic cell, solute diffusibility and the plant water economy. *Plant Physiology*, **33**, 264–271.

Pitman M.G. (1965) The location of the Donnan free space in disks of beetroot tissue. *Australian Journal of Biological Sciences*, **18**, 547–553.

Richter H. (1978) Water relations of single drying leaves: Evaluation with a dewpoint hygrometer. *Journal of Experimental Botany*, **29**, 277–280.

Robinson R.A. & Stokes R.H. (1959) *Electrolyte Solutions*. Butterworths, London.

Sanchez-Diaz M.F. & Kramer P.J. (1971) Behaviour of corn and sorghum under water stress and during recovery. *Plant Physiology*, **48**, 613–616.

Sanchez-Diaz M.F. & Kramer P.J. (1973) Turgor differences and water stress in maize and sorghum leaves during drought and recovery. *Journal of Experimental Botany*, **24**, 511–515.

Slatyer R.O. (1963) Climatic control of plant water relations. *Environmental Control of Plant Growth* (Ed. L.T. Evans), pp. 33–54. Academic Press, New York.

Slatyer R.O. (1967) *Plant–Water Relationships*. Academic Press, London.

Spanner D.C. (1973) The components of the water potential in plants and soils. *Journal of Experimental Botany*, **24**, 816–819.

Steudle E., Zimmermann U. & Lüttge U. (1977) Effects of turgor pressure and cell size on the wall elasticity of plant cells. *Plant Physiology*, **59**, 285–289.

Stout D.G. & Simpson G.H. (1978) Drought resistance of *Sorghum bicolor*. I. Drought avoidance mechanisms related to leaf water status. *Canadian Journal of Plant Science*, **58**, 213–224.

Talbot A.J.B., Tyree M.T. & Dainty J. (1975) Some notes concerning the measurement of water potential of leaf tissue with specific reference to *Tsuga canadensis* and *Picea abies*. *Canadian Journal of Botany*, **53**, 784–788.

Turner N.C., Begg J.E., Rawson H.M., English S.D., Hearn A.B. (1978) Agronomic and physiological responses of soybean and sorghum crops to water deficits. III. Components of leaf water potential, leaf conductance, $^{14}CO_2$ photosynthesis and adaptation to water deficits. *Australian Journal of Plant Physiology*, **5**, 179–194.

Tyree M.T. (1972) Ion exchange and transport of water and solutes in Characean cell walls. Ph.D. thesis, Cambridge University, England.

Tyree M.T. (1976a) Negative turgor pressure in plant cells. Fact or fallacy? *Canadian Journal of Botany*, **54**, 2738–2746.

Tyree M.T. (1976b) Physical parameters of the soil–plant–atmosphere system: Breeding for drought resistance characteristics that might improve wood yield. *Tree Physiology and Yield Improvement* (Ed. M.G.R. Cannell and F.T. Last), pp. 329–348. Academic Press, London.

Tyree M.T., Cheung Y.N.S., MacGregor M.E., & Talbot A.J.B. (1978) The characteristics of seasonal and ontogenetic changes in the tissue-water relations of *Acer*, *Populus*, *Tsuga* and *Picea*. *Canadian Journal of Botany*, **56**, 635–647.

Tyree M.T., Dainty J. & Hunter D.M. (1974) The water relations of hemlock (*Tsuga canadensis*). IV. The dependence of the balance pressure on the temperature as measured by the pressure-bomb technique. *Canadian Journal of Botany*, **52**, 973–978.

Tyree M.T. & Hammel H.T. (1972) The measurement of the turgor pressure and the water relations of plants by the pressure-bomb technique. *Journal of Experimental Botany* **23**, 267–282.

Vinters H., Dainty J. & Tyree M.T. (1977) Cell wall elastic properties of *Chara corallina*. *Canadian Journal of Botany*, **55**, 1933–1939.

Warren Wilson J. (1967a) The components of leaf water potential. I. Osmotic and matric potentials. *Australian Journal of Biological Sciences*, **20**, 329–347.

Warren Wilson J. (1967b) The components of leaf water potential. II. Pressure potential and water potential. *Australian Journal of Biological Sciences*, **20**, 349–357.

Warren Wilson J. (1967c) The components of leaf water potential. III. Effects of tissue characteristics and relative water content on water potential. *Australian Journal of Biological Sciences*, **20**, 359–367.

Weatherley P.E. (1970) Some aspects of water relations. *Advances in Botanical Research*, **3**, 171–206.

Wenkert W., Lemon E.R. & Sinclair T.R. (1978) Water content–potential relationship in soybean: changes in component potentials for mature and immature leaves under field conditions. *Annals of Botany*, **42**, 295–307.

Zimmermann U. (1978) Physics of turgor- and osmo-regulation. *Annual Review of Plant Physiology*, **29**, 121–148.

15. SOME LIMITATIONS AND APPLICATIONS OF THE PRESSURE–VOLUME CURVE TECHNIQUE IN ECOPHYSIOLOGICAL RESEARCH

HANNO RICHTER,[1] FRIEDRICH DUHME,[2]
GERHARD GLATZEL,[3] THOMAS M. HINCKLEY,[4]
AND HEIDRUN KARLIC[1]

[1]*Botanisches Institut der Universität für Bodenkultur, A-1180
Wien, Austria*
[2]*Lehrstuhl für Landschaftsökologie der TU München, D-8050
Freising-Weihenstephan, Federal Republic of Germany*
[3]*Institut für Forstökologie der Universität für Bodenkultur,
A-1180 Wien, Austria*
[4]*School of Forestry, Fisheries and Wildlife, University of
Missouri, Columbia, MO 65211, USA*

SUMMARY

1. The relationship between total water potential and volume of water in leaves can be analysed in several different ways to derive values of parameters such as the osmotic potential at saturation and the turgor loss point.

2. The values of the parameters obtained depend on the goodness of fit of the data to the assumed hyperbolic relationship and on the particular transformation used to obtain a linear regression.

3. The parameter variously called 'bound' or 'apoplastic' water is ambiguous and its interpretation unclear at present.

4. Other parameters such as the osmotic potential at the turgor loss point appear to be useful in defining acclimation and adaptation to environment, when combined with measurements in the field of total water potential and stomatal conductance.

PRESSURE–VOLUME CURVES

Graphical display and analysis of data relating changes in water potential to changes in cell, tissue or organ water content has become known as the pressure–volume (or *PV*) curve technique. The data for the analysis may be obtained from a number of different methods for the estimation of water potential and water content. The *PV* curve technique was first developed for the analysis of measurements made on leaves and twigs with the pressure

263

chamber (Scholander *et al.* 1965; Tyree & Hammel 1972) and has been extensively applied to such measurements. In such a case, total water potential ψ_t is equivalent to the equilibrium chamber pressure, and changes in water content are derived from the volume of xylem sap expressed from the leaves.

However, other methods may be used for the determination of ψ_t, such as thermocouple psychrometry (Talbot, Tyree & Dainty 1975), thermocouple hygrometry (Richter 1978a), refractometry of solutions with which the tissue has exchanged water (Kyriakopoulos & Richter 1977) or equilibration of cells with solutions of known potential (Dupre & Hempling 1978; Gerdenitsch 1979). Furthermore, osmotic potentials (ψ_o) determined on the sap of killed leaves can be analysed using the same type of graph (Kyriakopoulos & Richter 1977). The corresponding estimates of water content may also be obtained from measurements of transpirational water loss (Talbot, Tyree & Dainty 1975) or from microscopic measurement of the volume of single cells (Gerdenitsch 1979). Dupre & Hempling (1978) estimated volume changes of Ehrlich ascites tumor cells in hyperosmotic solutions by means of a Coulter Counter.

DERIVATION OF PARAMETERS

The basic relationship between osmotic potential and volume of solution, as stated by Boyle's Law and as applied by Van't Hoff, is:

$$\psi_o \cdot V = \text{constant} \tag{15.1}$$

This equation describes a rectangular hyperbola (Noy-Meir & Ginzburg 1967) which is, therefore, the primary pressure–volume curve. Two transformations are widely used to facilitate analysis of the curve:

$$\psi_o = 1/V \cdot \text{constant} \tag{15.2}$$

and

$$V = 1/\psi_o \cdot \text{constant} \tag{15.3}$$

Graphs using these transformations have a very useful feature. Curves drawn through data points for ψ_t show two clearly differing portions. Data points obtained from turgid cells, for which ψ_t is less negative than ψ_o, fall on a curved line, while the points obtained from flaccid cells or organs, for which $\psi_t = \psi_o$, fall on a straight line (Fig. 15.1a, b) (Tyree & Hammel 1972; Richter 1978b). The water potential at the turgor loss point, which occurs at the intersection of the two lines, can be estimated exactly from such a graph while there is no theoretically correct way to calculate this important parameter.

Pressure–volume curves may be drawn in two ways. The potentials (ψ_t and ψ_o) may be treated as the ordinate and this has been the standard

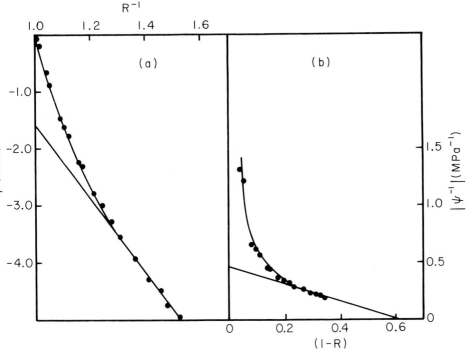

FIG. 15.1. Two types of PV curves for the same set of data from *Sorbus aria* L., 18.7.1978.
(a) ψ as a function of R^{-1}
   $\psi_0 = 4\cdot85 - 6\cdot46\ R^{-1}$; $\psi_{0(R=1)} = -1\cdot61$ MPa
(b) ψ^{-1} as a function of $(1 - R)$
   $\psi_0^{-1} = 0\cdot45 - 0\cdot73\ (1 - R)$; $\psi_{0(R=1)} = -2\cdot22$ MPa

procedure so far and is used in our figures. Alternatively the potentials may be treated as the abscissa and an estimate for water content as the ordinate. The only published example of the latter seems to be the work by Dupre & Hempling (1978). It may be shown that all four diagrams thus obtained give identical estimates of the main parameters of the curves, i.e. the osmotic potentials at full saturation $\psi_{o(sat)}$ and at the turgor loss point $\psi_{o(TLP)}$, as long as two conditions are fulfilled:

1. The linear part of the curves must strictly obey the theoretical relationships stated in equations 15.1, 15.2 and 15.3.

2. The correlation coefficient for these lines must be 1, that is, the data points must not be scattered around them.

However, these two conditions are seldom fully met, with the consequence that different values of the parameters are derived from the two transformations and the choice of independent variable. Data from living leaves (and less frequently also from expressed sap obtained after killing the leaves) give curves the linear portions of which often deviate pronouncedly from the theoretical behaviour of ideal solutions, i.e. the product $\psi_o \cdot V$ is not constant. This is evident in the slopes of the linear portions of the graph which are either too steep or too flat so that the intercepts are displaced along the axes. Furthermore, even if such a line has a correlation coefficient of 1·00 in one of the graphs, the fit will be less perfect when the reciprocal transformation is used, and the estimated values of the main parameters will change. This is not a purely statistical phenomenon, but rather one connected with systematic errors in the data (Fig. 15.1a, b). However, there is another ambiguity which is purely statistical. Scatter in the data leads to differences in the resulting parameters for the same transformation, depending on the choice of the independent variable, since regression lines are always calculated for the least sum of squares of deviation in the y-direction (Steel & Torrie 1960).

In view of these ambiguities the question arises as to why these transformations are used at all and why we do not analyse the untransformed data. The main reason is that the transformations provide a superior means for estimating the turgor loss point. The heavier weighting of data in the region where the PV curve departs from linearity facilitates the decision as to whether a data point should be included in the calculation of osmotic potentials or not. Furthermore, linear regression techniques (which do not require much statistical training) can be employed. It must be borne in mind, however, that these transformations do not improve the accuracy of the estimated parameters. Lengthy extrapolation of the linear part of the curves towards extreme values may result in large changes in the estimates of the parameters when the data points change only slightly.

THE BOUND WATER CONCEPT

As already mentioned, PV curves from living leaves show linear portions with a slope which does not always follow the theoretical prediction. Theoretically the linear portion of the curve in Fig. 15.1b should intercept the abscissa at a relative water content (R) of 0, or $(1 - R) = 1$; i.e. an infinite osmotic potential is reached when the last trace of water is removed from the tissue. As one can see from many curves (e.g. Fig. 15.1b), this is not the case. Many such straight line extrapolations intercept the abscissa somewhere between $(1 - R) = 0.7$ and $(1 - R) = 1$. This is the behaviour one would expect from the presence of water which is not part of the osmotically active solution but which is sequestered in a different compartment, such as the cell wall, or which is tightly bound to macromolecules (Tyree & Hammel 1972; Hellkvist, Richards & Jarvis 1974; Roberts & Knoerr 1977). The existence of such non-osmotic water (interchangeably called 'apoplastic' or 'bound' water) has

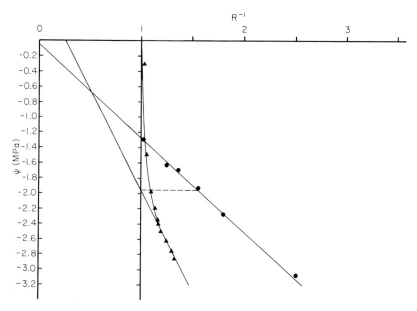

FIG. 15.2. *Prunus laurocerasus* L. (8.11.1978). A comparison between *PV* curves obtained on (a) living leaves and (b) expressed sap from killed leaves. The measurements are made with a thermocouple hygrometer.
(a) ▲: living leaves, $\psi_0 = 0.76 - 2.71\ R^{-1}$
(b) ●: expressed sap, $\psi_0 = 0.07 - 2.05\ R^{-1}$
The dashed line gives the estimate for 'apoplastic water'.

been well established (Gaff & Carr 1961). The influence of such water may be seen in *PV* curves when psychrometric measurements on living leaves and osmotic potentials of expressed sap from killed leaves are plotted together against *R* (Kyriakopoulos & Richter 1977). Fig. 15.2 shows that at a particular value of *R* values of ψ_0 derived from expressed sap are less negative than the values derived from living leaves. This difference may be considered to indicate the presence of some relatively pure water which is kept separate from the vacuolar sap as long as the membranes are intact. On killing a fully saturated leaf, mixing of the contents of the different compartments occurs. No change in the water content of the whole leaf is caused by this process, but the saturated value of ψ_0 becomes less negative. Water has to be removed from the expressed sap to obtain an osmotic potential identical with that found in the living leaves at full saturation; this surplus water is presumably not part of the symplasmic solution. The amount of water contained in the apoplast may be read from the graph as the length of the dashed line in Fig. 15.2. Clearly some simplifying assumptions enter into this reasoning. The presence of solutes in the apoplast is not taken into account, and changes in vacuolar solute content, such as precipitation, adsorption or

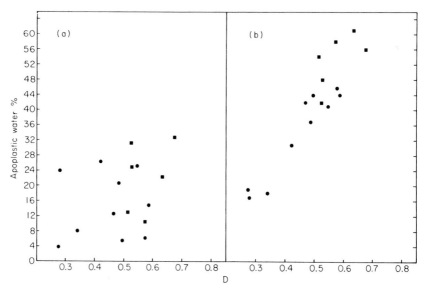

FIG. 15.3. Estimates of apoplastic water (in %) in relation to relative dry weight (*D*) determined by:
(a) the 'intercept method' (values estimated from the intercept with the water volume axis on curves from living leaves), and (b) the 'difference method' (values estimated from the distance between regression lines as shown in Fig. 15.2).
●: *Prunus laurocerasus* L., ■: *Ilex aquifolium* L., June to August 1978.

enzymatic degradation, after mixing with substances contained in the protoplasm and the cell wall, are disregarded.

Thus, we have two independent ways of estimating the amount of apoplastic water: the method of deriving it from the intercept of *PV* curves with the volume axis may be called the 'intercept method', while the technique of establishing it from parallel measurements on living and dead material, the 'difference method'. We may test these approaches against an independent criterion, using the following argument (Hellkvist, Richards & Jarvis 1974). The amount of apoplastic water should be related to the volume of the apoplast, and this is mainly composed of the cell walls. An easy if crude estimate of the amount of water bound to cell wall materials and kept in microcapillaries may be obtained from the dry weight of the organ, since this should be reasonably well correlated with the weight of cell wall materials. Since all weights in our graphs are referred to water content at full saturation, we introduce the term 'relative dry weight' (*D*) which we define as follows:

$$D = \text{(dry weight)}/\text{(saturated weight} - \text{dry weight)} \qquad (15.4)$$

Plotting the estimates of apoplastic water from both the 'intercept' and 'difference' methods against *D*, we obtain Fig. 15.3a and b. This direct

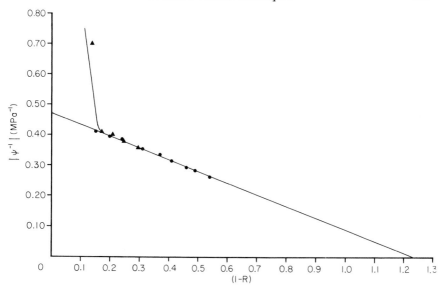

FIG. 15.4. A negative estimate of apoplastic water obtained with the intercept method from *Phragmites communis* Trin., 30.5.1978. Data from pressure chamber (▲) and thermocouple hygrometer (●).
$\psi_0^{-1} = 0.48 - 0.39 (1 - R)$; $\psi_{0(R=1)} = -2.08$ MPa; intercept at $R = -0.23$.

comparison indicates that the 'difference' method is probably a more effective way of estimating the amount of apoplastic water in an organ than the 'intercept' method. This is supported by the observation, made now and then, of rather improbable values of apoplastic water by the 'intercept' method applied to a graph of ψ^{-1} against R. There are examples of intercepts near or above $(1 - R) = 1$ even for leaves of large dry weight (Fig. 15.4). Since negative values of apoplastic water are meaningless, we should discuss other influences on the slope of the linear part of the graph in addition to (but not instead of) the existence of apoplastic water. A more neutral term to describe an intercept which is not zero on the volume axis, such as 'percentage of displacement', might help in further discussion.

ECOLOGICAL CONSEQUENCES

It seems most appropriate to concentrate on those parameters of the curve which are immediately relevant to the behaviour of plants in the field and may help in the interpretation of additional data, like diurnal patterns of total water potential or stomatal conductance. The most important of these parameters are the osmotic potential at full saturation, and the osmotic potential at the turgor loss point.

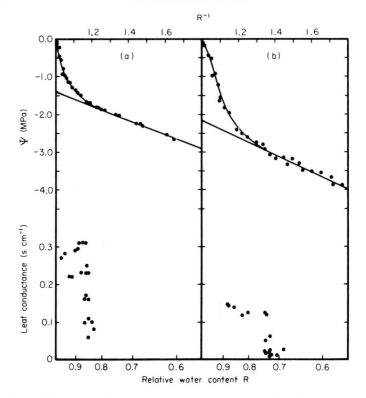

FIG. 15.5. *PV* curves for *Cornus sanguinea* L. (above), and leaf conductance measured with a diffusion porometer (below). On (a) 24.5.1979 and (b) 31.8.1978.

In Fig. 15.5 stomatal conductances at different relative water contents of the drought-hardy shrub *Cornus sanguinea* L. from a dry site near Vienna is compared with *PV* curves of leaves from the same tree. Values of R for the leaves in the field were estimated from the *PV* curves, using parallel pressure chamber measurements of total water potential. Comparison between Fig. 15.5a and b shows that there was an appreciable shift in the parameters of the *PV* curve as the leaves matured. The value of ψ_o at saturation changed from -1.38 to -2.15 MPa, and the value of ψ_o at the turgor loss point changed from -1.76 to -3.00 MPa with corresponding values of R of 0.83 and 0.74 respectively. The leaf conductance appears to drop sharply as the relative water content approaches the turgor loss point. The *PV* curve is therefore valuable for:

(a) The estimation of the relative water content and turgor relations of leaves from diurnal measurements of leaf water potential, and

(b) The demonstration of a relationship between the water status of the leaves and stomatal reaction during the day.

Although the *PV* curves in Figs. 15.1, 15.2 and 15.5 are quite different in detail, their general form is the same; that is, a steep curved portion meets a straight line, typically at a relative water content above 0·7. In contrast, *PV* curves from injured leaves are different. Salt damage in *Acer platanoides* L. as well as frost injury in several non-hardy, evergreen shrubs (*Aucuba japonica* Thunb., *Ilex aquifolium* L., *Prunus laurocerasus* L.) changed the shape of the curves in comparison with those of healthy leaves. The decline in potential with water loss was very small at first, increased gradually after the removal of more than 10 % of the volume of water, and was not followed by a clearly linear portion. This occurred even when the leaves showed only slight visible symptoms of injury at the macroscopic level. An effort is currently being made to investigate this phenomenon more closely and to assess the relationship between microscopical evidence of frost damage and characteristics of the *PV* curve.

ACKNOWLEDGEMENTS

This research was made possible through funds provided by Fonds zur Förderung der wissenschaftlichen Forschung, Vienna (Projekt 1465) and the Graduate School of the University of Missouri, Columbia, USA. Ernst Scharfetter drew the figures. Critical comments by Professor P.G. Jarvis are appreciated.

REFERENCES

Dupre A.M. & Hempling H.G. (1978) Osmotic properties of Ehrlich ascites tumor cells during the cell cycle. *Journal of Cellular Physiology*, **97**, 381–395.

Gaff D.F. & Carr D.J. (1961) The quantity of water in the cell wall and its significance. *Australian Journal of Biological Sciences*, **14**, 299–311.

Gerdenitsch W. (1979) Mikroskopische Beiträge zum Druck-Volumen—Schema des Zellwasserhaushaltes anhand von Einzelzellen und Zellfäden. *Protoplasma*, **99**, 79–97.

Hellkvist J., Richards G.P. & Jarvis P.G. (1974) Vertical gradients of water potential and tissue water relations in Sitka spruce trees measured with the pressure chamber. *Journal of Applied Ecology*, **11**, 637–668.

Kyriakopoulos E. & Richter H. (1977) A comparison of methods for the determination of water status in *Quercus ilex* L. *Zeitschrift für Pflanzenphysiologie*, **82**, 14–27.

Noy-Meir I. & Ginzburg B.Z. (1967) An analysis of the water potential isotherm in plant tissue. I. The theory. *Australian Journal of Biological Sciences*, **20**, 695–721.

Richter H. (1978a) Water relations of single drying leaves: evaluation with a dew point hygrometer. *Journal of Experimental Botany*, **29**, 277–280.

Richter H. (1978b) A diagram for the description of water relations in plant cells and organs. *Journal of Experimental Botany*, **29**, 1197–1203.

Roberts S.W. & Knoerr K.R. (1977) Components of water potential estimated from xylem pressure measurements in five tree species. *Oecologia*, **28**, 191–202.

Scholander P.F., Hammel H.T., Bradstreet E.D. & Hemmingsen E.A. (1965) Sap pressure in vascular plants. *Science*, **148**, 339–346.

Steel R.G.D. & Torrie J.H. (1960) *Principles and Procedures of Statistics.* McGraw-Hill, New York.

Talbot A.J.B., Tyree M.T. & Dainty J. (1975) Some notes concerning the measurement of water potentials of leaf tissue with specific reference to *Tsuga canadensis* and *Picea abies. Canadian Journal of Botany*, **53**, 784–788.

Tyree M.T. & Hammel H.T. (1972) The measurement of the turgor pressure and the water relations of plants by the pressure-bomb technique. *Journal of Experimental Botany*, **23**, 267–282.

16. THE RESPONSE OF PHOTOSYNTHESIS
TO TEMPERATURE*

OLLE BJÖRKMAN

*Department of Plant Biology, Carnegie Institution of Washington,
Stanford, California 94305*

SUMMARY

1. Transplant experiments with coastal and desert species in gardens on the California coast and in Death Valley have shown large species differences in the ability to survive and grow even when the plants were provided with adequate water, in a manner reflecting the temperature regimes of the native habitats.

2. Experiments in controlled environments showed that the widely different performances in the field were largely the result of intrinsic differences in the response of photosynthesis to temperature.

3. Both C_3 and C_4 coastal species had high rates of photosynthesis when grown at low temperatures and low rates when grown at high temperatures and had a low potential for acclimation to high temperatures. In contrast, the high temperature-adapted, C_4 species, *Tidestromia oblongifolia*, from Death Valley was unable to acclimate to low temperatures. A much higher acclimation potential was found in a number of evergreen desert species from habitats in which much larger seasonal variations in temperature occur. In the Old World shrub, *Nerium oleander*, photosynthesis of mature leaves possessed a high acclimation potental.

4. In C_3 species CO_2 was more limiting at high than at low temperatures, presumably because of the kinetics of the RuP_2 carboxylase/oxygenase. In C_4 species, CO_2 had only a small effect on the temperature dependence of photosynthesis. Changes in stomatal conductance with temperature could not account for the large differences in photosynthetic capacity of high and low temperature-adapted species when acclimated to different temperature regimes.

5. Adaptation and acclimation of photosynthesis to low temperature is correlated with increases in RuP_2 carboxylase in several C_4 species and with $FruP_2$ phosphatase in *N. oleander*.

6. Adaptation and acclimation to high temperatures involves an enhanced heat stability of the photosynthetic system, probably through increase in the viscosity of the thylakoid membranes. Inactivation at high temperatures may result from a functional dissociation of the light harvesting chlorophyll-protein complexes in the thylakoid membranes from the Photosystem 2 reaction centres.

7. The C_4 pathway improves photosynthetic performance in normal air at high temperatures because of the way it concentrates CO_2 at the RuP_2 carboxylase enabling the Calvin cycle to operate at near-saturating CO_2 levels, whereas in C_3 plants photosynthesis is severely CO_2-limited at high temperatures.

* Carnegie Institution of Washington–Department of Biology Publication No. 691.

INTRODUCTION

Together with the limited water supply, the intense summer heat that characterizes many of the world's arid regions imposes exceptional demands on the native vegetation. Frequently the plants are exposed to temperatures that would cause a cessation of photosynthesis and also result in severe and irreversible cellular damage to plants of cool temperate origin. It would thus appear that plants inhabiting such harsh environments possess special adaptations which enable them to cope successfully with extremely high temperatures.

In this paper I shall review some of the attempts by the Carnegie–Stanford group of investigators to determine the extent and mechanistic basis of photosynthetic response and adaptation of higher plants to contrasting temperature regimes. Our approach makes a special effort to bridge the gap that usually and unfortunately exists between ecological investigations of the overall performance of whole plant stands in their native habitats and detailed laboratory studies of specific processes at the level of the single leaf, isolated organelle or enzyme.

In our investigations we have made extensive use of plants occurring on the floor of Death Valley, California, one of the hottest and most arid habitats on earth. Transplant gardens in Death Valley and at Bodega Head on the cool Pacific coast of Northern California have enabled us to carry

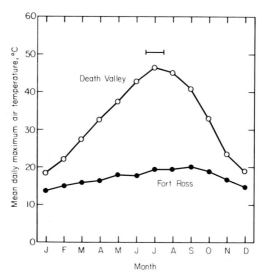

Fig. 16.1. Annual course of mean daily maximum air temperature in Death Valley, California and Fort Ross, on the Pacific coast of northern California. Graphs are based on official, long-term U.S. Weather Bureau records.

out comparative field studies of the performance of plants native to contrasting climates in reciprocal transplantation experiments, and under irrigated as well as unirrigated conditions. Fig. 16.1 shows the seasonal course of the mean daily maximum air temperatures for the location of our Death Valley transplant garden and for Fort Ross on the Californian coast, the closest meteorological station to Bodega Head for which long term official records are available. In this oceanic climate, mean daily maximum temperature shows only a small seasonal variation ranging from a minimum of about 13 °C in January to a maximum of about 18 °C in September. In contrast, the annual variation in mean daily maximum temperature for Death Valley ranges from 18 °C in January to over 46 °C in July. The average range of air temperature experienced by an evergreen plant in this habitat during a typical year extends from $-3 \cdot 1$ °C to $50 \cdot 6$ °C and the lowest and highest meteorological air temperatures ever recorded in Death Valley are $-9 \cdot 8$ °C and $56 \cdot 7$ °C.

FIELD TRANSPLANT EXPERIMENTS

Reciprocal transplant experiments with ample water supply, involving many species native to contrasting climates, have revealed striking differences in their ability to grow and survive in the two transplant gardens, in a manner reflecting the temperature regimes of the native habitats (Björkman *et al.* 1974a). The relative performance of the different species in the cool Bodega Head garden was inversely related to their performance during the summer in the Death Valley garden. For example, the C_3 species *Atriplex glabriuscula* Edmon., and the C_4 species *Atriplex sabulosa* Rouy., both of which are native to cool habitats on the Atlantic coasts of North Europe and northern North America, and *Atriplex triangularis* Willd., a cosmopolitan coastal species, in this case from the coast of Northern California, exhibited the highest relative growth rates and produced abundant seed when grown in the Bodega Head garden. However, these coastal species were unable to survive the summer in Death Valley even when supplied with ample water.

At the other extreme is the response of *Tidestromia oblongifolia* Wats. (Standl.), a summer-active and essentially winter-dormant C_4 perennial, native to the Death Valley floor and adjacent low desert areas. In Death Valley, little growth occurs until May but during June and July vegetative growth is extremely rapid, and at peak growth rate doubling of dry matter occurs in about three days. This plant maintains a very high productivity throughout the summer when temperatures approach or even exceed 50 °C, and a prodiguous seed crop is produced in late autumn. This thermophilic species was the only one of the species studied which completely failed to grow in the coastal garden at any time of the year.

Evergreen species native to the Death Valley floor showed an intermediate response. For example, *Atriplex hymenelytra* (Torr.) Wats. (C$_4$), *Atriplex lentiformis* (Torr.) Wats. (C$_4$), and *Larrea divaricata* Cav. (C$_3$) remained photosynthetically active throughout the year in Death Valley when supplied with water but peak productivity occurred during the milder seasons, and summer productivity was far lower than for *T. oblongifolia*. Although these evergreen species were able to produce some growth during the summer in the Bodega Head garden, growth rate was very slow and none produced ripe seed.

CONTROLLED-GROWTH EXPERIMENTS

In addition to their contrasting temperature regimes, the environments of the two transplant gardens inevitably differ in other important variables, notably air humidity and edaphic factors. Subsequent growth experiments were therefore conducted in the laboratory under controlled temperature regimes, simulating those of Death Valley and Bodega Head during the summer, while other conditions were kept constant. Air humidity and the supply of water and nutrients were maintained at high and non-limiting levels, and other environmental stresses that may differ between the coastal and desert habitats were completely eliminated.

These experiments confirmed that the widely different performances among species in the two natural habitats are largely attributable to intrinsic differences in their response to temperature. The overall growth responses

TABLE 16.1. Effect of temperature on daily relative growth rate on a total dry weight basis, allocation and daily growth rate on a leaf dry weight basis, for species from cool and from hot environments (Data from Björkman *et al.* 1974b).

Species	Habitat of origin	Daily relative growth rate per unit dry weight		Per cent dry matter allocated to leaves		Daily relative growth rate per unit leaf weight	
		Cool*	Hot*	Cool	Hot	Cool	Hot
Atriplex glabriuscula	coastal C$_3$ annual	0·15	0	55	—	0·26	0
Atriplex sabulosa	coastal C$_4$ annual	0·14	0	53	—	0·26	0
Atriplex hymenelytra	desert evergreen C$_4$	0·13	0·09	65	60	0·19	0·15
Tidestromia oblongifolia	desert summer-active C$_4$	<0·04	0·22	67	43	<0·05	0·50

* 'Cool' and 'Hot' regimes were 16 °C day/11 °C night, and 45 °C day/31 °C night, respectively.

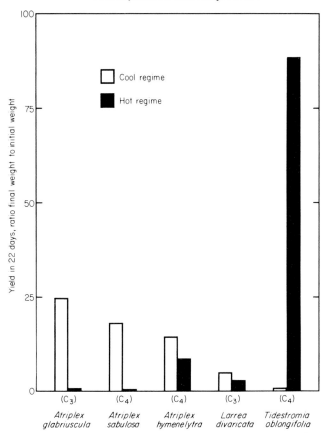

FIG. 16.2. Ratios of the yield of dry matter for coastal and desert species grown under controlled, contrasting temperature regimes. Cool regime: 16 °C day/11 °C night temperature; hot regime: 45 °C day/31 °C night temperature (from Björkman *et al.* 1974b and O. Björkman, unpublished).

closely resembled those obtained in the two transplant gardens (Björkman *et al.* 1974b). As shown in Fig. 16.2, the two coastal species, *A. glabriuscula* (C_3) and *A. sabulosa* (C_4) both produced high yields in the cool regime but died in the hot regime, whereas the thermophilic desert species *T. oblongifolia* (C_4) produced very high yields in the hot regime but failed to show significant growth in the cool regime. The desert evergreens *L. divaricata* (C_3) and *A. hymenelytra* (C_4) were able to grow under both regimes but their yields (Fig. 16.2) and relative growth rates (Table 16.1) were lower than those of the coastal species in the cool regime and much lower than those of *T. oblongifolia* in the hot regime.

The effect of temperature on the dry matter distribution to the leaves relative to other organs (Table 16.1) was surprisingly similar among the different species and the ratio of total dry matter allocated to the leaves in the cool regime was *smaller* in the coastal species which have the highest relative growth rates and total yields than in the two desert C_4 species which have low growth rates. Consequently, daily growth rates in the cool regime expressed on the basis of *leaf* dry weight, show even larger differences between the coastal and the desert species than do growth rates expressed on a total dry weight basis. As is also shown in Table 16.1, *T. oblongifolia* allocated a considerably smaller fraction to the leaves when grown in the hot regime than in the cool regime. Moreover, the distribution of dry matter to the leaves in this species under the hot regime was smaller than in the two evergreen desert species, yet the relative growth rate and yield were much superior in *T. oblongifolia*. It follows that the contrasting responses of productivity of the different species must primarily be a result of intrinsic differences in the temperature dependence of primary growth processes, notably photosynthesis.

PHOTOSYNTHETIC GAS EXCHANGE CHARACTERISTICS

Adaptation to high temperature

Positive evidence supporting the conclusion that the response characteristics of photosynthesis to temperature play an important, perhaps paramount, role in the different abilities of the species studied to grow and survive in contrasting thermal regimes has been obtained from direct measurements of photosynthesis both in laboratory studies on plants grown under controlled conditions and in field investigations using a mobile laboratory unit (Björkman *et al.* 1972; Björkman 1975; Björkman, Mooney & Ehleringer 1975; Björkman, Boynton & Berry 1976; Mooney *et al.* 1976; Mooney, Björkman & Collatz 1978; Pearcy *et al.* 1974; Pearcy 1976, 1977). *In situ* studies in Death Valley (Fig. 16.3) have shown that *T. oblongifolia* possesses record rates of CO_2 uptake (~ 5 nmol cm^{-2} s^{-1}) during the hottest time of year; the optimum temperature for photosynthesis occurs at 45 to 47 °C and substantial rates can be sustained even at temperatures in excess of 50 °C. On the other hand, photosynthesis shows a steep decline below 35 °C and the photosynthetic performance below 20 °C is very poor. This contrasts with the response of photosynthesis in coastal species such as *A. glabriuscula* (Fig. 16.3) which has a much higher photosynthetic capacity at low temperatures but fails to increase its photosynthetic rate as the temperature is increased above 20 to 25 °C; furthermore in this species photosynthesis is completely inactivated at the same temperatures that result in optimum performance in *T. oblongifolia*. In their native habitats the evergreen species

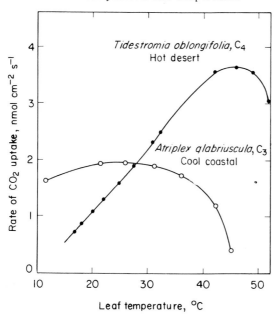

FIG. 16.3. Comparison of the temperature dependences of photosynthesis by whole plants of *Tidestromia oblongifolia* during July in Death Valley and coastal *Atriplex glabriuscula*, grown under a temperature regime simulating that for the month of July at Bodega Head (from Björkman, Mooney & Ehleringer 1975).

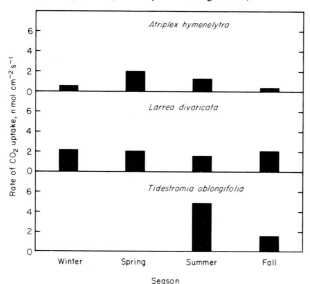

FIG. 16.4 Seasonal course of photosynthetic capacity of three species native to the floor of Death Valley. Rates were determined on naturally-occurring plants. All measurements were made at high light, in normal air and at a leaf temperature of 30 °C for *Atriplex hymenelytra* and *Larrea divaricata* and 35 °C for *Tidestromia oblongifolia* (from Mooney *et al.* 1976).

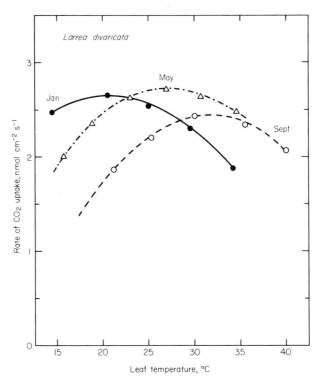

FIG. 16.5.   Seasonal change in the temperature dependence of photosynthesis by leaves of *Larrea divaricata* in Death Valley, California. Measurements were made *in situ* on watered plants under saturating light and normal air (from Mooney, Björkman & Collatz 1977).

A. hymenelytra and *L. divaricata* are incapable of attaining the high maximum photosynthetic rates of *T. oblongifolia* (Pearcy *et al.* 1974; Mooney, Björkman & Collatz 1977) but unlike this species they are able to sustain a moderately high photosynthetic activity throughout the year (Fig. 16.4). Moreover, as shown in Fig. 16.5, *L. divaricata* exhibited a seasonal change in the temperature dependence of photosynthesis in concert with the prevailing ambient temperature regime. Similar seasonal adjustments in the field have also been found to occur with *A. lentiformis*, a C_4 evergreen from Death Valley and the hot desert at Thermal, California (Pearcy & Harrison 1974).

Acclimation to high temperature

The response of photosynthesis to temperature when the growth temperature is changed shows considerable and important differences among species. This is illustrated in Fig. 16.6 which summarizes the temperature dependence of photosynthesis of the species studied when grown under contrasting

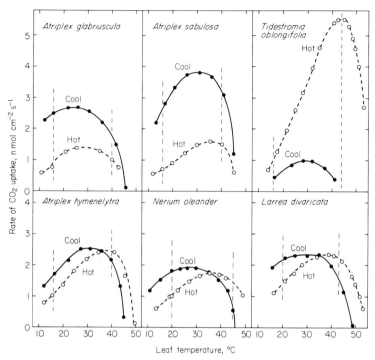

FIG. 16.6. Effect of the temperature during growth on the rate and temperature dependence of light-saturated photosynthesis for a number of species native to habitats with contrasting thermal regimes. The vertical broken lines indicate the daytime temperatures of the 'cool' and 'hot' growth regimes for each species. (Data from Björkman, Mooney & Ehleringer 1975; Mooney, Björkman & Collatz 1977; Björkman, Badger & Armond 1978; O. Björkman, unpublished.)

temperature regimes in controlled conditions. In addition to the species discussed above, Fig. 16.6 also includes the response of *Nerium oleander* L., an evergreen C_3 shrub native to the desert areas of Southwest Asia and North Africa and widely planted in many parts of the world, especially in California where it is commonly found in climatic zones ranging from mild coastal to hot desert areas. When supplied with water it is highly successful on the floor of Death Valley and, like the native *L. divaricata*, it remains active throughout the year. *N. oleander* has been widely used in our investigations primarily because of its unusual suitability for study with a wide range of experimental techniques, especially biochemical ones.

As shown in Fig. 16.6, growth temperature had profound effects on the photosynthetic characteristics of all the species. However, the mode of the response obtained was strongly species dependent. When *A. glabriuscula* was grown under a cool regime, simulating its native habitat, the photosynthetic capacity measured at the daytime growth temperature of 16 °C was quite high and reached about 95% of the maximum value which occurred at about 25 °C. Growing this species under a hot regime, in this case 40 °C day/25 °C night, resulted in a strong reduction in photosynthetic activity at all measurement temperatures. Although an upward shift in the temperature optimum of photosynthesis occurred when these plants were grown under a hot regime, there were considerable counteracting reductions in photosynthetic capacity, resulting in an inferior rather than a superior performance at high temperatures, in comparison with plants grown in the cool regime.

The coastal C_4 species *A. sabulosa* was capable of at least as high photosynthetic rates at low temperatures as the C_3 species *A. glabriuscula*, showing that C_4 plants need not be inferior to C_3 plants in cool climates, a conclusion also supported by the growth analysis experiments discussed earlier. However, since the presence of C_4 photosynthesis permits *A. sabulosa* to operate at or near CO_2 saturation even in normal air, maximum photosynthetic rates are reached at higher temperatures than in the C_3 species in which photosynthesis becomes increasingly CO_2 limited as temperature is increased. The role of C_4 photosynthesis in the temperature dependence of photosynthesis is further discussed in the following section.

As with *A. glabriuscula*, growth of *A. sabulosa* in a hot regime caused a severe reduction in photosynthetic rate at all measurement temperatures (Fig. 16.6). Thus, neither of these two cool-coastal species are capable of acclimating to high temperatures. In contrast to these coastal species, the summer-active, desert, C_4 species *T. oblongifolia* was capable of extremely high photosynthetic rates at high measurement temperatures but was a very poor performer at low temperatures (Fig. 16.6). Although this species is capable of some degree of photosynthetic acclimation when grown at temperatures in the 30 to 50 °C range (data not shown), its acclimation potential is very limited and it completely lacks the ability to acclimate to temperatures below 20 °C. When this species is maintained under a cool growth regime subsequent photosynthesis, measured at any temperature, is drastically inhibited.

It thus appears that in plants from environments which are subject to only a moderate temperature variation during the period of active growth, the potential for photosynthetic temperature acclimation is quite limited, regardless of whether the prevailing temperatures are low or high. A considerably greater acclimation potential is found in the evergreen desert species which experience large seasonal changes in temperature regime. Excellent examples of this type of response are the C_3 species *L. divaricata* (Mooney, Björkman & Collatz 1977), *N. oleander* (Björkman, Badger & Armond 1978; Björkman & Badger 1979), and *Heliotropum curassavicum* L., the last-named a herbaceous perennial ranging in distribution from hot deserts to mild coastal regimes (Mooney 1979). Desert forms of C_4 species such as *A. lentiformis* (Pearcy 1977) and to a lesser degree *A. hymenelytra*

(Björkman, Mooney & Ehleringer 1975) likewise show a wide photosynthetic acclimation potential. Interestingly, coastal forms of *A. lentiformis* have a very limited acclimation potential indicating the presence of ecotypic differentiation in photosynthetic characteristics in this species (Pearcy 1976).

The effect of growth temperature regime on the subsequent temperature dependence of some of these species is illustrated in the lower part of Fig. 16.6. In all cases growth under the cool regime resulted in a substantially increased photosynthetic capacity at low measurement temperatures in comparison with the same genotype grown under the hot regime. Conversely, growth under the hot regime resulted in a superior performance at high temperatures. The rate of photosynthesis at the respective optimum temperatures remained essentially unaffected. It is noteworthy, however, that those species which have a wide acclimation potential do not reach the photosynthetic rates achieved by the coastal *Atriplex* species at low temperatures nor those achieved by *T. oblongifolia* at high temperatures.

The changes in temperature dependence of photosynthesis effected by growth under different controlled temperature regimes in *L. divaricata*, *A. lentiformis*, and *A. hymenelytra*, closely resemble those observed at different seasons in the field. It is thus clear that growth temperature alone is sufficient to bring about the seasonal temperature acclimation.

For the laboratory studies reported above the experimental plants had been continuously grown under the different temperature regimes for several

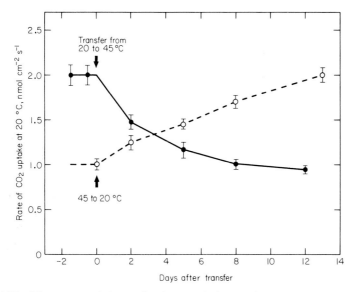

FIG. 16.7. Time course of changes in photosynthetic capacity at 20 °C of *Nerium oleander* following transfer of plants grown at 45 °C and 20 °C to the opposite growth regime (from Björkman & Badger 1979).

months and the leaves used for the measurements had developed entirely under the designated regime. Leaf longevity in many evergreen species may be considerable (in *N. oleander* it usually exceeds an entire year), and the time scale over which developing and mature leaves are capable of acclimating to a *change* in growth temperature is therefore an important factor. Fig. 16.7 shows the time course of photosynthetic capacity measured at 20 °C of fully mature *N. oleander* leaves following transfer of plants, grown at either 45 °C or 20 °C, to the reciprocal regime (Björkman & Badger 1979). Surprisingly the photosynthetic capacity of mature leaves, developed at the hot regime, increased after transfer to the cool regime so that 12–14 days after transfer this capacity was indistinguishable from that of leaves that had developed and been continuously maintained under the cool regime. Conversely, within 12–14 days after transfer of plants grown under the cool regime, the photosynthetic capacity gradually declined to the level found in the plants continuously maintained at 45 °C. Similarly, the optimum temperature for photosynthesis shifted upward by 6–9 °C within 10 days or less after the plants were moved from the cool to the hot regime, while transfer in the opposite direction caused a gradual downward shift. These results demonstrate that at least in *N. oleander*, even in mature leaves photosynthesis has a high potential for temperature acclimation, permitting full adjustment to the new regime in spite of the fact that the plants were subjected to quite abrupt and drastic changes in temperature. Thus, the much more gradual seasonal changes occurring in the field should allow the leaves to acclimate completely to the thermal regime existing at any time of the year.

MECHANISMS OF PHOTOSYNTHETIC ADAPTATION AND ACCLIMATION TO TEMPERATURE

With the exception of early photochemical events which are intrinsically temperature-independent, the rate of every step of the photosynthetic process can be expected to be directly, immediately, and reversibly affected by temperature. This includes processes as diverse as photosynthetic electron transport, photophosphorylation and CO_2 fixation as well as stomatal control and other aspects of diffusive transport of CO_2 into the leaf. In addition, the temperature range within which the integrity of the photosynthetic system can be maintained is limited, and exposure to too low or too high temperatures may cause inactivation and even irreversible damage to various parts of the system. It is clear therefore that the temperature response of overall photosynthesis is the result of diverse and complex effects on various parts of the photosynthetic process.

   Let us first consider the temperature range within which the functional integrity of the photosynthetic system is maintained. In the presence of normal atmospheric CO_2 and O_2 the intercellular CO_2 concentration exerts a strong limiting influence on photosynthesis in C_3 plants. It is a widely held view that the primary cause of this is the relatively low affinity for CO_2 of the carboxylation enzyme, ribulose-1,5-bis-phosphate (RuP_2) carboxylase, together with the fact that this same enzyme also catalyses the reaction of the CO_2 acceptor, RuP_2, with O_2. When RuP_2 reacts with CO_2 two molecules of 3-phosphoglyceric acid (PGA) are formed, but when RuP_2 reacts with O_2 it is split into one molecule of PGA and one molecule of phosphoglycolic acid with no carbon gain. Phosphoglycolic acid is metabolized in the photo-respiratory pathway and one-third of the carbon contained in this compound is released as CO_2 in the process, thus counteracting photosynthetic CO_2 uptake and causing a reduction in net photosynthetic rate. In this view the oxygenase activity of RuP_2-carboxylase causes competition between CO_2 and O_2 for RuP_2 and because of the very high ratio of O_2/CO_2 in the atmosphere a substantial fraction of the RuP_2 pool reacts with O_2 rather than CO_2, resulting in a reduced photosynthetic rate. Increasing the CO_2 concentration to several times the normal atmospheric value increases the net photosynthetic rate for two reasons: it enables the carboxylase to operate closer to maximum activity and at the same time suppresses the oxygenase reaction and sub-sequent photorespiratory CO_2 release.

This has important implications bearing upon the temperature dependence of photosynthesis. Because the affinity for CO_2 of RuP_2 carboxylase declines sharply with increasing temperature, CO_2 becomes progressively rate-limiting as leaf temperature is increased; moreover, because the decrease in the affinity for O_2 with temperature is smaller than that for CO_2 the fraction of RuP_2 reacting with O_2 rather than CO_2 increases as the temperature is raised. Hence, the limitation of photosynthesis in C_3 plants by lack of CO_2 is much more pronounced at high temperatures than at low temperatures. This is illustrated in Fig. 16.8, which shows the temperature dependence of photosynthesis of low-temperature-grown *A. glabriuscula* and high-temperature-grown *L. divaricata* at normal and at saturating CO_2 levels. Similarly Fig. 16.10 shows the effect of CO_2 on the temperature dependence of photosynthesis for *N. oleander* and *L. divaricata*, grown at high and low temperatures. In each of these C_3 species the shape of the temperature dependence curves is strongly influenced by CO_2 concentration. Rates of photosynthesis at high temperatures are much more affected by CO_2 than those at low temperature resulting in an upward shift of the temperature optimum when the CO_2 concentration is increased. It follows that stomatal conductance, because of its influence on the rate of diffusive transport of CO_2 into the leaf and hence on the CO_2 concentration inside the leaf would be expected to affect the temperature response curve of photosynthesis. I shall briefly return to this aspect later.

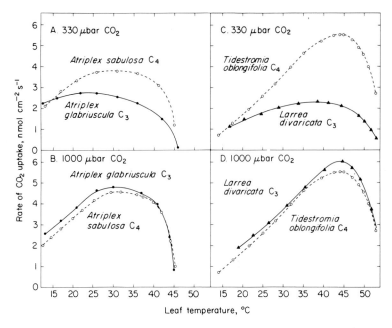

FIG. 16.8. Temperature dependence of photosynthesis of cold-adapted and heat-adapted C_3 and C_4 species at normal (330 μbar) CO_2 and high (1000 μbar) CO_2 partial pressures. The two *Atriplex* species were grown at 16 °C, the other two species were grown at 45 °C daytime temperature. (Based on data of Björkman, Mooney & Ehleringer 1975; Mooney, Björkman & Collatz 1977 and O. Björkman, unpublished.)

Role of C_4 photosynthesis

It is now widely agreed that the main function of the C_4 pathway of CO_2 fixation is to serve as a CO_2 concentrating mechanism for the Calvin cycle. Atmospheric CO_2 is initially fixed by phosphoenolpyruvate carboxylase, which has a high affinity for CO_2 and is unaffected by high O_2, to form C_4 dicarboxylic acids, in the mesophyll cells. These C_4 products are subsequently transported to the bundle sheath cells which contain RuP_2 carboxylase and the other enzymes of the Calvin cycle. The C_4 dicarboxylic acids are decarboxylated in these cells, yielding a high local concentration of CO_2 which is refixed by the Calvin cycle in a manner similar to that occurring in C_3 plants. In this way, RuP_2 carboxylase operates closer to its maximum rate and the reaction of RuP_2 with O_2 is suppressed. In other words, a C_4 plant in normal air may be expected to respond like a C_3 plant would respond under elevated ambient CO_2 levels.

This is shown in Fig. 16.8, which compares the photosynthetic temperature dependence of low-temperature-grown *A. sabulosa* (C_4) and *A. glabriuscula*

(C_3) as well as of high-temperature-grown *T. oblongifolia* (C_4) and *L. divaricata* (C_3), at normal and at high CO_2. In contrast to the response of the two C_3 species, CO_2 enrichment has only a small effect on the temperature dependence of the two C_4 species, indicating that for the latter plants, normal atmospheric CO_2 concentration is close to being saturating at all temperatures. At elevated CO_2, the coastal C_3 and C_4 species have almost identical temperature response curves (Fig. 16.8B) and so do the desert C_3 and C_4 species (Fig. 16.8D). However, the difference in the photosynthetic temperature dependence curves between the coastal and desert species remains as striking when compared at high or at normal CO_2 levels.

Influence of stomatal factors

As was mentioned earlier, because of its effect on intercellular CO_2 concentration stomatal conductance may have an important influence on the photosynthetic response curve under conditions where CO_2 concentration is less than saturating. One would predict that in an atmosphere of normal CO_2 and O_2 content, a low stomatal conductance would depress the photosynthetic rate to a larger extent at high than at low temperature, especially in C_3 plants which in normal air are almost always limited by CO_2. This flattening effect on the temperature dependence curve would of course be especially striking if the stomata tend to close with increasing temperature, but low stomatal conductance would have a pronounced flattening influence even if the conductance remained constant when the temperature was increased.

Although these considerations fully apply to the photosynthetic behaviour of all of the species shown in Figs. 16.3 to 16.8, simultaneous measurements of photosynthesis and stomatal conductance show that the stomata cannot account for the differences in photosynthetic capacity between different species, between low- and high-temperature-grown plants of the same genotype, or in the shapes of their temperature dependence curves. Evidence for this conclusion is presented in Tables 16.2 and 16.3 and Fig. 16.9. In the comparison between the contrasting C_4 species shown in Table 16.2, it is evident that there tends to be a correlation between photosynthetic rate and stomatal conductance. However, it is also evident that differences in photosynthetic capacity are not caused by differences in intercellular CO_2 concentration. For example, at low temperature the 6- to 7-fold superiority of *A. sabulosa* over *T. oblongifolia* occurs in spite of the fact that the intercellular CO_2 partial pressure in *A. sabulosa* is less than half of the value for *T. oblongifolia*. Conversely, at high temperature, *T. oblongifolia* has a more than 3-fold higher photosynthetic rate at only half the intercellular CO_2 partial pressure of *A. sabulosa*. Similarly, the photosynthetic acclimation to growth temperature regime which occurs in the C_3 species *L. divaricata* and *N. oleander* is unrelated to stomatal conductance (Table 16.3). In both of these

TABLE 16.2. Photosynthetic rate (F), stomatal conductance to water vapour transfer (g_s) and intercellular CO_2 partial pressure (C_i) in three contrasting C_4 species, grown at a 16/11 °C and a 40/30 °C regime. (Data from Björkman, Mooney & Ehleringer 1975; O. Björkman, unpublished).

Species	Grown and measured at 16 °C			Grown and measured at 40 °C		
	F $nmol\,cm^{-2}\,s^{-1}$	g_s $cm\,s^{-1}$	C_i μbar	F $nmol\,cm^{-2}\,s^{-1}$	g_s $cm\,s^{-1}$	C_i μbar
Atriplex sabulosa	2·80	0·51	117	1·54	0·36	161
Atriplex hymenelytra	1·62	0·31	127	2·41	0·41	98
Tidestromia oblongifolia	0·42	0·29	271	5·00	0·79	85

TABLE 16.3. Rate of photosynthesis, stomatal conductance to water vapour transfer and intercellular CO_2 partial pressure at two leaf temperatures on leaves of *Larrea divaricata* and *Nerium oleander*, grown under a 20/15 °C and a 45/31 °C regime (data from Mooney, Björkman & Collatz 1977; Björkman, Badger & Armond 1978; O. Björkman, unpublished).

	Larrea divaricata				*Nerium oleander*			
Daytime growth temperature, °C	20		45		20		45	
Measurement temperature, °C	20	45	20	45	20	45	20	45
Photosynthesis, $nmol\,cm^{-2}\,s^{-1}$	2·10	1·10	1·00	2·00	2·02	0·80	1·04	1·40
Stomatal conductance, $cm\,s^{-1}$	1·10	0·95	0·65	0·95	0·60	0·70	0·80	0·80
Intercellular CO_2, μbar	252	281	266	245	201	286	280	263

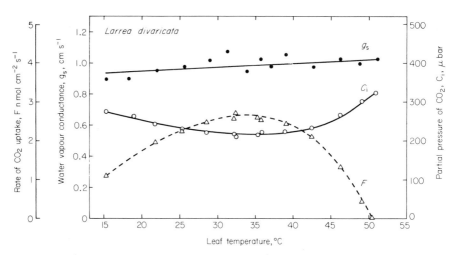

FIG. 16.9. Temperature dependence of stomatal conductance, photosynthesis and intercellular CO_2 partial pressure in *Larrea divaricata* leaves. Measurements were conducted in normal air (325 μbar CO_2) and saturating light. The plant was grown at a daytime temperature of 30 °C (from Björkman, Mooney & Collatz 1977).

species plants grown at $20\,°C$ have twice as high photosynthetic capacity at $20\,°C$ as the plants grown at $45\,°C$ but the intercellular CO_2 concentrations are lower in the plants grown at $20\,°C$. Conversely, at $45\,°C$ the plants grown at $45\,°C$ have about twice the photosynthetic rate of the plants grown at $20\,°C$ but they do not have a higher intercellular CO_2 concentration. Finally, as shown in Fig. 16.9, the temperature dependence curve for photosynthesis of *L. divaricata* merely mirrors that for intercellular CO_2 concentration. The decline in photosynthetic rate at either side of the optimum occurs in spite of an increase in intercellular CO_2 concentration.

Adaptation and acclimation to low temperature

The results presented above demonstrate that cool-adapted species such as the C_4 species *A. sabulosa* have considerably higher capacities for photosynthesis than heat-adapted species such as C_4 *T. oblongifolia*. A similar difference is also found between cool-acclimated and heat-acclimated plants of the same genotype in species which possess a wide acclimation potential, such as the C_3 species, *L. divaricata* and *N. oleander*. As shown in Table 16.4, this superiority of cold-adapted plants is as least as pronounced at saturating CO_2 as it is in normal air, thus providing further support for the conclusion that the differences are not attributable to factors involved in the diffusive transport of CO_2 but reflect intrinsic differences in photosynthetic capacity. One would expect that an increased photosynthetic capacity could be achieved simply by increase in the amount of photosynthetic systems per unit leaf area. However, determinations of the amount of leaf dry matter, chlorophyll and total leaf protein per unit leaf area (Table 16.4) show that such changes cannot account for the differences in photosynthetic capacity. This indicates that both adaptation and acclimation to low temperature must involve increases in particular rate-limiting catalysts.

In a comparison between the C_4 species *A. sabulosa* and *T. oblongifolia*, Björkman and Badger (1977) determined the activities of 14 enzymes involved in photosynthesis and carbon metabolism. Of these enzymes, only RuP_2 carboxylase showed differences between the two species that were similar in extent to the difference in photosynthetic capacity (Table 16.4). The increased activity of this enzyme was related to corresponding increases in the amount of Fraction-I protein and no qualitative differences in the catalytic properties of the enzyme isolated from the two species could be detected. The low content of RuP_2 carboxylase protein in *T. oblongifolia* accounted for the small amount of soluble leaf protein and to some extent also for the small amount of total leaf protein. Calculations show that the activity of RuP_2 carboxylase is just sufficient to support the measured rates of photosynthetic CO_2 fixation in both species. This lends further support to the hypothesis

Table 16.4. Photosynthetic capacity at 20 °C, specific leaf weight, chlorophyll and protein contents of cold-acclimated and heat-acclimated plants of cold- and heat-adapted species. (Data from Björkman, Mooney & Ehleringer 1975; Björkman & Badger 1977; Mooney, Björkman, & Collatz 1977; Björkman, Badger & Armond 1978; O. Björkman, unpublished.)

	Atriplex sabulosa cold-adapted	Tidestromia oblongifolia heat-adapted	Nerium oleander cold-acclimated	Nerium oleander heat-acclimated	Larrea divaricata cold-acclimated	Larrea divaricata heat-acclimated
Growth temperature, °C	21/15	40/32	20/15	45/32	20/15	44/32
Photosynthetic capacity at 20 °C $nmol\ cm^{-2}\ s^{-1}$						
Normal air	3·20	1·59	2·02	1·04	2·10	1·30
Saturating CO_2	3·30	1·59	3·02	1·17	3·50	2·01
Specific leaf dry weight, $mg\ cm^{-2}$	6·76	6·36	12·60	15·4	6·54	7·94
Chlorophyll content, $\mu g\ cm^{-2}$	50	45	76	65	40	45
Total protein content, $mg\ cm^{-2}$	1·69	1·18	1·41	1·65	1·50	1·38
Soluble protein content, $mg\ cm^{-2}$	0·79	0·41	0·67	0·64		
Fraction I protein, $mg\ cm^{-2}$	0·16	0·064				
RuP_2 carboxylase activity, $nmol\ cm^{-2}\ s^{-1}$	4·08	1·72	3·62	2·66		

that RuP_2 carboxylase level is an important rate-limiting factor at low temperature and may account for the difference in photosynthetic capacity between these two C_4 species. Similar conclusions may be drawn from Pearcy's (1977) comparisons of leaves of *A. lentiformis*, acclimated to two contrasting temperature regimes.

However, differences in RuP_2 carboxylase level do not explain the difference in photosynthetic capacity between *N. oleander* plants grown at 20 °C and 45 °C. Calculations based on the CO_2-saturated activity, the affinity for CO_2 and the temperature dependence of these kinetic properties of *N. oleander* RuP_2 carboxylase, suggest that this enzyme may well be limiting photosynthesis in the 20 °C-grown plants both at normal and saturating CO_2, especially at temperatures below 30 °C. Nevertheless, these calculations also indicate that it is very unlikely that RuP_2 carboxylase limits photosynthesis in the 45 °C-grown plants in the 10–30 °C range. In addition, although these plants have a somewhat lower RuP_2 carboxylase level than the 20 °C-grown plants (Table 16.4), this difference is small in comparison with the difference in photosynthetic capacity. Extensive comparisons of the activities of photosynthetic electron transport, photophosphorylation and of ten different enzymes of photosynthetic carbon metabolism revealed no differences that could explain the difference in photosynthetic capacity between cool- and heat-adapted *N. oleander* leaves with one exception—$FruP_2$ phosphatase. This is another Calvin cycle enzyme which has been implicated in the literature potentially to limit photosynthetic CO_2 fixation under certain conditions. As shown in Table 16.5, the difference in $FruP_2$ phosphatase level between the 20 °C-grown and the 45 °C-grown *N. oleander* plants closely matched the difference in photosynthetic capacity. Moreover, the increase in photosynthetic capacity (measured at 20 °C) that took place following transfer of 45 °C-grown plants to the 20 °C growth regime (see Fig. 16.7) was at all times

TABLE 16.5. Activities of enzymes of photosynthetic carbon metabolism from leaves of *Nerium oleander* expressed per unit fresh weight before and after transfer of plants for 12–14 days to contrasting growth temperature regimes (data from Björkman & Badger 1979).

Daytime growth temperatures °C	Enzyme activity, nmol g^{-1} s^{-1}				
	$FruP_2$ phosphatase	RuP_2 carboxylase	3-PGA kinase	P-hexose isomerase	P-gluco mutase
45 → 20	1·50 → 3·12 (3.33)*	5·10 → 7·77 (8·38)	51·7 → 64·5 (63·9)	5·32 → 6·55 (5·85)	4·17 → 5·27 (4·43)
20 → 45	2·52 → 1·56 (1·47)	8·06 → 5·50 (3·47)	63·1 → 59·5 (43·3)	5·43 → 5·00 (4·45)	4·53 → 3·67 (3·00)

* Values in parentheses are for new leaves developed after transfer.

closely correlated with increases in the level of $FruP_2$ phosphatase. Conversely, the decline in photosynthetic capacity that occurred in the reciprocal transfer experiments was matched by a similar decline in $FruP_2$ activity (Table 16.5). The changes in RuP_2 carboxylase activity were less pronounced than the changes in photosynthetic capacity and $FruP_2$ phosphatase activity. The other enzymes examined, some of which are shown in Table 16.5, have very similar activities in leaves grown under the two temperature regimes and they show only small changes after transfer of the plants from one growth regime to the other.

The close relationship between the changes in photosynthetic capacity and in $FruP_2$ phosphatase activity in this C_3 species and the correlation between photosynthesis and RuP_2 carboxylase level in the C_4 species mentioned earlier, support the concept that these enzymes may be rate limiting to photosynthesis at low and moderate temperatures and that regulation of their synthesis may be a key factor in the acclimation and adaptation of photosynthesis to temperature.

HIGH-TEMPERATURE INACTIVATION

As the measurement temperature is increased, a point is eventually reached beyond which photosynthesis starts to decline steeply (see Figs 16.8 and 16.10). In the presence of saturating CO_2 (i.e. in the absence of diffusional limitations to CO_2 uptake and of a competitive interaction of O_2 with CO_2 in the carboxylation reaction and associated photorespiratory CO_2 release) the decline in photosynthesis at high temperatures is caused by an inactivation of processes at the chloroplast level. At a temperature only a few degrees beyond the optimum (indicated by arrows in Fig. 16.10), photosynthesis becomes time-dependent and the longer the leaf is kept at such a supra-optimal measurement temperature the larger is the decline in photosynthetic rate, indicating that the leaf is no longer able to maintain functional integrity of the photosynthetic system. This is further supported by the fact that the inactivation of photosynthesis is in part irreversible. (This does not mean that the system, given sufficient time at non-inhibiting temperatures, is not subject to repair; see Bauer 1978.)

The irreversible nature of the inactivation is illustrated in Fig. 16.11. In these experiments the photosynthetic rate was measured at a constant, high, but non-inhibitory, temperature. The intact, attached leaf was then subjected to the temperature indicated on the abscissa for 10 min. Finally, the leaf was returned to the original temperature and the rate was again measured as soon as a steady-state level had been reached. The data points depict this second rate expressed as a percentage of the original rate. It is apparent from Fig. 16.11 that an abrupt inactivation of the photosynthetic machinery

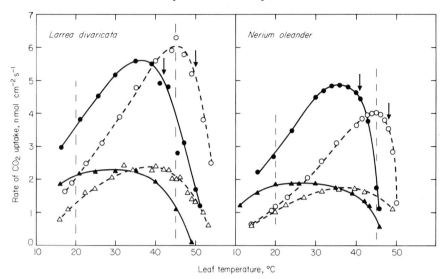

FIG. 16.10. Effect of CO_2 concentration on the temperature dependence of photosynthesis in leaves of *Larrea divaricata* and *Nerium oleander*, grown at 20 °C (filled symbols) and 45 °C (open symbols). Triangles and circles depict measurements at normal (330 μbar) and saturating (1000 μbar) partial pressures, CO_2, respectively. Arrows indicate the temperatures above which photosynthesis became irreversibly inactivated (from Mooney, Björkman & Collatz 1977; Björkman, Badger & Armond 1978).

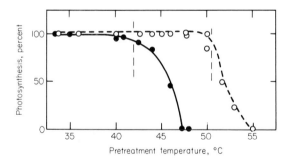

FIG. 16.11. Effect of previous leaf temperature on photosynthesis in *Atriplex sabulosa* (solid line) and *Tidestromia oblongifolia* (broken line) leaves. Photosynthesis measurements were made at a constant leaf temperature (see text) (from Björkman & Badger 1977).

occurs above 40 °C in the cold-adapted *A. sabulosa* and above 50 °C in the heat-adapted *T. oblongifolia*. As shown in Table 16.6, similar differences exist between *N. oleander* leaves that had developed and acclimated at low and high growth temperatures.

High-temperature inactivation of photosynthesis could be an indirect consequence of a heat-induced general disturbance of cellular integrity that would result if the semipermeability of the cell's partitioning membranes were lost. This would be likely to cause adverse changes in the immediate environment of the chloroplasts, and thus to inhibit photosynthesis. However, measurements of the temperature at which the leaf cells begin to leak ions (Table 16.6) show that for each species or growth regime, loss of semipermeability occurs at much higher temperatures than does the inactivation of photosynthesis. Also, no adverse effects on respiratory activity could be detected until temperatures were reached that caused complete inhibition of photosynthesis. This strongly suggests that heat inactivation of photosynthetic activity is primarily the result of a direct effect of temperature on the photosynthetic system itself, a conclusion also supported by the results of Krause & Santarius (1975). It also follows that the differences in photosynthetic performance at high temperatures between different species, and between individuals of the same species acclimated to contrasting temperature regimes, are largely attributable to intrinsic differences in the heat stability of one or several components of their photosynthetic systems.

Fig. 16.12 and Table 16.6 summarize some of the results of our experiments, designed to identify these components. It is evident that activities

TABLE 16.6. Comparison of the heat sensitivity *in vivo* of whole leaf photosynthesis (CO_2 and light saturated), respiration, cell membrane semipermeability, photosynthetic electron transport, photophosphorylation, F_0 fluorescence, soluble leaf protein, and selected enzymes of photosynthetic carbon metabolism in leaves of *Atriplex sabulosa*, *Tidestromia oblongifolia*, and *Nerium oleander* grown at high and low temperatures. The indicated temperatures depict the onset of irreversible inactivation or other damage. (Data from Björkman, Boynton & Berry 1976; Björkman & Badger 1977; Schreiber & Berry 1977; Björkman, Badger & Armond 1978; Raison & Berry 1979.)

	Atriplex sabulosa	*Tidestromia oblongifolia*	*Nerium oleander*	
Growth temperature, °C	21/15	40/32	20/15	45/32
Onset of damage temperature (°C) for:				
Photosynthesis	41	49	41	48
Dark respiration	49	54	—	—
Ion leakage	52	55	52	58
Loss of soluble leaf protein	42	50	45	49
PS1-driven electron transport	52	52	50	50
PS2-driven electron transport	41	49	44	49
Photophosphorylation	—	—	40	49
F_0-fluorescence rise	41	49	43	52
NADP-G3P dehydrogenase	42	50	41	49
Ru5P kinase	43	51	42	49
Adenylate kinase	47	49	47	49
$FruP_2$ phosphatase	—	54	53	54
RuP_2 carboxylase	49	56	49	56

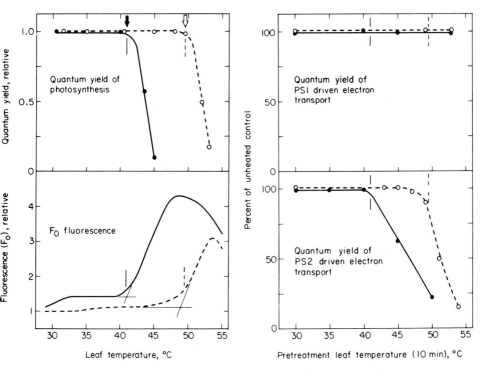

Fig. 16.12. Effect of leaf temperature on the quantum yield for photosynthesis by intact leaves of *Atriplex sabulosa* (solid line) and *Tidestromia oblongifolia* (broken line), and of pretreating illuminated leaves for 10 min at different temperatures on the quantum yields for photosynthetic electron transport by chloroplasts isolated from the leaves (from Björkman, Boynton & Berry 1976). Also shown is the effect of leaf temperature on the fluorescence yield of detached leaves (from Schreiber & Berry 1977).

residing in the thylakoid membranes of the chloroplasts are especially sensitive to heat inactivation. As shown in Fig. 16.12, upper left, the light-limited quantum yield of photosynthesis by *A. sabulosa* and *T. oblongifolia* leaves exhibits a steep decline at approximately the same temperatures as those which cause an irreversible inhibition of light-saturated photosynthesis (see Fig. 16.11), indicating that the generation of NADPH and /or ATP, driven by the photochemical reactions, is inactivated by high temperature. This is in agreement with the temperature dependence of F_0 fluorescence, emitted primarily by the chlorophyll associated with photosystem 2 (Fig. 16.12, lower left). At low and moderate temperatures most of the excitation light energy is used to drive electron transport and photophosphorylation, and the fluorescence intensity is low. The sudden increase in fluorescence that occurs at the same threshold temperatures as those which inhibit overall photosynthesis is interpreted as being caused by a reduction in the transfer of excitation energy to the photosystem 2 reaction centres. Further evidence for such impairment in the function of photosystem 2 is shown in Fig. 16.12, lower right. In this experiment the activity of photosystem 2-driven electron transport was determined on chloroplasts isolated from *A. sabulosa* and *T. oblongifolia* leaves that had been exposed for 10 min to the temperatures indicated on the abscissa. Photosystem 1 activity proved to possess a much higher heat stability in both species than photosystem 2 activity and photosynthesis by the intact leaf (Fig. 16.12, upper left). Similar results were also obtained with intact leaves and chloroplast preparations from *L. divaricata* (Armond, Schreiber & Björkman 1978) and from *N. oleander*, grown under 20 °C and 45 °C regimes (Table 16.6).

These and other results demonstrate that heat-induced damage to the photosynthetic system involves a functional dissociation of the light-harvesting pigment complex from the photosystem 2 reaction centres. Analyses of chloroplast membranes from *N. oleander* leaves by freeze-fracture electron microscopy further show that this heat activation involves a physical as well as a functional dissociation of these complexes in the thylakoid membrane (Armond, Björkman & Staehelin 1979). Adaptation and acclimation to high temperature obviously involves an increased heat stability of the thylakoid membranes. Although it is possible that the proteins of these membranes are themselves altered so that the stability of the pigment-protein complexes is enhanced, it now seems probable that the increased heat stability can be accounted for by changes in the lipid matrix in which the pigment-protein complexes are embedded. Using electron spin label techniques, Raison & Berry (1979) provided experimental evidence that the apparent viscosity of the thylakoid membrane lipids, isolated from *N. oleander* leaves, is increased when the plants are grown at high temperature, and they suggested that this increased viscosity strengthens the stability of the membrane by improving the balance of the hydrophobic and the hydrophilic interactions at high temperatures. Recent work (Berry *et al.*, unpublished) suggests that these changes in viscosity are caused by modifications in the composition of the chloroplast polar lipids.

While it seems evident that the difference in the heat stability of photosynthesis between plants adapted to low and to high temperatures can be explained by changes in the properties of the thylakoid membranes, it is interesting to note that these plants also differ in the heat stability *in vivo* of soluble enzymes considered to be located outside the thylakoid membrane, either in the stroma region of the chloroplast or in the cytoplasm. As shown in Table 16.6, a decline in the amount of soluble leaf protein, normally extractable with aqueous buffers, can be detected at temperatures that initiate inhibition of photosynthesis. This loss of protein presumably occurs through some form of denaturation and aggregation. Further studies (Björkman & Badger 1977) have shown that the soluble proteins being lost are in the less than 100 000 dalton category. The identity of this protein fraction has not yet been determined. The loss cannot be adequately explained by denaturation of any of the 12 soluble enzymes studied so far. These include the major soluble chloroplast enzymes and should account for at least 50 % of the total soluble leaf protein.

Fig. 16.13 shows the effect of treating illuminated leaves of *A. sabulosa* and *T. oblongifolia* for 10 min at different temperatures on the activities of a number of soluble enzymes of photosynthetic carbon metabolism, subsequently extracted from these leaves. The temperatures at which heat inactivation of some of these enzymes can be detected in 20 °C-grown and 45 °C-grown *N. oleander* leaves are listed in Table 16.6. It is clear that there are marked differences in the heat stabilities among the different enzymes. Enzymes such as RuP_2 carboxylase, $FruP_2$ phosphatase, phosphohexose isomerase, phosphoglucomutase and NAD malate dehydrogenase exhibit high heat stability in all cases, although for most of them, the heat stability is appreciably greater in the plants acclimated to high temperatures rather than low temperatures. We do not know whether these differences in heat stability *in situ* are caused by a change in the protein or to an alteration of the immediate environment in which the enzyme exists. In any event, the heat stabilities of these enzymes in each plant are too high to account for the heat inactivation of photosynthesis. Other enzymes such as NADP glyceraldehyde-3P dehydrogenase,

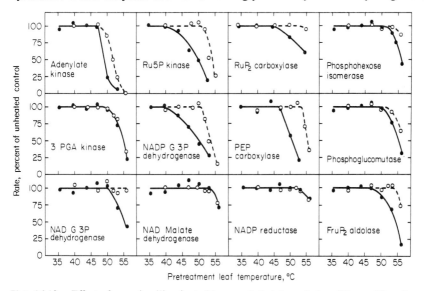

FIG. 16.13. Effect of exposing illuminated leaves of *Atriplex sabulosa* (●——●) and *Tidestromia oblongifolia* (○——○) to different temperatures for 10 min on the subsequent activities of a number of soluble enzymes of photosynthetic carbon metabolism, subsequently extracted from the leaves. Measurements of enzyme activities were made at 30 °C (from Björkman & Badger 1977).

NADP malate dehydrogenase, Ru5P kinase, and adenylate kinase show apparent heat stabilities which in most cases are similar to those of photosynthetic activity. However, with the exception of adenylate kinase these enzymes present a special case since they are known to require light for their activation. There is evidence that their activation state is dependent on reducing equivalents ultimately provided by the photochemical reactions (Wolosiuk & Buchanan 1977). Hence, the observed decline in the activity of these 'light-activated' enzymes at high temperature could well be a consequence of the heat inactivation of photosystem 2 activity discussed earlier but the possibility that this decline reflects a low heat stability of the enzymes themselves cannot be ruled out. In any event, these enzymes are not involved in the heat-induced loss of soluble leaf protein discussed earlier, for full activity can be restored by addition to the enzyme solutions of appropriate sulfhydryl-containing systems which are capable of activating these enzymes *in vitro*.

CONCLUSIONS

It is evident that the temperature response characteristics of photosynthesis play an important role in the ability of plants to grow and survive in habitats with contrasting thermal regimes. Plants exhibit large differences in the potential for photosynthetic acclimation to temperature in a manner apparently reflecting the temperature regimes of their native habitats. Plants from cool coastal environments, exhibiting only a moderate variation in temperature during the period of vegetative growth, such as *A. sabulosa* and *A. glabriuscula*, have high photosynthetic rates at low and moderate temperatures but their ability to acclimate to high temperatures is quite limited. Conversely, the desert species *T. oblongifolia* whose active growth period is restricted to the very hot summer months exhibits an unusually superior photosynthetic performance at high temperatures, but it lacks the ability to adjust its photosynthetic characteristics in a manner that would permit an adequate performance at low temperatures. Evergreen desert species, such as *A. lentiformis*, *L. divaricata* and *N. oleander*, which experience a wide range of seasonal variation in temperature regime, possess a much higher acclimation potential. Growing these plants under a cool regime results in a superior photosynthetic performance at low temperatures while growth under a warm regime improves their performance at high temperatures.

Photosynthetic adaptation and acclimation to low and to high temperature may be regarded as two, apparently separate, phenomena. The superior performance at low temperatures of cold-adapted plants is attributable to an increased photosynthetic capacity. This increased capacity is not related to changes in stomatal or other factors which influence the diffusive transport of CO_2 into the leaf, nor can it be accounted for by a general increase in the amount of photosynthetic systems per leaf area. The close correlation found to exist between photosynthetic capacity and the amount and activity of RuP_2 carboxylase in the C_4 species examined, and between photosynthetic capacity and $FruP_2$ phosphatase activity in the C_3 species *N. oleander*, indicate that changes in the level of these two enzymes may be a key factor in photosynthetic adaptation and acclimation to low temperature.

Photosynthesis is one of the most heat-sensitive aspects of growth. Excessively high temperatures cause an irreversible inactivation of photosynthesis. Several lines of experimental evidence show that this heat damage involves both a functional and a physical dissociation of the chlorophyll–protein complexes in the thylakoid membranes of the chloroplast, resulting in an inactivation of photosystem 2 activity and hence also of overall photosynthesis. In addition to the damage caused to these membrane-bound processes, high temperature also adversely affects some of the soluble enzymes located outside the thylakoid membranes, presumably by some form of denaturation of aggregation of the native protein.

Adaptation and acclimation to high temperature involve an enhanced heat stability of the photosynthetic apparatus. Experiments show that one of the key underlying factors is an increase in the apparent viscosity of the thylakoid membrane lipids in which the chlorophyll–protein complexes are embedded. The heat stability of certain chloroplast proteins may also be enhanced.

C_4 plants are not necessarily inferior to C_3 plants at low temperatures nor do they possess greater tolerance to heat damage than C_3 plants. However, the presence of the C_4 pathway greatly improves the photosynthetic performance in normal air at high (but non-damaging) temperatures. In C_3 plants the rate at which the carboxylation reaction can proceed is limited by CO_2 at normal levels of CO_2, and this CO_2 limitation increases with increasing temperature. Because of its function as a metabolic CO_2 concentrating mechanism, the C_4 pathway allows the Calvin cycle to operate near saturating CO_2 levels over a wide temperature range. C_4 photosynthesis together with high heat stability of the photosynthetic system should therefore result in an especially superior performance at high temperatures, as is indeed found in the thermophilic C_4 species, *T. oblongifolia*.

REFERENCES

Armond P.A., Björkman O. & Staehelin L.A. (1979) Dissociation of supramolecular complexes in chloroplast membranes: a manifestation of heat damage to the photosynthetic apparatus. *Carnegie Institution Year Book*, **78**, 153–157.

Armond P.A., Schreiber U. & Björkman O. (1978) Photosynthetic acclimation to temperature in the desert shrub, *Larrea divaricata*. II. Light-harvesting efficiency and electron transport. *Plant Physiology*, **61**, 411–415.

Bauer H. (1978) Photosynthesis of ivy leaves (*Hedera helix*) after heat stress. *Physiologia Plantarum*, **44**, 400–406.

Björkman O. (1975) Thermal stability of the photosynthetic apparatus in intact leaves. *Carnegie Institution Year Book*, **74**, 748–751.

Björkman, O. & Badger M. (1977) Thermal stability of photosynthetic enzymes in heat- and cool-adapted C₄ species. *Carnegie Institution Year Book*, **76**, 346–354.

Björkman O. & Badger M. (1979) Time course of thermal acclimation of the photosynthetic apparatus in *Nerium oleander*. *Carnegie Institution Year Book*, **78**, 145–148.

Björkman O., Badger M. & Armond P.A. (1978) Thermal acclimation of photosynthesis: effect of growth temperature on photosynthetic characteristics and components of the photosynthetic apparatus in *Nerium oleander*. *Carnegie Institution Year Book*, **77**, 262–282.

Björkman O., Boynton J. & Berry J. (1976) Comparison of the heat stability of photosynthesis, chloroplast membrane reactions, photosynthetic enzymes, and soluble protein in leaves of heat-adapted and cool-adapted C₄ species. *Carnegie Institution Year Book*, **75**, 400–407.

Björkman O., Mahall B., Nobs M.A., Ward W., Nicholson F. & Mooney H.A. (1974b) Growth responses of plants from contrasting thermal environments. An analysis of the temperature dependence of growth under controlled conditions. *Carnegie Institution Year Book*, **73**, 757–767.

Björkman O., Mooney H.A. & Ehleringer J. (1975) Photosynthetic responses of plants from habitats with contrasting thermal environments: comparison of photosynthetic characteristics of intact plants. *Carnegie Institution Year Book*, **74**, 743–748.

Björkman O., Nobs M., Mooney H.A., Troughton J., Berry J., Nicholson F. & Ward W.L. (1974a) Growth responses of plants from habitats with contrasting thermal environments. Transplant studies in the Death Valley and the Bodega Head experimental gardens. *Carnegie Institution Year Book*, **73**, 748–757.

Björkman O., Pearcy R.W., Harrison A.T. & Mooney H.A. (1972) Photosynthetic adaptation to high temperatures: a field study in Death Valley, California. *Science*, **172**, 786–789.

Krause G.H. & Santarius K.A. (1975) Relative thermostability of the chloroplast envelope. *Planta*, **127**, 285–299.

Mooney H.A. (1979) Seasonality and gradients in study of stress adaptation. *Adaptations of Plants to Water and High Temperature Stress* (Ed. by P. Kramer and N. Turner). Wiley-Interscience, New York.

Mooney H.A., Björkman O. & Collatz G.J. (1977) Photosynthetic acclimation to temperature and water stress in the desert shrub, *Larrea divaricata*. *Carnegie Institution Year Book*, **76**, 328–335.

Mooney H.A., Björkman O., Ehleringer J. & Berry J.A. (1976) Photosynthetic capacity of *in situ* Death Valley plants. *Carnegie Institution Year Book*, **75**, 410–413.

Pearcy R.W. (1976) Temperature effects on growth and CO₂ exchange rates in coastal and desert races of *Atriplex lentiformis*. *Oecologia*, **26**, 245–255.

Pearcy R.W. (1977) Acclimation of photosynthetic and respiratory CO₂ exchange to growth temperatures in *Atriplex lentiformis* (Torr.) Wats. *Plant Physiology*, **59**, 795–799.

Pearcy R.W. & Harrison A.T. (1974) Comparative photosynthetic and respiratory gas exchange characteristics of *Atriplex lentiformis* (Torr.) Wats. in coastal and desert habitats. *Ecology*, **55**, 1104–1111.

Pearcy R.W., Harrison A.T., Mooney H.A. & Björkman O. (1974) Seasonal changes in net photosynthesis of *Atriplex hymenelytra* shrubs growing in Death Valley, California. *Oecologia*, **17**, 111–121.

Raison J. & Berry J.A. (1978). The physical properties of membrane lipids in relation to the adaptation of higher plants and algae to contrasting thermal regimes. *Carnegie Institution Year Book*, **77**, 276–282.

Raison J. & Berry J.A. (1979) Viscotropic denaturation of chloroplast membranes and acclimation to temperature by adjustment of lipid viscosity. *Carnegie Institution Year Book*, **78**, 149–152.

Schreiber U. & Berry J.A. (1977) Heat-induced changes of chlorophyll fluorescence in intact leaves correlated with damage of the photosynthetic apparatus. *Planta*, **136**, 233–238.

Wolosiuk R.A. & Buchanan B.B. (1977) Thioredoxin and gluthathione regulate photosynthesis in chloroplasts. *Nature*, **266**, 565–567.

17. THE EFFECT OF TEMPERATURE ON THE GROWTH AND PHOTOSYNTHESIS OF THE TEMPERATE C₄ GRASS *SPARTINA TOWNSENDII*

R. DUNN,[1] S.P. LONG[1] AND S.M. THOMAS[2]

[1]*Department of Biology, University of Essex, Colchester, CO4 3SQ, and*
[2]*Department of Botany, Rothamsted Experimental Station, Harpenden, AL5 2JQ*

SUMMARY

The responses of growth and photosynthesis to temperature in the cool temperate C_4 grass *Spartina townsendii* were measured in field and controlled environments, to enable a comparison with native C_3 grasses. At monthly intervals from March 1977 to April 1978 leaf photosynthetic rates were measured in an open gas-exchange system. Simultaneously, changes in above-ground biomass were determined by harvesting plants from quadrats and changes in below-ground biomass determined from soil cores. In addition leaf and stem growth were measured on plants in permanently positioned quadrats. Controlled temperature environments were used for a comparison of *S. townsendii* with the C_3 grass *Lolium perenne* using growth analysis procedures. This comparison was conducted at temperatures from 10–25 °C.

Peak standing crop was $1.2\ kg\ m^{-2}$ at Southport, Lancashire and $0.3\ kg\ m^{-2}$ at Manningtree, Essex. Below-ground biomass accounted for 75% of total plant dry weight at Manningtree. The increase in biomass and leaf growth were strongly correlated with increasing mean air temperature. The field data indicated a marked reduction in photosynthetic capacity during the months when the mean air temperature dropped below 10°C. Thus the reduction in CO_2 assimilation correlated well with the poor leaf growth and biomass production during this period. The controlled environment studies demonstrate that at the warmer temperatures mean Unit Leaf Rate ($U\bar{L}R$) of *S. townsendii* was higher than that of *L. perenne*. The lower $U\bar{L}R$ is more than compensated for by a higher leaf area in the C_3 species. However, at 10°C both the $U\bar{L}R$ and leaf area production of *S. townsendii* were lower than in *L. perenne*.

The results suggest that whilst *S. townsendii* can achieve a comparable biomass to native C_3 grasses in the field, this must be achieved through high photosynthetic efficiencies in the warmest months of the year compensating for inferior efficiencies during the colder months, particularly the early spring.

INTRODUCTION

Tropical and sub-tropical C_4 grasses are usually unable to survive, or are severely damaged by exposure to low day-time temperatures (Ludlow & Wilson 1971; Taylor & Rowley 1971). In contrast *Spartina townsendii* (*sensu lato*) and some dicotyledonous species of the genera *Atriplex* and *Salsola* survive in habitats where such low temperatures are common (Long, Incoll & Woolhouse 1975; Caldwell, Osmond & Nott 1977). Thus, inability to survive at low air temperatures is not an inherent feature of all C_4 plants. The features that contribute to the acclimatization of cool-tolerant C_4-species are not understood. Nor do we know how well such plants compare with native C_3-species in their growth, productivity and photosynthesis in the field. Controlled environment studies of plants grown in 16 °C/13 °C day/night temperatures suggested that *S. townsendii* can attain mean leaf photosynthetic rates at 5 °C and 10 °C equal to those of temperate C_3 grasses, whilst rates at 15 °C and above exceeded those of the C_3 grasses (Long & Woolhouse 1978a). However this superiority might not hold in the field situation. For example, Caldwell, Osmond & Nott (1977) demonstrated in cool-tolerant *Atriplex* species that photosynthetic rates at 10 °C and 4 °C were higher in plants that had been grown at 20 °C than in those grown at 10 °C. Furthermore leaf photosynthetic rates of field grown *S. townsendii* were significantly lower during the spring than during the summer even when the leaf temperatures at the times of measurement were the same (Long & Incoll 1979). Thus a reduction in photosynthetic capacity of leaves grown at low air temperatures may contribute to the acclimatization of these C_4 species. As leaf photosynthetic rate is one of the factors that determines the productivity of a stand of vegetation, this acclimatization may have a detrimental effect on productivity relative to C_3 grasses and may offset the higher rates of photosynthesis of the C_4 grass at higher temperatures.

This study aims to assess the productivity of *S. townsendii* under the temperature conditions of a cool temperate climate, and to compare it with a C_3 grass. We report the productivity, growth and photosynthetic CO_2 assimilation of *S. townsendii* grown under field conditions, and the growth of *S. townsendii* and *Lolium perenne* in a range of constant temperatures.

PHOTOSYNTHETIC CO_2 EXCHANGE

Plants of *S. townsendii* were collected from a field site on the Stour estuary, Suffolk, at monthly intervals, from March 1977 to April 1978. Rates of CO_2 assimilation under controlled light and temperature conditions were determined on individual attached leaves within 48 h of collection, using an open gas exchange system as described by Long & Woolhouse (1978b). Only leaves which showed no visible evidence of senescence and in which the ligule had emerged were used.

Under the same light and temperature conditions, rates of CO_2 assimilation per unit leaf area (F_c) were similar for plants collected from June to December, but markedly lower in plants collected from January to May. This

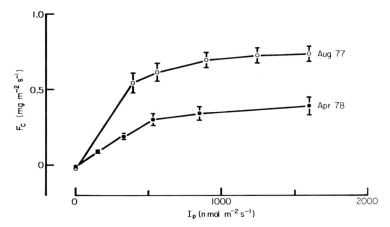

FIG. 17.1. The relationship between leaf photosynthetic rate, ± its standard error, of *S. townsendii*, (F_c) and photon flux density (I_p) for a random sample of leaves present in April 1978 (●) and August 1977 (○). Each point is the mean of 6 measurements.

difference is shown in a comparison of the mean response of F_c to light at 15 °C, for a random sample of leaves from plants collected in April and August (Fig. 17.1). F_c was lower at all photon flux densities in the April collection.

The lower rates of CO_2 assimilation in spring were not the result of any qualitative changes in the biochemistry of assimilatory carbon metabolism (Thomas & Long 1978). Furthermore, measurements of CO_2 compensation points in different months gave values of between 0–20 mg m^{-3} of CO_2 in air, even for leaves that had been exposed to several degrees of frost, as in February and March 1978. These values were typical of C_4 species. However, leaves formed from January to May showed strong red pigmentation; probably an accumulation of anthocyanin in the vacuoles of the epidermal cells. This must have reduced the amount of light reaching the underlying chlorenchyma and may account for the apparent decrease in the initial slope of the response of F_c to I_p (Fig. 17.1). Spring leaves may also have had a higher stomatal resistance, contributing to the reduction of F_c. Unfortunately this cannot be tested by water vapour diffusion porometry since *S. townsendii* plants growing in saline soils actively secrete an aqueous salt solution from salt glands on both leaf surfaces (Skelding & Winterbotham 1939).

FIELD STUDIES OF GROWTH AND DRY MATTER PRODUCTION

Field studies were conducted between March 1977 and April 1978, on a salt marsh on the Stour estuary. This contained more than 100 ha of a nearly pure and continuous stand of *S. townsendii*. Growth was assessed non-

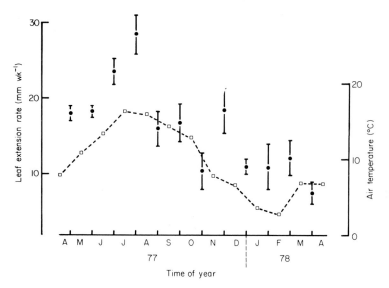

FIG. 17.2. Extension rate ± its standard error, of leaves of *S. townsendii* over a 12 month period. Mean leaf extension rate (●) has been calculated from monthly measurements of the lengths of the two uppermost leaves on all stems present in permanently positioned 0·01 m² quadrats at the field site. The mean air temperatures (□) for each month are from Meteorological Office records for Clacton, Essex.

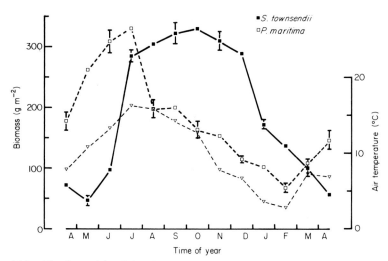

FIG. 17.3. The dry weight of the above-ground live material, ± its standard error, of two salt marsh grasses over a 12 month period. Dry weight of *S. townsendii* (■) at Manningtree, Essex, has been determined by harvesting 20 × 0·1 m² quadrats at monthly intervals. Data for *Puccinellia maritima* (□) at Colne Point, Essex, from Hussey (unpublished data). Mean air temperature (▽) from Meteorological Office records for Clacton, Essex.

destructively by measuring changes in stem length, leaf length and leaf height of plants in 20×0.01 m^2 permanently positioned quadrats. Aerial biomass was harvested from 20×0.1 m^2 quadrats at monthly intervals. In addition 10 soil cores of 0.0087 m^2 cross-sectional area were taken for the estimation of the weight of live roots and rhizomes. All quadrats were positioned according to a restricted randomized design.

Leaf elongation occurred throughout the year (Fig. 17.2), and was most rapid in June and July and slowest from January to March, corresponding to the highest and lowest mean monthly air temperatures, respectively. This contrasts with typical perennial grasses of cool temperate regions which show a burst of growth in April and May, as can be seen in the C$_3$ species *Puccinellia maritima* studied at Colne Point; a coastal marsh in north-east Essex (Fig. 17.3).

The biomass data (Fig. 17.3) shows that a significant increase in the live above-ground material of *S. townsendii* did not occur until June. This corresponded with the first month in which the mean air temperature exceeded 10 °C. Total above and below-ground live dry weight increased rapidly in July, reaching a peak of 1.2 kg m^{-2} in October. The below-ground fraction, i.e. roots and rhizomes, were a high proportion of the total, amounting to about 75 % (Table 17.1).

TABLE 17.1. The biomass of *S. townsendii* present in October 1978 at two sites. Dry weight in kg m^{-2}, $\pm$ standard error.

	Above-ground	Below-ground
Manningtree Essex	0.33 ± 0.03	0.90 ± 0.09
Southport, Lancs.	1.20 ± 0.11	0.94 ± 0.09

The maximum above-ground biomass of 0.3 kg m^{-2} is low by comparison to C$_3$ forage crops, where 1.0 kg m^{-2} can be expected from a well fertilized *L. perenne* hay crop whilst a total dry-matter yield of 2.9 kg m^{-2} has been obtained from *L. perenne* under an optimum cutting regime (Cooper 1970). However, production of *S. townsendii* on other marshes in Britain could be considerably higher. The above-ground biomass recorded on a marsh at Southport, Lancashire, was 1.2 kg m^{-2} (Table 17.1), and comparable in quantity to the yield of a good hay crop.

CONTROLLED ENVIRONMENT GROWTH STUDIES

To isolate the effects of temperature from other climatic variables the growth of *S. townsendii* at a range of day-time air temperatures from 10 °C to 25 °C was compared with the C_3 perennial grass *Lolium perenne* in constant environment rooms.

Seed of *S. townsendii* and *L. perenne* cv S24 were germinated at 20 °C. Seedlings were planted individually in 6-inch pots of potting compost (EFF Ltd.) and placed in four controlled environment rooms (Ford & Thorne 1975) at 20 °C. Temperatures were gradually adjusted over a 2-week period to the final growth temperatures to allow for temperature acclimatization. The day-time air temperatures in the four rooms were 25, 20, 15 and 10 °C and night-time air temperatures 3 °C less. Each room had a photoperiod of 14 hours, a photon flux density of 500 μmol m^{-2} s^{-1} at plant pot height and a water vapour pressure deficit of 0·5 kPa. Growth was measured by harvesting 6 plants of each species from each room using a randomized block design which incorporated a grading and pairing system (Ford & Thorne 1975). Harvests were carried out at weekly intervals at 10 °C and at 3–4 day intervals at the higher temperatures. From the harvests, mean growth analysis parameters were calculated (Evans 1972).

At all temperatures, plants of *L. perenne* had both a greater dry weight and leaf area than plants of *S. townsendii*. Table 17.2 compares the growth of seedlings for 36 days at 20 °C and at 10 °C. Although at the warmer temperature the leaf area of *S. townsendii* was lower than that of *L. perenne*, its mean rate of dry weight accumulation per unit leaf area ($U\bar{L}R$) was higher, suggesting that its lower dry weight was due to a lower rate of leaf growth. The amount of leaf area *S. townsendii* produced at 10 °C in the 36 days following the acclimatization period was markedly less than at 20 °C (Table 17.2). At 10 °C both $U\bar{L}R$ and leaf area of *S. townsendii* were lower than those of *L. perenne*. The $U\bar{L}R$ of *S. townsendii* declined steadily with time of exposure to 10 °C, from a value greater than that of *L. perenne* after 7 days to a much lower value after 63 days at 10 °C (Table 17.3). By this time, more than 60 % of the leaves on the *S. townsendii* plants had expanded at 10 °C.

The controlled environment studies suggest that prolonged exposure of *S. townsendii* to low temperature resulted in the production of physiologically

TABLE 17.2. Comparison of the growth of seedlings after 36 days in 20 °C and 10 °C.

Species	Temperature (°C)	Mean dry weight per pot (g)	Mean leaf area per pot (m²)	Mean leaf area ratio (m² kg⁻¹)	Mean unit leaf rate (g m⁻² d⁻¹)
L. perenne	10	4·0 ± 0·3	0·02 ± 0·002	5·0 ± 0·1	12·1 ± 2·0
	20	11·0 ± 1·2	0·09 ± 0·004	9·0 ± 0·4	12·8 ± 2·0
S. townsendii	10	0·5 ± 0·04	0·001 ± 0·0001	4·0 ± 0·2	4·1 ± 3·7
	20	2·4 ± 0·3	0·01 ± 0·002	6·1 ± 0·2	15·0 ± 3·8

Mean figures calculated by pairing plants between consecutive harvests, ± standard error.

TABLE 17.3. Mean unit leaf rate ($U\bar{L}R$), $\pm$ its standard error.

| Days of exposure to 10 °C | $U\bar{L}R$. (g m^{-2} day^{-1}) $\pm$ standard error | |
	S. townsendii	*L. perenne*
7	$11\cdot0 \pm 2\cdot0$	$9\cdot4 \pm 1\cdot3$
36	$4\cdot1 \pm 3\cdot7$	$12\cdot1 \pm 2\cdot0$
63	$2\cdot9 \pm 0\cdot3$	$15\cdot0 \pm 4\cdot7$

Mean figures calculated by pairing plants between consecutive harvests.

TABLE 17.4. Unit leaf rates $U\bar{L}R \pm$ its standard error, at 20 °C following low temperature exposure.

| | $U\bar{L}R$ (g m^{-2} day^{-1}) | |
	Control	Treatment
L. perenne	$12\cdot0 \pm 1\cdot4$	$12\cdot0 \pm 1\cdot7$
S. townsendii	$15\cdot0 \pm 2\cdot6$	$10\cdot6 \pm 3\cdot6$

Mean figures calculated by pairing plants from consecutive harvests, $n = 6$ for each treatment combination.

different leaves from those developed at higher temperatures. These leaves had lower mean unit leaf rates and expanded more slowly. They were comparable to the leaves developed in the field in early spring which had low rates of CO_2 assimilation and expansion. The controlled environment leaves developed at 10 °C were also deeply pigmented with anthocyanin as were the field grown winter and spring leaves.

To examine the effect of a relatively short exposure to low temperature on the growth of *S. townsendii* and *L. perenne*, plants of these species were grown in a controlled environment room at a day-time temperature of 20 °C for 25 days, then transferred to a similar room with a day-time temperature of 10 °C for 3 days, and finally returned to the higher temperature. Unit leaf rates of the plants, one week after the low temperature treatment are compared in Table 17.4 with a control batch of plants which remained at 20 °C throughout. The results suggest that although long exposure to low air temperatures result in reduced $U\bar{L}R$ in *S. townsendii*, short exposures did not reduce it significantly.

CONCLUSION

The field studies demonstrated that *S. townsendii* was capable of leaf extension growth and the maintenance of photosynthetically active leaves throughout the year, even during the coldest months. Peak biomass for *S. townsendii* at the Southport site was comparable to that for intensively managed C_3 grass crops, such as *Lolium perenne*. The controlled environment studies showed that *S. townsendii* had a higher $U\bar{L}R$ at 20 °C and 25 °C than *L. perenne*, which was offset by a lower leaf area. The lower leaf area of *S. townsendii* probably reflected its different growth form. *S. townsendii* has a deep anchor root system and underground storage rhizomes, which are not present in *L. perenne*. Therefore, a greater proportion of assimilate is likely to be invested in non-photosynthetic organs. Indeed, 75% of the biomass of *S. townsendii* at Manningtree was below-ground. At 10 °C and below, field and controlled environment studies of *S. townsendii* showed a marked reduction in photosynthetic rate and $U\bar{L}R$. The amount of reduction increased with duration of exposure to the lower temperature. Thus, whilst biomass accumulation in *S. townsendii* may approach that of native C_3 grasses, this could only be achieved through high photosynthetic efficiencies in the warmest months of the year compensating for the inferior efficiencies of this species during the colder months. This is reflected in the slow rise in *S. townsendii* biomass which peaks 2–3 months after native C_3 grasses.

ACKNOWLEDGEMENTS

We thank Drs A.J. Keys and G.N. Thorne for many helpful discussions relating to this work and for their critical comments of the draft manuscript. We also thank H. Gogay, R.W. Soffe and J.C. Turner for their technical assistance with the controlled environment studies and I. Meek for his assistance with the field studies. One of us (R.D.) was supported by an S.R.C.-C.A.S.E. Studentship during the course of this work.

REFERENCES

Caldwell M.M., Osmond C.B. & Nott D.L. (1977) C_4 pathway photosynthesis at low temperatures in cold-tolerant *Atriplex* species. *Plant Physiology*, **60**, 157–164.

Cooper J.P. (1970) Potential production and energy conversion in temperate and tropical grasses. *Herbage Abstracts*, **40**, 1–15.

Evans G.C. (1972) *The Quantitative Analysis of Plant Growth*. Vol. 1. Blackwell Scientific Publications, Oxford.

Ford M.A. & Thorne G.N. (1975) Effects of variation in temperature and light intensity at different times on growth and yield of spring wheat. *Annals of Applied Biology*, **80**, 283–291.

Long S.P., Incoll L.D. & Woolhouse H.W. (1975) C_4 photosynthesis in plants from cool temperature regions, with particular reference to *Spartina townsendii*. *Nature*, **257**, 622–624.

Long S.P. & Woolhouse H.W. (1978a) The response of net photosynthesis to vapour pressure deficit and CO_2 concentration in *Spartina townsendii* (*sensu lato*), a C_4 species from a cool temperate climate. *Journal of Experimental Botany*, **29**, 567–577.

Long S.P. & Woolhouse H.W. (1978b) The response of net photosynthesis to light and temperature in *Spartina townsendii* (*sensu lato*), a C_4 species from a cool temperate climate. *Journal of Experimental Botany*, **29**, 803–814.

Long S.P. & Incoll L.D. (1979) The prediction and measurement of photosynthetic rate of *Spartina townsendii* (*sensu lato*) in the field. *Journal of Applied Ecology*, **16**, 879–891.

Ludlow M.M. & Wilson G.L. (1971) Photosynthesis of tropical pasture plants. I. Illuminance, carbon dioxide concentration, leaf temperature, and leaf-air vapour pressure difference. *Australian Journal of Biological Sciences*, **24**, 449–470.

Skelding A.D. & Winterbotham J. (1939) The structure and development of the hydrathodes of *Spartina townsendii* Groves. *New Phytologist*, **38**, 69–79.

Taylor A.O. & Rowley J.A. (1971) Plants under climatic stress. I. Low temperature, high light effects on photosynthesis. *Plant Physiology*, **47**, 713–718.

Thomas S.M. & Long S.P. (1978) C_4 photosynthesis in *Spartina townsendii* (*sensu lato*). At high and low temperatures. *Planta*, **142**, 171–174.

18. PLANT GROWTH IN RESPONSE TO VARIATIONS IN TEMPERATURE: FIELD AND LABORATORY STUDIES

I. H. RORISON

Unit of Comparative Plant Ecology (NERC), Department of Botany, The University, Sheffield, S10 2TN

SUMMARY

Diurnal and seasonal fluctuations of temperature are usually greater in range and amplitude, and more variable, in the field than in laboratory experiments. The ways in which different temperature regimes affect germination, seedling establishment and phenological pattern are explored with reference to growth and nutrient uptake in the field and in controlled environments.

INTRODUCTION

This chapter is written from the viewpoint of an investigator with an interest in both chemical and physical aspects of the environment and in the function of the root as well as the leaf. It is broadly concerned with the ecological aspects of plant temperature relations, seeking to link the aerial with the soil environment and the shoot with the root. As a result, some of the complexities of both field and laboratory experiments and some current over-simplifications are exposed. The aim is to provide an ecological basis for the contributions from the crop physiologists, whose relatively simple systems allow more precise modelling and measurement of variables. Some of the techniques used in the analysis of plant performance in the natural environment are examined; some of the questions which arise from such investigations are posed; and some indication of how these have been tackled using controlled environments is given.

Basic response

Temperature is one of the easier factors of the environment to measure in the field and to control in the laboratory. The response of plants both inter- and intra-specifically follows a basic pattern (Fig. 4.1 in Sutcliffe 1977) but both the temperature range over which growth occurs and the optimum rate of growth achieved can vary widely (see Fig. 16.6). Also the optimum temperature for growth may vary between different organs on the same plant and even between two sides of the same organ. Roots may have a lower temperature optimum than shoots and tolerate a narrower range (Nielsen & Humphries 1966).

There are potentially large and variable diurnal and seasonal fluctuations of temperature in the field and something of their effect on the plant was indicated by Went's (1945) earlier work showing the benefit of lowered night temperatures on the growth of tomato plants. There is a gradient down through the canopy of vegetation and into the soil profile, with the soil being better buffered against fluctuation and with a greater capacity to store heat. There is an added complication in certain environments in that the aerial parts of plants achieve temperatures which may be either considerably above or below those of the surrounding air (Watts 1975; Regehr & Bazzaz 1976), depending on radiation load, life form, leafiness, rate of air movement and transpiration rate (Gates 1968).

In view of these differences it is surprising to an ecologist that in most published experimental work dealing with the effects of root and shoot temperature, it is the air temperature that has been kept constant and the root temperature that has been varied (Cooper 1973). There are some technical advantages (Davidson 1969) but such experiments bear little relation to reality.

Ecological implications

Since temperature can have a marked effect both directly and indirectly on plant growth, adaptation to a particular pattern may well play a vital part in the survival of plants in natural vegetation. The importance of temperature may be judged at different levels of organization as when vegetation is classified in relation to the major climatic zones of the world (Walter 1977), and the distribution of individuals and of populations is considered in relation to local microclimates. The ability to grow at low and/or high temperatures may extend the growing season of both annuals and perennials.

The problem arises of how to distinguish the climatic from the edaphic and biotic influences on plant growth and, among the climatic influences to distinguish between the effects of radiation, temperature and moisture.

Little of the landscape left to natural vegetation in Britain is flat, and since slope and aspect have large influences on conditions of temperature they can be of ecological importance. Valleys, hills and industrial tips often provide a range of slope and climate within an area accessible to a variety of plants and offer an outdoor laboratory in which to make preliminary measurements.

FIELD MEASUREMENTS IN NATIVE GRASSLANDS

Such measurements have been carried out for several years in Lathkilldale, Derbyshire (Nat. Grid. reference SK177654), the long axis of which runs roughly east-west. It is reasonably uniform topographically and has areas of ungrazed, semi-natural grassland on both north-* and south-slopes. These, together with the plateau grasslands, have been monitored for soil and air temperatures, soil moisture, and for nutrient availability. Solar radiation is measured only on the plateau (Rorison & Sutton 1976). Measurements have also been made of the germination, seedling growth and phenology of a number of species. From these a picture of seasonal change and of the range and degree of fluctuation of the climatic factors has emerged.

Two observations were of particular interest: the persistence of green leaves in many species throughout the winter and some 'unseasonal' flushes of nutrient availability during milder winter weather. These led to the hypothesis that some of the perennial herbaceous species might grow and take up nutrients when they were thought previously to be dormant; that they might, in other words, respond during winter, to some degree, like winter annuals and this was tested in controlled environment rooms. The potential survival value of this behaviour as a response to fluctuating temperature of the amplitude found under field conditions was also assessed.

Temperature measurements

Small matched silicon bead thermistors were connected to compact data loggers from four levels in the community: 2 and 4 cm below the soil surface, 2 cm above it, and from a moveable position just above canopy height. The 1 mm diameter sensors were unshielded, since preliminary tests had shown shielding to be unnecessary, and measurements at each level were replicated four times within the experimental area. The annual mean daily air temperatures vary barely 1 °C between north and south, i.e. the overall mean for both soil and air on the south-slope was 9·2 °C and for the north air was 8·2 °C and soil 8·1 °C. But the amplitudes of daily fluctuations may be considerable (cf. Figs 18.2 and 18.3). Contrary to expectation, the south-slope was found to be less favourable for plant growth than the north-slope, in both winter and summer.

In winter, freezing and thawing may increase mineralization in the soil but they also cause heaving and solifluction on the south-slope and subsequent disturbance to plants which on the north-slope are unmoved and are

---

* Used in this paper to mean a north-*facing* slope following the current German use of *Nordhang*.

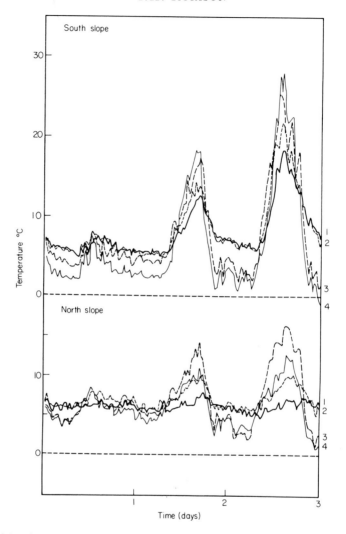

FIG. 18.1.   A computer trace showing temperatures on three consecutive days with increasing solar radiation on the north- and south-slopes of Lathkilldale (14–16 April 1976):
1. 4 cm below the soil surface
2. 2 cm below the soil surface
3. 2 cm above the soil surface
4. immediately above the plant canopy
   (a height of *c*. 15 cm on both slopes at this time)

more often insulated from extremely low temperatures by a layer of snow (Rorison & Sutton 1976, Fig. 15.5).

   In summer, the shallow soils of the south-slope quickly dry out and soil temperatures near the surface fluctuate widely, e.g. the temperature range in °C 2 cm above and below the soil surface on sunny days in June and July is:

	Below	Above
South	10–35	5–40
North	8–18	5–22

The vegetation of the south-slope tends to be more stunted, open and xeromorphic than that of the north-slope and a comparison of standing crop data (Sutton & Rorison 1970) confirms the low productivity under relatively dry mid-summer conditions. In spring and autumn, however, standing crop is larger on the south- than on the north-slopes (Sydes 1980).

In spring both air and soil temperatures increase more rapidly during the morning on the south-slope than on the north-slope (Fig. 18.1). Again there is a tendency for wide temperature fluctuations both in space and time, but, provided there is adequate soil moisture, the rise in temperature favours plant growth and germination on the south-slope at this time of year.

Plant response

Climatic change is reflected in several plant processes, such as germination and emergence, dry weight increase per plant (relative growth rate) or per unit area (phenology). Examples of each of these are considered below.

Germination

Measurements of seedling emergence and of climate following seedfall over a number of years can give useful indications of field conditions which are either suitable or unsuitable for the germination and survival of species.

In order to condense the time-scale of the operation and using species whose viability remains constant over the experimental period, a series of sowings can be made over a single season. For example in Lathkilldale surface sowings were made at 5-weekly intervals for one year with five replicates for each species at each sowing (1974–75). Results showed the response of each species to a number of combinations of soil temperature and moisture, many of which could occur at time of natural seedfall.

Responses of two species are considered here. They are *Arrhenatherum elatius* and *Centaurea nigra**, which, although showing no aspect preferences in their general distribution, occurred in Lathkilldale mainly on the south- and on the north-slopes respectively. Seed of both remains viable throughout the year and no chilling is required for germination.

The seed of *Arrhenatherum elatius* was shed locally during July and August

* Nomenclature follows Clapham, Tutin & Warburg (1962).

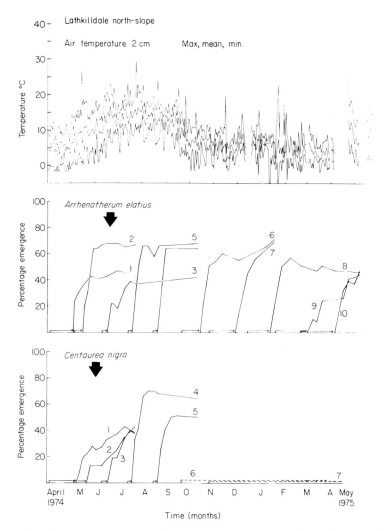

FIG. 18.2.   Seedling emergence in relation to air temperature on the north-slope of
Lathkilldale.

Upper: A computer trace of daily maximum, mean and minimum air temperatures at
2 cm above ground. Centre: *Arrhenatherum elatius*. Horizontal arrowed lines at the base
indicate the length of time between sowing and emergence. Numbered lines indicate the
percentage of seedlings surviving from the different sowings during the year. The solid
downward pointing arrow indicates a period of soil moisture stress of *c.* −0.2 MPa at
2 cm depth.

The time taken from sowing to emergence is significantly correlated ($P = 0.05$) with the
mean daily air temperature during the thirty days following sowing. Lower: *Centaurea
nigra*. (Details as for *Arrhenatherum elatius*.)

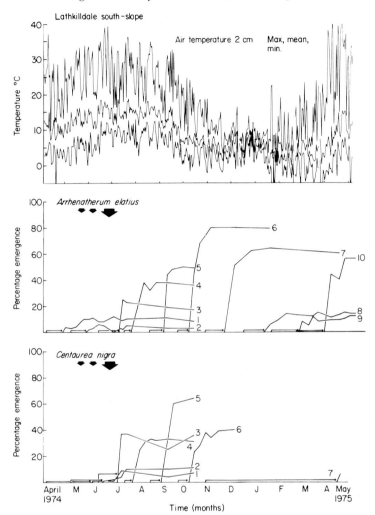

FIG. 18.3. Seedling emergence in relation to air temperature on the south-slope of Lathkilldale.

Upper: A computer trace of daily maximum, mean and minimum air temperatures at 2 cm above ground. Centre: *Arrhenatherum elatius*. Horizontal arrowed lines at the base indicate the length of time between sowing and emergence. Numbered lines indicate the percentage of seedlings surviving from the different sowings during the year. The three solid downward-pointing arrows indicate periods of soil moisture stress of *c.* −0.4 and −0.5 MPa respectively.

The time taken from sowing to emergence is significantly correlated ($P = 0.05$) with the mean daily air temperatures during the thirty days before sowing and with the daily mean and minimum soil temperatures both before sowing and during the time taken to emergence. Lower: *Centaurea nigra*. (Details as for *Arrhenatherum elatius* except that there is no correlation between emergence and air or soil temperatures.)

and that of *Centaurea nigra* during September and October. Both species were affected adversely by the dry conditions on the south-slope compared with the north-slope but only *A. elatius* germinated throughout the year (Figs 18.2 and 18.3). No *C. nigra* germinated between September and April, suggesting that a 'basic' mean soil temperature of *c.* 10°C is required for emergence (Figs 18.2 and 18.3).

In standardized tests on temperature gradient bars *A. elatius* germinated at constant temperatures between 0–34°C and most rapidly at *c.* 13°C (within 2·5 days) while *C. nigra* germinated between 9–38°C and most rapidly (2 days) at 21°C (Grime, Curtis, Neal & Rodman, personal communication). *A. elatius* germinated freely in the light and dark, whereas *C. nigra* was inhibited in the dark (Grime & Jarvis 1976).

Diurnal fluctuations of temperature are also of ecological importance in initiating or accelerating germination of certain species. The effectiveness of the stimulus varies according to the amplitude of fluctuation and the presence or absence of light. Such effects have been invoked by Thompson, Grime & Mason (1977) to explain why the buried seed of species which require fluctuating temperatures to germinate in the dark are activated by the creation of gaps in, and the dispersal of litter from, existing vegetation. Removal of such insulating materials allows an increase in the fluctuation of soil temperatures which would also be influenced by aspect.

Seedling growth

Once seedlings have emerged their rates of growth may be measured as they pass through the phase of establishment. A suitable experiment is easy to plan but not to execute. The vagaries of climate and predation can play havoc with replicated plots in the field, as occurred with the sequential sowings (p. 317).

To safeguard plants to a greater degree a possible alternative is a pot experiment which ensures equal volumes of soil per plant, if not of natural root penetration. The experiment can also be designed to allow a distinction to be made between the role of temperature and some of the other environmental variables.

In one such experiment soil from randomized and replicated sites was potted and seeded in a controlled environment room (Rorison 1964) and after fourteen days groups of pots were deployed both in the field and laboratory, as indicated in Table 18.1. It was thus possible to indicate the direct importance of the temperature regime and to distinguish its role in relation to radiation and to soil moisture regimes. Soil water potential was measured using small resistance blocks (Lloyd 1968). Growth curves were fitted to the data from six five-weekly harvests (Fig. 18.4) using the method of Hunt & Parsons (1974).

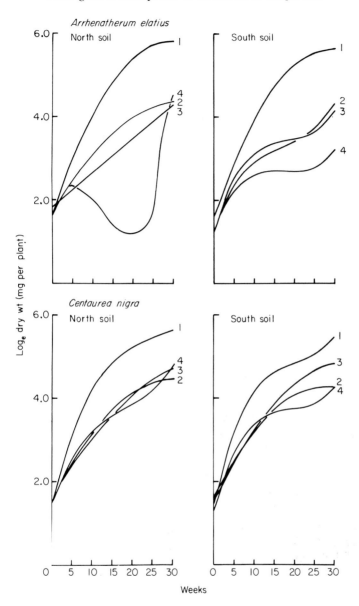

FIG. 18.4. Growth curves for seedlings of *Arrhenatherum elatius* and *Centaurea nigra* grown in pots from April until November 1974. Data are transformed to $\log_e$ and curves are fitted by computer program (Hunt & Parsons 1974). The span of the 95% confidence limits is normally within 20% of the fitted values. Numbers indicate the location of the pots during the experiment: 1 in the controlled environment room with 20 °C day 15 °C night; 2 on the north-slope; 3 on the south-slope (watered to soil water potential of soil on north-slope); 4 on the south-slope.

TABLE 18.1. Experimental sequence for a pot experiment indicating the treatments aimed to distinguish between effects on growth of temperature, soil moisture and soil type.

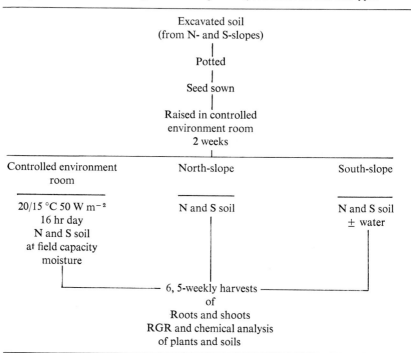

Plants of both species grew significantly better in the controlled environment room than under any of the field conditions, indicating, *inter alia*, that there was no limitation of soil volume in the field. *A. elatius* was adversely affected to a significant degree by the soil moisture stress experienced on the south-slope. In the pots on the south-slope, maintained at the soil moisture regime of the north-slope, growth was not significantly depressed, indicating that soil and air temperatures were not limiting.

In contrast, the growth of seedlings of *C. nigra* was affected by soil moisture stress to a far smaller degree and was generally less sensitive to both low spring temperatures and to early summer droughts than the germination of its seeds—an indication of the importance of studying different phases of a plant's life-cycle.

Phenological pattern

The cycle of phenological phenomena such as bud break, flowering, fruiting and senescence, are traditionally linked with seasonal changes in weather (Wells 1971, Fig. 9; Larcher 1975, Fig. 141). It is possible to monitor seasonal

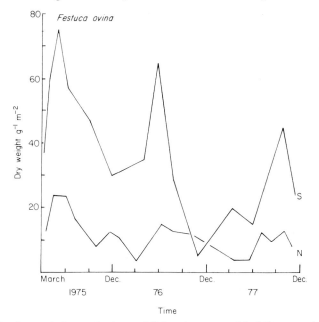

FIG. 18.5. Seasonal changes in the total living shoot material of *Festuca ovina* from the south- (S) and north- (N) slopes of Lathkilldale from March 1975 until December 1977. Note seasonal changes.

growth pattern and to assess the contribution of a species to the community by regular sampling and separation of component vegetation.

The growth of shoots is relatively easy to measure but not so the growth of roots. Hoffman (1972) reports that in a range of tree species roots begin to grow earlier and finish later in the season than shoot growth, i.e., they have lower temperature minima than shoots. He states that while root growth is regulated primarily by external environmental factors, shoot growth proceeds by successive spurts separated by endogenously fixed pauses in trees such as oak, beech and pine, while in birch, poplar and larch, shoot growth is synchronized by daylength and temperature changes. Grasses have also been shown to have differing patterns of root and shoot growth (Stuckey 1941; Davidson 1969; Sauer 1978).

In order to detect any extension of the growing season and to monitor growth under 'unfavourable' climatic conditions, it is necessary to make measurements throughout the year and, if possible, for more than one season (Fig. 18.5). These can indicate which species retain a proportion of green leaves throughout the winter and which die back to dormant buds (Al-Mufti *et al.* 1977). The terms 'evergreen' and 'deciduous' are not applicable to herbaceous species in the same way that they are to woody plants. Probably of all grasses native to Britain only *Molinea caerulea* is truly deciduous.

Leaves of the majority of native grasses die back gradually, usually after a few months of life and those which begin to grow in late summer often survive until the following spring. Examination of plants of *A. elatius*, *Festuca ovina* and *Dactylis glomerata* from Lathkilldale, in February 1969, showed that these species had 38, 57 and 65 per cent of their total above-ground dry weight in the form of green leaf material which was protected to varying degrees by the remaining weight of dead leaf material (the late P.S. Lloyd, personal communication). These amounts tend to vary according to the severity of the winters.

The formation of tufts, like cushion forms of some arctic plants, is an advantage to plants subject in winter to low temperatures and strong winds (Salisbury & Spomer 1964). Even more protected is the overwintering rosette. This form enables some winter annuals to attain leaf temperatures adequate for photosynthesis during the winter and thus to build up reserves for mobilization in spring (Regehr & Bazzaz 1976). Several perennial rosette plants, e.g. *Scabiosa columbaria*, *Succisa pratensis* and *Hieracium pilosella*, also retain basal leaves during winter and the following question arises. Do they, and the grasses which die back only partially in winter, maintain the capacity to photosynthesize and grow (Peacock 1976) whenever micro-clir atic conditions permit?

The advantage of such a mechanism would be greatest on relatively open and south-slopes which, in Britain, receive direct solar radiation in winter.

Seasonal variation in nutrient supply

Another environmental factor that is influenced by changes in temperature and which is then an influence on plant survival is nutrient availability. Seasonal changes have been reported particularly in undisturbed and relatively highly organic soils in which increased levels of available phosphorus and nitrogen can be detected at different times of the year.

Plants respond to changes in the external phosphorus supply particularly sensitively in changing root : shoot ratios (Rorison & Gupta 1974). Thus soil 'flushes', like additional photosynthate, may well be of potential advantage in the survival of some plants, and the relation between occasional photosynthetic activity and nutrient uptake should be explored. Knowledge of the range of root temperature over which a plant can take up nutrients compared with the range of its leaf temperatures over which it is capable of photosynthesizing is particularly relevant in understanding the survival of native species in infertile and unfertilized environments.

Having explored some ways of detecting in the field the influence of temperature on plant growth, is it possible to extend knowledge by means of simple simulation in the laboratory?

SIMULATION IN THE LABORATORY

The control of temperature and allied factors

The technical and biological limitations of controlled environment rooms have been discussed in the past by Evans (1963) and Van Bavel (1973).

Of the controllable variables temperature creates the fewest problems.

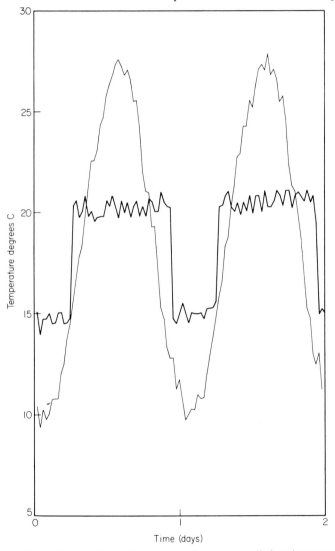

FIG. 18.6.   Diurnal fluctuations of temperature in two controlled environment rooms. A computer trace of air temperatures recorded on two successive days, one set for 20/15 °C and the other for a curve extending from *c.* 28 °C to *c.* 8 °C.

The temperature of the air can be maintained to within one degree Celsius without great expense. However, between the air and parts of the plant such as leaves, basal growing points and roots, there are gradients of temperature which by careful control can be maintained, extended or minimized as required (Watts 1975; Sheehy, Woodward, Farrow, Cooper & Windram, personal communication). Diurnal fluctuations, once achieved in controlled environment rooms in Sheffield by means of a cam system (Sutton & Rorison 1968) are now more easily reproduced using data from a punched code paper-tape. An interface produces a variable resistance which is related to the output from a tape reader and offsets the Wheatstone bridge circuit of a temperature controller (Sutton & Rorison 1980). Any paper-tape program is quickly and easily produced at negligible cost either manually or as an output from a computer program. It may be a continuous diurnal loop (Fig. 18.6) or may contain data for a succession of days (Fig. 18.7), each with time steps sufficiently short to produce a reasonably smooth temperature curve.

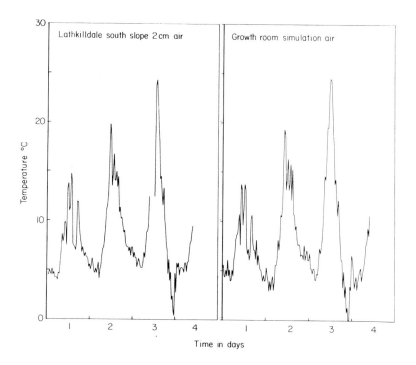

FIG. 18.7. Computer traces of (left) air temperature measured by compact logger system in the field and (right) the simulation achieved in a controlled environment room by means of a punch encoded paper-tape (Sutton & Rorison 1980).

Plant response to fluctuations of ecological and physiological amplitude

Many experiments have been done using a combination of 20 °C during daytime and a simple step down to 15 °C at night. In the field, however, there are many days during the growing season when plants are exposed to a diurnal fluctuation in air temperature of more than 20 °C (see Fig. 18.1) and on south-slopes at least the upper soil temperatures reflect closely the air temperature.

The questions arise: are certain species adapted to fluctuating temperatures of the magnitude experienced in the field and are the processes measured by metabolic physiologists at constant and high temperatures, such as 25 °C, relevant to the plant in the field and to its survival? There is one major difference between field and laboratory conditions. In the field, plant temperatures are the result of the energy balance and may differ substantially from air temperatures. In a controlled environment room, on the other hand, temperatures are controlled by air conditioning and plant and air temperatures are closely similar. There is therefore a very substantial conceptual and practical difference between plant responses to air temperature in the field and in the growth room.

Diurnal fluctuations

To test the response of a number of plants of differing phenologies to contrasting amplitudes of temperature, replicated groups of plants were grown in two rooms, one programmed for a single diurnal step of 20/15 °C and the other for a sine-wave extending from *c.* 8 °C to *c.* 28 °C. In both rooms the mean daily temperature was 18·3 °C (Fig. 18.6).

In order to minimize any effects of differing saturation deficits, humidistats were set to provide sufficient moisture vapour for a constant saturation deficit of *c.* 7 mbars (700 Pascal). In the event this was achieved in the 20/15 °C room but fluctuations of between 4–10 mbars were recorded at the extremes of the range 8–28 °C. The combination of an, albeit short, exposure to 28 °C and a saturation deficit of 10 mbars may have affected some species more strongly than others. But all seven species examined, ranging from the fast-growing *Holcus lanatus* to slower-growing *Deschampsia flexuosa*, grew as well in the 20/15 °C regime as in the 28/8 °C one for six weeks after emerging.

It is reported by Evans (1972) that a range of saturation deficits from 2–6·4 mbars had no significant effect on the growth of *Impatiens parviflora*.

Some physiologists are rightly strong advocates of constant saturation deficits in experiments where temperature is the variable but, apart from the comments of Evans (1972) above, no record of work designed to distinguish the effect of these two factors on the growth of plants has been found. There is obvious scope for comparative experiments since in the field plants experience fluctuations in both saturation deficit and temperature (Woodward & Pigott 1975).

Seasonal fluctuations

Possibly more important for plant growth than diurnal temperature variation is the seasonal pattern of temperature. In order to test this hypothesis a simple simulation of autumn–winter–spring temperatures has been carried

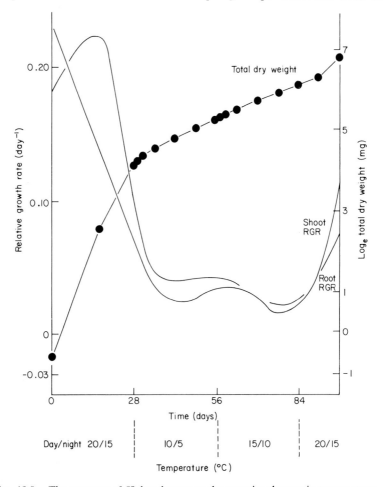

FIG. 18.8.   The response of *Holcus lanatus* to the stepwise changes in temperature indicated below the abscissa. Root and shoot were maintained at the same temperature. The progress curve for whole-plant dry weight was fitted by means of splined regressions (Hunt 1978). Also shown are instantaneous relative growth rates of roots and shoots, derived from these fitted curves.

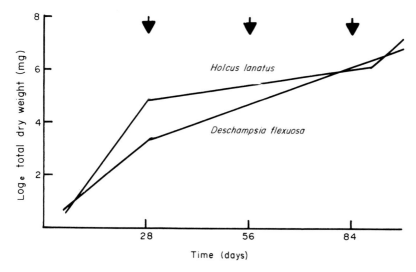

F<small>IG</small>. 18.9. Dry weight curves for *Holcus lanatus* and *Deschampsia flexuosa* compared. Downward-pointing arrows indicate temperature changes (see Fig. 18.8).

out. At first a series of abrupt steps (Fig. 18.8) were made each with a plateau of four weeks to allow measurement of relative growth rates and specific absorption rates after each change.

It was of particular interest to consider the relationship between root and shoot response when temperature regimes were similar to those frequently experienced in the field.

Two temperature regimes were applied. In the first, called the growth-room regime, roots and shoots were exposed to the same diurnal fluctuation of temperature involving a night depression of 5 °C, whereas in the second, hereafter called the cold bath regime, an equal number of plants had their roots held constantly at the lower temperature while their shoots were exposed to the diurnal fluctuation.

The response of two species is considered here. *Holcus lanatus* is a fast-growing species whose green leaf material dies back by at least 50 per cent in the winter and *Deschampsia flexuosa* is a slower-growing species which retains a majority of green leaves during the winter.

In *H. lanatus* (Fig. 18.8) the weight of both root and shoot increased at approximately the same rate. The curve of total plant dry weight shows a marked response to the initial drop to low temperatures and a much smaller response on return to higher temperatures. This is reflected in the fitted curves for relative growth rate of both root and shoot. In the growth-room regime it was noticeable that the shoot responded slightly more quickly than the root to a return to higher temperatures. In the cold bath regime the response to the initial drop in temperature was seen most immediately in the root and response to a return to higher temperatures was less than in the growth-room regime.

In *D. flexuosa* there were similar responses except that the relative growth rate of the roots was more strongly affected by the constant low temperature than that of the shoots. The response as temperatures rose did not match the original relative growth rate and this may reflect the ageing of the plant and also an ability or inability to overcome the 'pre-conditioning' effects of the previous cold spell. The continued growth of both species throughout the whole experiment indicated a lack of any induced dormancy.

When the dry weight curves of the two species are plotted together an interesting trend is revealed concerning their ability to grow at low temperatures (Fig. 18.9). For the first four weeks the growth curves diverge as expected. In the next four weeks the rate of growth falls, following the drop in temperature. However, there is a bigger fall for *H. lanatus* than for *D. flexuosa* and as a result the absolute weight of *D. flexuosa* overtakes that of *H. lanatus* by the twelfth week. From then on, back at 20/15 °C the relative growth rate of *H. lanatus* increases more quickly than that of *D. flexuosa* which is then overtaken. The significance of these responses to temperature change must be considered subsequently in the light of the overall complex of environmental factors including day length.

CONCLUSIONS

Some species will respond immediately and measurably to changes in environmental factors such as temperature or nutrient availability. Others do so more slowly.

This response is interesting not only when it takes place during, but also if it extends beyond, the 'normal' growing season since it may well have a bearing on the survival of the plant.

Many questions remain unanswered. For example: How far does daily amplitude in temperature as opposed to mean temperature affect the growth and survival of different species? Do species have different temperature minima for photosynthesis to take place which allow some activity during the winter? If leaf temperatures of some species become adequate for photosynthesis is the benefit immediate in the form of new growth or is it in the form of energy reserves for use in the spring? Does this facility influence the species' chance of survival in a habitat?

Is it related to or influenced by responses to root temperature and mineral nutrient uptake over the same period?

Measurement of the response of the whole plant in the natural environment is not easy but techniques of simple simulation and of measurement are currently available. These tend to provide an integration both of conditions and of plant responses and so long as the time scale is carefully chosen the results can be as ecologically meaningful as rapid, unequilibrated reactions to instantaneous change. The problem of precise evaluation of the significance of natural fluctuations remains.

REFERENCES

Al-Mufti M.M., Sydes C.L., Furness S.B., Grime J.P. & Band S.R. (1977) A quantitative analysis of shoot phenology and dominance in herbaceous vegetation. *Journal of Ecology*, **65**, 759–791.

Clapham A.R., Tutin T.G. & Warburg E.F. (1962) *Flora of the British Isles*, 2nd edn. Cambridge University Press, London.

Cooper A.J. (1973) *Root Temperature and Plant Growth*. Research Review, 4, Commonwealth Bureau of Horticultural and Plantation Crops, East Malling, Kent.

Davidson R.L. (1969) Effect of root/leaf temperature differentials on root/shoot ratios in some pasture grasses and clover. *Annals of Botany*, **33**, 561–569.

Evans G.C. (1972) *The Quantitative Analysis of Plant Growth*. Studies in Ecology, Volume 1. Blackwell, Oxford.

Evans L.T. (1963) Extrapolation from controlled environments to the field. *Environmental Control of Plant Growth* (Ed. by L.T. Evans), pp. 421–437. Proceedings of a Symposium held at Canberra, Australia 1962. Academic Press, New York.

Gates D.M. (1968) Transpiration and leaf temperature. *Annual Review of Plant Physiology*, **19**, 211–238.

Grime J.P. & Jarvis B.C. (1976) Shade avoidance and shade tolerance in flowering plants. II. Effects of light on the germination of species of contrasted ecology. *Light as an Ecological Factor II*, pp. 525–532. Symposia of the British Ecological Society 16. (Ed. by I.H. Rorison). Blackwell, Oxford.

Hoffman G. (1972) Growth rhythms of root and shoot axes in forest trees. *Flora*, **161**, 303–319.

Hunt R. (1978) Analysis of complex growth curves. *UCPE Annual Report 1978*, pp. 14–17.

Hunt R. & Parsons I.T. (1974) A computer program for deriving growth-functions in plant growth analysis. *Journal of Applied Ecology*, **11**, 297–307.

Larcher W. (1975) *Physiological Plant Ecology*. Springer-Verlag, New York. Translated by M.A. Biedermann-Thorson.

Lloyd P.S. (1968) A miniature gypsum resistance block for measuring soil moisture tensions. *The Measurement of Environmental Factors in Terrestrial Ecology*. Symposia of the British Ecological Society 8. (Ed. by R.M. Wadsworth), p. 273. Blackwell, Oxford.

Nielsen K.F. & Humphries E.C. (1966) Effects of root temperature on plant growth. *Soils and Fertilizers*, **29**, 1–7.

Peacock T.J. (1976) Temperature and leaf growth in four grass species. *Journal of Applied Ecology*, **13**, 225–232.

Regehr D.L. & Bazzaz F.A. (1976) Low temperature photosynthesis in successional winter annuals. *Ecology*, **57**, 1297–1303.

Rorison I.H. (1964) A double shell plant growth cabinet. *New Phytologist*, **63**, 358–362.

Rorison I.H. & Gupta P.L. (1974) The growth of seedlings in response to variable phosphorus supply. *Plant Analysis and Fertilizer Problems*, pp. 373–382. Proceedings of 7th International Colloquium, Hanover 1974.

Rorison I.H. & Sutton F. (1976) Climate, topography and germination. *Light as an Ecological Factor II.* Symposia of the British Ecological Society 16. (Ed. by G.C. Evans, R. Bainbridge & O. Rackham), pp. 361–383. Blackwell, Oxford.

Salisbury F.B. & Spomer G.G. (1964) Leaf temperatures of alpine plants in the field. *Planta,* **60,** 497–505.

Sauer R.H. (1978) A simulation model for grassland primary producer phenology and biomass dynamics. *Grassland Simulation Model* (Ed. by G.S. Innis), pp. 55–87. Ecological Studies no. 26. Springer-Verlag, New York.

Stuckey I.H. (1941) Seasonal growth of grass roots. *American Journal of Botany,* **28,** 486–491.

Sutcliffe J. (1977) *Plants and Temperature.* Studies in Biology no. 86. Edward Arnold, London.

Sutton F. & Rorison I.H. (1968) Programmed control of plant growth-room climate. *New Phytologist,* **67,** 973–975.

Sutton F. & Rorison I.H. (1970) The modification of a data logger for the recording of temperatures in the field, using thermistor sensors. *Journal of Applied Ecology,* **7,** 321–329.

Sutton F. & Rorison I.H. (1980) An interface, allowing a commercial paper-tape reader to be used in the continuous programming of growth-room temperatures. *Journal of Experimental Botany,* **31,** 691–696.

Sydes C.L. (1980) Some aspects of competition and coexistence in various types of herbaceous vegetation. Ph.D. Thesis, University of Sheffield.

Thompson K., Grime J.P. & Mason G. (1977) Seed germination in response to diurnal fluctuations of temperature. *Nature,* **267,** 147–149.

Van Bavel C.H.M. (1973) Towards realistic simulation of the natural plant climate. *Plant Response to Climatic Factors* (Ed. by R.O. Slatyer), pp. 441–447. Proceedings of the Uppsala Symposium 1970. UNESCO, Paris.

Walter H. (1977) *Vegetationszonen und Klima* (3rd edn) *Die Ökologische Gliederung der Biogeosphäre.* Verlag Eugen Ulmer, Stuttgart.

Watts W.R. (1975) Air and soil temperature differences in controlled environments, as a consequence of high radiant flux densities and of day/night temperature changes. *Plant and Soil,* **42,** 299–303.

Wells T.C.E. (1971) A comparison of the effects of sheep grazing and mechanical cutting on the structure and botanical composition of chalk grassland. *The Scientific Management of Animal and Plant Communities for Conservation.* Symposia of the British Ecological Society 11. (Ed. by E. Duffey and A.S. Watt), pp. 497–515. Blackwell, Oxford.

Went F.W. (1945) Plant growth under controlled conditions. V. Relation between age, variety and thermoperiodicity of tomatoes. *American Journal of Botany,* **32,** 469–479.

Woodward F.I. & Pigott C.D. (1975) The climatic control of the altitudinal distribution of *Sedum rosea* (L.) Scop. and *S. telephium* L. I. Field observations. *New Phytologist,* **74,** 323–334.

19. ASSESSING PLANT RESPONSE
TO THE WEATHER

E.D. FORD AND R. MILNE

Institute of Terrestrial Ecology, Bush Estate,
Penicuik, Midlothian, EH26 0QB

SUMMARY

Some problems which the ecologist faces in making an analysis of how conditions in the atmospheric environment may influence the growth and distribution of plants are reviewed.

(1) Variation in the weather has a complex structure: different elements are correlated to varying degrees, variation occurs at a range of frequencies, from daily to periods of several decades all of which can have an observed effect on growth. The significance of these complexities is discussed through a review of analyses which have been made of field observations of growth in relation to variation in the environment.

(2) The growth of a plant is the result of interaction between its different parts. From attempts made to describe the growth in weight of the above ground parts by simple equations, complex models have evolved. These describe growth in terms of environmental influences on the different parts of the plant and how these parts interact. However the complexity of these models renders them of little direct use to the ecologist studying growth in a constantly changing environment.

An example is presented of the application of time series analysis to the study of plant response to environmental change. This technique facilitates the stepwise analysis of the interrelations between environmental fluctuation, physiological processes and growth itself, with a clearly stated error.

INTRODUCTION

The papers in this volume demonstrate how different factors of the aerial environment operate through a range of microclimatological and physiological processes to influence plant function. However, a decision to frame an hypothesis on the control of plant growth or distribution in terms of one particular set of processes, e.g. those concerned with water relations, involves a degree of subjectivity that may not be justified even in extreme environments, e.g. deserts (see Björkman, this volume). The ecologist interested in the control of plant growth and plant distribution must be prepared to make an assessment of the relative importance of different environmental influences operating through different processes.

One approach, which has come to prominence over the last decade, is to attempt a synthesis of information from measurements of the different components of plant function, i.e. photosynthesis, water balance, nutrient uptake and the carbohydrate economy (see Landsberg, this volume). An obvious technical difficulty which prohibits this approach to many ecologists lies in obtaining such a range of measurements for even a single species over a range of conditions. A more fundamental problem involves welding such diverse information into a model so that interactions under different environmental conditions can be assessed within an acceptable probability. The more complex the model the more difficult it is to assess how the errors involved in estimating a range of different parameters may combine.

Traditionally ecologists have employed two approaches to the direct analysis of the influence which the environment may have on plant growth and, by implication, on plant distribution: (i) the parallel observation of growth and environment in time or over a geographical gradient where relationships have been sought by various forms of correlation analysis; and (ii) application of the techniques of plant growth analysis (Evans 1972) both in controlled environments and in the field.

In this paper we review some of the contributions which such studies have made to the understanding of environmental influences on plant growth. In these studies the aerial environment has been revealed to demonstrate complex variation, in time and space, which interacts with the soil environment to affect plant performance. The plant has also been revealed as a dynamic system—both the processes which comprise growth and their interactions change markedly over time. Thus to further our understanding of the environmental influences on plant performance there is a requirement to take account of both the fluctuating nature of the environment and the dynamic nature of plant response. We present an example of an application of the technique of time series analysis (see Kendall 1973 for an introduction) to modelling, with a specified error, the response of a single plant process to environmental change.

FIELD OBSERVATION OF DIFFERENCES IN PLANT GROWTH

Observation of differences in growth of a species either between years or between geographical regions stimulates attempts to examine their correlation with environmental variation. However the ecologist is immediately faced with the problem of deciding which aspects of the environment he should measure. The number of environmental variables which can be included in any form of analysis based on correlation must be limited to many less than the number of data points available, e.g. years or sites. Furthermore it must be recognized that environmental variables are themselves correlated.

R.A. Fisher discussed these problems in reporting an investigation into the relationship between rainfall and the yield of wheat over a 61-year period (Fisher 1924). Prior to his investigation it had frequently been suggested that yields could be depressed by increased amounts of rainfall, largely, it was thought, because nitrates were leached from the soil. However, there had been controversy, particularly over the effect of seasonal differences in rainfall amount. Fisher recognized that yields were likely to be affected by weather variables other than rainfall and that rainfall itself might be partially correlated with other aspects of the weather. He appreciated that by restricting his study to yield solely in relation to rainfall his was necessarily a preliminary investigation into the factors which do control wheat yields. His investigation was a critical examination of an established hypothesis though not a test.

Fisher's data were values of daily rainfall and the annual yields of wheat, in bushels (1 bushel = 36·4 l), from 13 plots on Broadbalk field at Rothamsted Experimental Station, Harpenden, U.K., each of which had been continuously maintained under a different fertilizer regime over the period 1854–1918. His objective was to build up an equation of partial regressions of annual yield on seasonal rainfall and this required that each annual period of rain (September 1–August 31) and annual yield be considered as independent samples, i.e. the value in one year should not be correlated with values before or after. However the values of yield *were* correlated. Fisher observed (i) a decline in yield over the whole period on plots receiving inorganic fertilizers though not on those receiving farmyard manure, which he attributed to a slow diminution of some nutrients; (ii) a cyclic variation in yield on all plots which he ascribed to variations in weed infestation over the whole field. He extracted both of these scales of variation by fitting a 5th order polynomial separately to the data for each plot and making his further calculations on the observed deviations from these fitted curves which he considered as independent. Cyclic variation in annual rainfall was also apparent. Over the 61 years there were two periods when mean annual rainfall rose to 32 inches (813 mm) separated by periods during which it fell to 27 inches (686 mm) but this cyclic variation was small, 6·7% of the total variance, in comparison with the year to year variation, 93·3% of the total variance, and he ignored it in the analysis.

To construct a regression equation of annual yield on seasonal rainfall Fisher had available 61 values of yield and a rainfall value for each day. It was not possible to treat each of the 365 daily values of rainfall as a separate weather variable, as there would be five times as many variables (365) as sample points (61)! Furthermore Fisher noted that any 'reasonable' aggregation of daily values e.g. into weeks (52 variables) or perhaps fortnights (26 variables), would still give too many weather 'variables' for their effect to be estimated by direct regression analysis. He also pointed out that even if there were sufficient yield values to permit direct regression analysis this would ignore any serial correlation which existed between rainfall periods within a year.

His solution was to solve for a regression integral

$$\bar{w} = c + \int_0^t ar \cdot dt \qquad (19.1)$$

where $\bar{w}$ is yield, c is a constant, r is rainfall in the period dt and a is the average benefit to the crop in bushels/acre/inch of rain that fell in the period dt. He approximated the course of annual rainfall $r \cdot dt$, by taking mean values for 5-day periods and then fitting a polynomial with 6 terms for each year thereby reducing the number of 'variables' to 6 whilst preserving accuracy in the representation of seasonal variation. The values of yield were used to estimate multipliers for each of the polynomial coefficients calculated separately for each plot, and which themselves could be represented in a single curve for the seasonal effect equivalent to $a \cdot dt$. Differently shaped curves were obtained for different fertilizer treatments (Fig. 19.1), which demonstrated 'how the response of the crop to weather is intimately connected with the manurial condition of the soil' although the direct value of the regression equation in predicting yield was limited. Considering the 13

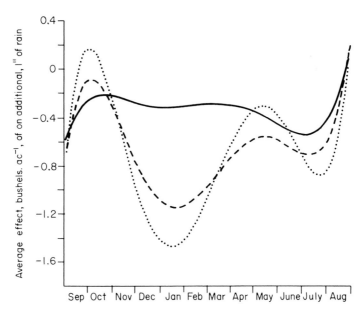

FIG. 19.1. The effect of rainfall at different seasons of the year on the yield of wheat for a low fertility soil which received no added nitrogen, ——; a soil receiving added nitrogen as NH_4 in spring and autumn, – – –; a soil receiving twice the total amount of nitrogen as the previous plot distributed 3 : 1, spring to autumn, . . . (Fisher 1924).

different fertilizer treatments the regression equation accounted for a maximum of 40% and a minimum of 11% of the variation in yield, i.e. after the trends ascribed to gradually decreasing soil fertility and cyclical weed infestations had been removed.

This early work demonstrated:

1. that more than one time scale of variation can exist in records of plant growth and that these must be isolated if correlation analyses are to be made;

2. the effect of a 'single' variable, in this case rainfall, can be altered by variation in other conditions, in this instance soil fertility;

3. some form of simplified description or classification of the environment is necessary before correlation analyses are feasible;

4. that a 'single' variable may not account for a high percentage of the observed variation in growth—even although other scales of variation have been extracted.

We now consider these four features in a broader ecological context.

Tree rings provide the longest and most widely spread records of plant growth and differences in these annual increments are particularly useful for exploring the interactions between time and place in relation to variations in growth. The sequence of variation in ring widths is frequently highly consistent from tree to tree at a site, particularly those with more extreme environments and this has long been used as a dating system for timber. It is generally assumed that variation in width of growth rings is related to variation in the weather.

Direct correlations between ring widths and environmental factors have been found to be largest in extreme environments though with different factors showing the largest correlations in different regions. Thus in arid zones, positive correlations have been found between ring width and environmental factors related to a favourable water balance for the tree (Julian & Fritts 1968) whilst in alpine regions narrow rings have been related to low temperatures and the advance of glaciers (LaMarche 1974a; LaMarche & Fritts 1971) and in northern latitudes with low summer temperatures (Mikola 1962). In less extreme environments, simple relationships may not be found and indeed dendrochronologists go so far as to classify sequences of ring width as 'sensitive' only where correlations with environmental factors are apparent (Fritts 1976).

Studies of tree rings have demonstrated that patterns of growth may vary in a complex way over comparatively short geographical distances. LaMarche (1974b) examined sequences of ring width of Bristle cone pine (*Pinus longaeva*) at four sites within 11 km of each other in the generally dry environment of eastern Nevada. These sites were distributed along an ecological gradient, one at the upper altitudinal limit of the forest, two within the main forest region and a fourth at the dry, lower altitudinal limit of the forest.

At each site 14–18 individual sequences were obtained from a minimum of nine trees. At the two extreme sites the standard deviation of ring width was larger than at the intermediate sites as was the (positive) correlation coefficient between individual trees within sites. A standardized ring width sequence for 109 years was prepared for each site. Those of the lowest and two intermediate altitudes were positively correlated between each other but that from the upper altitudinal limit was correlated with none of the other three. LaMarche noted that the variation in the ring widths comprised short term and long term trends—a similar structure to that found by Fisher (1924) in the sequence of annual wheat yields. By using reciprocal digital filters LaMarche was able to separate the variation within each sequence attributable to long term changes (low frequency variation) and short term changes (high frequency variation) (Fig. 19.2). Correlations between each of the four sites were then calculated separately for the high and low frequency variation.

There was a positive correlation between all sites for the high frequency variation. However, whilst the low frequency variation was positively correlated between each of the lower three sites, each in turn was negatively correlated with that of the upper site. LaMarche concluded that the apparent lack of correlation between the upper site and the other three for the unfiltered sequences was the result of a combination of the opposed effects of high and low frequency variation.

By comparing sequences of tree rings with patterns of variation from a single weather station LaMarche made general observations on which weather factors were important for tree growth in the region. He suggested that the high frequency variation, positively correlated between all sites, was related positively to rainfall amount in a similar manner over the whole of this dry region. He suggested that the low frequency variation was related to increased temperature during the summer months which had a positive effect on growth at the upper altitudinal limit but a negative effect at the other three sites, presumably, in this latter case, through a direct or indirect effect on the water balance.

The results from these analyses of sequences of tree rings in natural vegetation show similarities with the environmental influences on wheat yields. Patterns of variation with different time scales were apparent in both and the observable effect of variation in one variable was influenced by conditions in others. Additionally LaMarche's analysis of tree rings at these four sites demonstrates a measurable change in environmental influence over a short geographical distance. Such changes over short distances are not confined to dry regions like eastern Nevada where plants are under stress. Miller & Cooper (1976) observed patterns in tree ring widths of *Pinus sylvestris* taken from the Dee valley in east Scotland which correlated with periodicities in annual temperature and particularly with periodicities in

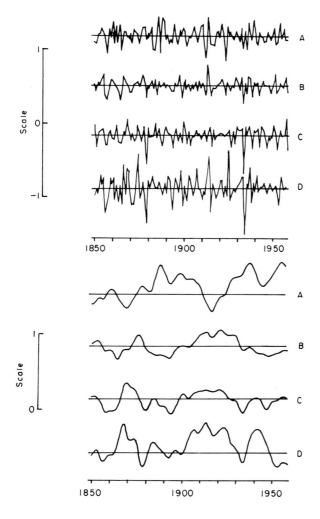

Fig. 19.2 The variation in annual ring width of the *Pinus longaeva* at high altitude, A, low altitude, D, and two intermediate sites, B and C, in eastern Nevada, separated into high (upper) and low (lower) frequency components (LaMarche 1974b).

May and June rainfall. Commenting that similar patterns had not been observed in other tree ring sequences from Scotland, they showed that the periodicities in the amounts of May and June rainfall were restricted to sites in the rain shadow of the Grampian mountains.

In all but extreme climates direct correlations between variation in growth and single factors of the environment are small and this has stimulated a consideration of multivariate influences. However limits are set to the number of variables which can be considered in any study by the number of independent values of yield or growth which are available. This, and the observation that many factors of the environment vary in similar ways (positively correlated) or in diametrically opposed ways (negatively correlated) has led to attempts to classify the environment of a region, e.g. by principal component analysis (Gittins 1969). If the environment can be specified in terms of a few orthogonal components this reduces the number of 'variables' to be considered when seeking quantitative relationships with plant growth. As with all classifications both the individual elements selected on which to make the classification and to what level of division the classification is allowed to proceed must rely on subjective decisions (Cormack 1971). In general such a classification can account for a high proportion of total variance and with a few components if the region is one of gradual environmental change, e.g. W. Australia (Patterson, Goodchild & Boyd 1978), rather than one with major orographic irregularities, e.g. Scotland (Malcolm 1970).

Malcolm (1970) surveyed the growth of Sitka spruce (*Picea sitchensis*) in 77 plantations distributed between four forests in different regions of Scotland —Glentress (Peeblesshire), Bin (Banffshire), Glengarry (Inverness-shire) and Inverliever (Argyllshire). At each, 41 environmental variables were measured, the majority being concerned with aspects of soil physics and chemistry; as is frequently the case reliable meteorological records were not available for each site and Malcolm was compelled to use general descriptions of the habitat such as altitude, aspect and topographic class. A principal components analysis of the environmental variables from all sites accounted for 78% of the variance by eight components (Table 19.1). When restricted to sites within a single forest the percentage variation accounted for increased, often with fewer components, e.g. 94% of the variation with six components at Glentress.

Having classified the environment in terms of its principal components Malcolm then regressed the mean increment of annual tree height for each plantation on its values of each component. The 3rd component in the classification accounted for the largest single reduction in variability, 26%, and in the complete multiple regression five components accounted for 58% of the variation. Components 1 and 2, i.e. the most effective in the classification of the environment, produced no significant reduction in the construction of the multiple regression equation. For individual forests larger reductions in variation were achieved in multiple regressions whilst employing fewer components of the classification. Glentress forest, a 91% reduction with three components, was the most efficient.

The major difficulty associated with this multivariate approach to the analysis of environmental influences on plant growth lies in the interpretation

TABLE 19.1. A principle components analysis of the environmental variables from all sites (Malcolm 1970).

Component	Vector	% of variability accounted for
1	Concentrations of Ca, K and Mg in the total soil profile, pH and concentrations of Ca, K and Mg in the A horizon, soil rootability index	25
2	Available water capacity of the soil, cation exchange capacity and nitrogen concentration of the total profile, soil rootability index	14
3	Rooting depth, slope, soil texture, phosphorus concentration in the LFH horizons	12
4	Phosphorus concentration in the LFH horizons	7
5	Site elevation	6
6	Topographic class	6
7	Phosphorus concentration in the total profile	4
8	Nitrogen and phosphorus concentrations in the LFH horizons, Mg concentration in the A horizons	4
		78%

of the components of the environment classification. Each component is a vector in multidimensional space and as such is specified in terms of elements of a number of different measured variables. Successive components are orthogonal, so it is permissible to regress a dependent variable on them, but this expression of the multivariate data set in terms of a series of 'independent' axes is frequently achieved at the expense of a blurred environmental definition of the components. The calculation that a component produces the largest reduction in variation in the multiple regression of mean annual height increment may not lead to a testable hypothesis as to how growth is controlled by the environment. Its value is limited by the initial choice and specification of the variables in the study. In Malcolm's study the importance of a number of soil variables was highlighted and this is typical of such forest productivity work (Ralston 1964) though little could be learnt of their interaction, if any, with climatic variables.

Multivariate studies can produce evidence of direct environmental influences on growth where (i) variation in the environment is more gradual than that measured by Malcolm; and (ii) selection of the environmental variables to be measured can be made on the basis of some prior evidence

about the control of plant growth. From a study of 40 years of wheat production in 58 administrative shires in W. Australia, Goodchild & Boyd
(1975) concluded that, unlike the productivity of the region's natural vegetation, annual yields were not obviously related to the amounts of rainfall nor
to any other major climatic variable except in extreme conditions. A principal
component analysis of these yields showed variation between years, 55·9%,
greater than variation between sites within years, 17·1%. Boyd et al. (1976)
attempted a more detailed analysis of climatic effects by (i) stratifying climate
in relation to the known growing requirements of wheat, i.e. length of growing season, rainfall amount, moisture requirements, evaporation, radiation
and temperature; (ii) representing each of these climatic factors by up to six
variables and conducting a principal component analysis on them—each
group was represented by two components which accounted for between 85
and 98% of the variation; (iii) calculating multiple correlation coefficients
between the principal components of these environmental classifications and
(a) the mean yield for each site and (b) the two principal components of
variation in yield, i.e. between years and between sites.

The correlations between the principal components of the environmental
classification and both mean yield and the between-year component of
variation in wheat yield were low. Those with the between-site component of
yield were high, particularly that for moisture. Boyd et al. (1976) concluded
that rainfall was of major importance for wheat yield, particularly its pattern
of distribution during the growing season, and that this was the result of an
interaction between region and season which they discussed in terms of the
climatic circulation pattern of W. Australia. During the growing season
there was fluctuation in the conditions favourable and unfavourable for
growth caused by the passage of depressions from west to east across the
country. Their rainfall measurements included no measure of these changes
in rainfall distribution during the growing season and so the variations from
year to year in wheat yields appeared to be random fluctuations.

The studies described in this section show that the environment varies in
a complex way both in time and space and that this complexity must be
described with some considerable precision if environmental influences on
growth are to be understood. Weather variables cannot be specified simply
as a succession of means over a certain time period; account must be taken
of the weather as a dynamic system. Conditions are likely to be produced
which vary in a manner critical to plant growth over short time periods and
small geographical distances and the effect which variation in the weather
can have on growth can be influenced by soil conditions. In general we can
conclude that attempts to seek an understanding of environment → plant
can only ever be partly successful when an integrated value of growth is
taken, e.g. annual yield. Since the weather is a dynamic process emphasis
must be placed on patterns of growth in relation to environmental change.

ANALYSIS OF PLANT GROWTH

Two rather different quantitative approaches have been made to analyse environmental influences on the process of plant growth. One is synthetic and involves a direct representation of the many physiological processes which constitute growth in a complex model. The advantages of this approach are discussed by Landsberg in this volume. A second approach has evolved from attempts made to fit simple equations, containing only a few parameters, to growth observed in different environments. The objective is to summarize different responses through comparison of a few parameters some of which may have a direct biological interpretation.

A number of different mathematical functions have been fitted to growth data and, excluding the use of polynomials, it has repeatedly been found that the formulae with best fit obtained depend upon (1) species, (2) whether the whole plant or part of it is measured and (3) the particular environment in which growth has taken place (see Erikson 1976 for a review). The single most valuable contribution made to the use of the exponential type of growth function was that of Richards (1959). He described an equation with four parameters flexible enough to fit a wide variety of asymptotic sigmoid type growth curves. This can be written as:

$$W = A(1 - be^{-kt})^{1/(1-m)} \qquad m < 1 \qquad (19.2)$$

$$W = A(1 + be^{-kt})^{1/(1-m)} \qquad m > 1 \qquad (19.3)$$

where W is size at time t, A is the maximum limiting value, k is a rate constant for the growth process. Four parameters completely define the growth curve; b has no biological information and indicates adjustment of the time scale, m indicates the basic shape of the curve, k/m is the mean relative growth rate and $Ak/(2m + 2)$ is a weighted mean growth rate.

Richards fitted his equation to data on the growth in length of the hypocotyl of *Cucumis melo* in the dark at each of a number of constant temperatures. All parameters changed between 15 and 25 °C, with both growth rate and relative growth rate increasing and the shape of the curve changing (Fig. 19.3). For higher temperatures, i.e. ≥ 30 °C, the shape of the curve stabilized and the relative growth rate did not change from that at 25 °C but both the absolute growth rate and maximum length achieved were maximal at 30 °C.

Richards' approach permits estimates of growth rate, relative growth rate and maximum size obtained under different constant conditions to be compared. However, problems have arisen in attempting to extend it from laboratory conditions to the more complex natural environment (Hunt 1979). The main difficulty encountered is that in the natural environment

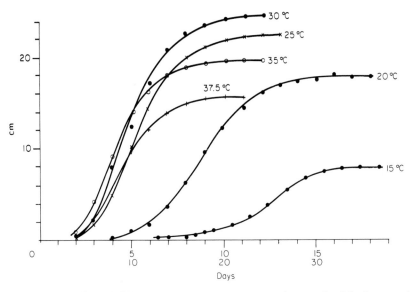

FIG. 19.3. The effect of different constant temperatures on the growth of the hypocotyl of *Cucumis melo* grown in the dark. The time axis for 15 °C, lower scale, is twice that for the other conditions (Richards 1959).

growth does not follow a curve represented by a single mathematical process. This is hardly surprising because the aerial environment fluctuates at a range of time scales. An example is provided by the attempt made by Austin, Nelder & Berry (1964) to examine the growth of carrots under four fertilizer regimes in each of three seasons. Leaf and tap root weights were measured separately and for both an attempt was made to fit a generalized form of the logistic equation for each fertilizer regime in each season.

$$\text{Relative growth rate} = R_w = \frac{1}{W} \cdot \frac{dW}{dt} = K \left[1 - \frac{W^\theta}{A} \right] \quad (19.4)$$

where t is time, W weight, and k, θ and A constants. K determines relative growth rate during the exponential phase of growth, K/A^θ determines the rate at which R_w declines with increasing plant size. Austin, Nelder & Berry (1964) were aware of the difficulties experienced in attempting to fit such equations (Nelder *et al.* 1960) and so attempted additional fitting where time was replaced by the integral of an environmental variable, E: $\tau = \int E \, dt$. Environment variables tested were solar radiation, day degrees and evaporation from an open water surface.

Equation 19.4, with a time base in days, did not fit the root data but did fit the leaf weight data for each individual treatment in each year. For the leaf weight data, values of K and θ were significantly different between fertilizer treatments but values between years for each treatment were not different. In contrast, the values of A differed more between years than between treatments. The use of integrated meteorological variables as a base gave larger error mean squares than the use of time in days.

Austin, Nelder & Berry (1964) suggested that their failure to fit equation 19.4 to the leaf weight data with non-significant differences between years within treatments may have been due to inadequate selection of a meteorological alternative to time in days. In particular they suggested that because differences in total leaf weight were correlated with total rainfall a scale based on calculated transpiration rather than evaporation from an open water surface might have made an improvement. They also suggested that the correspondence of K and θ between years was a reflection that these parameters are determined to a large extent by R_w during the early exponential phase of growth and that since the ground was always irrigated before sowing, differences in rainfall then would have little effect on K and θ.

The major drawback to their approach was the failure to fit this flexible equation to the root growth data. They considered that this was partly due to the reasons suggested for the difference between fits for the leaf data but also because tap root weight was dependent upon leaf weight. They suggested a pair of differential equations by which this dependance could be built into equation 19.4.

This inter-relationship between plant parts in the growth process and the effect which the environment may have on it has received considerable attention from plant scientists. Over the past 70 years (see Evans 1972 for a review) a model has been developed from the logistic equation for growth, expressing it in the following way.

Relative growth rate = Unit leaf rate × Specific leaf area × Leaf weight ratio.

$$\frac{1}{W}\cdot\frac{dW}{dt} = \frac{1}{L_A}\cdot\frac{dW}{dt} \times \frac{L_A}{L_W} \times \frac{L_W}{W} \quad (19.5)$$

where t is time, W dry weight, L_A leaf area and L_W leaf weight (Evans 1972). Obviously this model is not written specifically in terms of the physiological processes of photosynthesis and assimilate distribution, it takes no direct account of how these processes may be influenced by other physiological processes, e.g. nutrient uptake, the water balance, but it is of sufficient flexibility to act as a basis for studies analysing how different factors can influence various aspects of growth under steady state environments. Using equation 19.5 a comprehensive analysis has been made of environmental influences and particularly variation in light, on the growth of *Impatiens parviflora*, a plant of both woodland and grassland (Evans 1976). The species has been grown under a wide range of conditions, in shaded woodland and clearings, in controlled environments in the laboratory and partial controlled environments in the botanical garden. Measurements of the parameters in

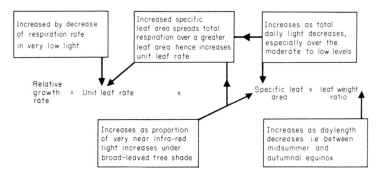

FIG. 19.4.   A summary of the effects which varying light climates have on the relative growth rate of *Impatiens parviflora* (Evans 1976).

equation 19.5 and detailed specification of the different environments led to a synthesis of the effects which *Impatiens parviflora* shows in its response to shade (Fig. 19.4). In this analysis the plant is considered as a dynamic system, and environmental influence, in this case shade, has effect on a component of the growth process, e.g. specific leaf area which in turn effects another component, e.g. unit leaf rate, and has a net effect on R_w.

Evans made his interpretation of changes in response to shade through the solution for the parameters of equation 19.5 in different environments. However, equation 19.5 itself does not incorporate the dynamic aspects of his description of the changes which take place in plant structure, i.e. the control of specific leaf area and leaf weight ratio are not specified. Thornley & Hurd (1974) have attempted analyses of the dynamic aspects of assimilate distribution by considering plant weight, W, in two parts, that of structural, W_G and storage, W_S, materials. They made a series of experiments (Hurd & Thornley 1974) where young tomato plants in water culture were exposed to different steady state conditions of daylength, total light received and atmospheric CO_2 concentration. As conditions were made more favourable for growth there was an increase in net assimilation rate, equivalent to the unit leaf rate of equation 19.5, accompanied by a decrease in leaf area ratio.

However with continuing increase in net assimilation rate the changes caused in R_W and leaf area ratio progressively decreased. In specifying a model to describe this system Thornley & Hurd (1974) made two assumptions about the structure and growth of tomato plants:

(i) L_A is proportional to the structural weight of the plant, W_G, i.e. $L_A = F_G \cdot W_G$ where F_G is a constant (i.e. specific leaf area of equation 19.5 is a constant).

(ii) Structural growth is dependent on the weight of storage material, W_S, i.e. leaf weight ratio is made dynamic.

They expressed these relationships, using the Δ notation to denote increase in unit time rather than the d notation of instantaneous rates

$$\frac{\Delta W_G}{\Delta t} = \frac{R_W \cdot W_G \cdot W_S}{W_S + bW_G} \quad (19.6)$$

Increase in storage material was specified as

$$\frac{\Delta W_S}{\Delta t} = L_A \cdot E_A - \frac{\Delta W_G}{\Delta t} \quad (19.7)$$

where E_A is net assimilation rate expressed on a leaf area basis.

The maximum value of growth rate is obtained when W_S is large, the maximum value of F_G occurs when there is no storage material. b specifies the way in which the growth of the plant depends upon the amount of storage substrate material present, when $b = 0$ the plant works at a constant rate independent of substrate level unless the substrate level itself is zero when the growth process works at the level at which substrate is produced. For $b = 1$ the growth process responds linearly to changes in substrate concentration. Thornley & Hurd (1974) suggested that for the young tomato plant a value of b between 0·5 and 1·0 was appropriate to their range of experiments but implied that this might change if temperature was varied.

A more complex model relating the parameters of net assimilation rate and relative growth rate to the physiology of both carbohydrate and nutrient uptake and utilization was developed by Cooper & Thornley (1976). This was applied in the analysis of experiments with young tomato plants grown in water culture and exposed to various levels of root temperature, day-lengths and light intensities. Measurements were made of per cent dry weight of non-structural carbohydrate, C, and per cent dry weight of Kjeldahl nitrogen, N. Plant relative growth rate was seen to increase as a function of the product of these, i.e. $C \cdot N$. Large values of C occurred with small values of N at low root temperatures whilst high values of N occurred with low C and at high temperatures. The fraction of plant dry weight occurring in the root decreased with time and was lower at high temperatures.

Their model described the plant in two parts, root and shoot. The rate of utilization of C and N for growth were functions of their concentrations in each part of the plant, and the fluxes of C to roots and of N to the shoot were considered to depend on their relative concentrations in the two organs. Plant growth is thus described as an interlocking network of rates (Fig. 19.5) and was written in six differential equations with six unknowns; the structural dry weight of root and shoot, W_{GR} and W_{GS}, the soluble carbohydrate concentrations of root and shoot, C_R and C_S and the nitrogen concentrations in root and shoot N_R and N_S.

Cooper & Thornley (1976) did not attempt rigorous optimization of the

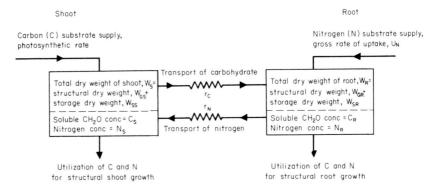

FIG. 19.5. Diagrammatic representation of a model for the partitioning of the growth of the tomato plant into root and shoot components. The transport of carbohydrate from shoot to root is opposed by the resistance r_C, the transport of nitrogen from root to shoot is opposed by the resistance r_N (Cooper & Thornley 1976).

parameters in their model but did obtain qualitative agreements between model and data for the ratio R_W : net assimilation rate and the relationship of whole plant N and C and root temperature. However they considered the predictions of root : shoot weight ratio by the model as less satisfactory and drew attention to the difference between what they termed 'fast' variables, i.e. C and N, which dominated the short term behaviour of the system, and 'slow' variables, e.g. root : shoot ratio, which dominated the long term behaviour. They considered that more needs to be known about the time dependence of these two types of variable.

The studies made by Thornley & Hurd (1974) and Cooper & Thornley (1976) demonstrate the potential for modelling growth directly in terms of basic physiological processes where it is required to understand the effect of different environmental regimes. However, their models, and indeed other models derived from the growth analysis equations, may have limited value for ecologists wishing to analyse environmental influence on plant growth for any purpose other than a comparison of two very different regimes, e.g. sun and shade.

Firstly, these models are in the form of systems of differential equations and whilst this is generally appropriate for dynamic systems their solution for an exhaustive series of steady state conditions is rarely possible in an ecological framework. In the field, a plant experiences the environment as a series of changes and the effect of these on growth can be recorded.

Secondly, although these models specify the dynamic relationships between physiological processes and growth in more detail than previously attempted they are still of restricted scope insofar as many plants growing in uncontrolled environments are concerned. Analyses of observed growth differences have demonstrated how the environmental factors with the greatest influence on growth may change over short time scales or geographical distances.

Thirdly, the link between 'short' term physiological and 'long' term growth processes has yet to be achieved.

Fourthly and perhaps most importantly, these models, based on systems of differential equations, take no explicit account of error.

ENVIRONMENT→PLANT GROWTH VIEWED AS A STOCHASTIC, DYNAMIC SYSTEM

Differences in plant growth between years can be related only imprecisely to environmental differences because in making integrated assessments of growth and environment separately the details of the within-season dynamic relationships are lost. Attempts have been made to describe the course of plant growth in different environments both in terms of logistic equations and by the inter-relationship between physiological processes. These approaches were discussed earlier and take account of some of the dynamic aspects of plant growth and are often expressed in terms of systems of differential equations. However two problems arise with this approach. Firstly, although it has been possible to specify fluctuations in some physiological processes with considerable detail, how these combine in the total growth process still remains to be solved. Secondly, systems of differential equations are most suited to the analysis of experiments under stable conditions or where a known and readily quantifiable environmental variation is applied (Milsum 1966). However, there seems little evidence, even for the growth of agricultural crops where there is considerable uniformity of soil and plants, that such conditions will occur over time spans of interest for plants grown in typical field situations.

Evans (1972) stated the problem, 'To understand in detail the growth of a plant in a natural environment at any particular time, it would thus appear to be necessary to record in detail and to interpret the results of the past environments which have contributed to its make-up. But in a climate such as that in Great Britain where substantial fluctuations of hot and cold, wet and dry, can succeed each other irregularly throughout the growing season, this is counsel of despair'.

One approach to this problem is to construct a deterministic model of plant growth containing a very detailed specification of physiological properties (Landsberg, this volume). Empirical relationships may be incorporated into such a model from a range of different experiments but necessarily, as the number of relationships specified increases, rigorous treatment of error becomes impossible (Fisher 1924).

An alternative approach is to continue the direct analysis of growth but at a time scale which corresponds to the time scale of environmental variation. There is evidence from work with trees that growth rates do change in response to short term changes in the environment. White (1972, 1974) recorded daily height increments between zero and 0·6 cm for young *Pinus sylvestris* and zero and 0·7 cm for young *Betula pubsecens* and found positive correlation between these rates and different sets of environmental variables at different times of the growing season (Fig. 19.6). During the early part of a growing season Ford, Robards & Piney (1978) measured daily cell production along the radial files of tracheids of *Picea sitchensis* between zero and 12. There was a correlation of $+0·69$ ($P = 0·01$) between number of cells produced and total solar radiation. Deans (1979) measured a concentration of 23·5 cm of fine root ($\leq 1·0$ cm diameter) cm^{-3} of soil in the fallen needle and decomposing turf surface horizons of a 14 yr old monoculture of *P. sitchensis* in mid June. During the following period of 10 days with no rain this decreased to 13 cm cm^{-3} whilst soil moisture tension decreased from 0·0 to $-0·26$ bars. Rain fell and in the following 10 days soil moisture tensions increased to $-0·13$ bars whilst root concentration increased to 20·5 cm cm^{-3}.

Direct analysis of growth responses requires acknowledgement that our assessment of influences may not be completely accurate. We must analyse growth as a dynamic process but also as a stochastic one, i.e. structure our model to account for uncertainty (Gold 1977).

The uncertainties are likely to arise because of incomplete specification of inputs, inadequate specification of relationship between environment and growth and inaccuracies in measurement in both environment and growth. Such an approach does not question that plant growth is the result of photosynthesis, translocation, nutrient uptake etc. but does question our present ability to specify how the environment might limit growth through these processes in particular conditions.

As well as the foregoing which influences the decision to consider environmental control of plant growth as a stochastic, dynamic system it is necessary to consider the following six points before deciding in detail which particular mathematical method to adopt.

1. The weather comprises non-random variation at a range of time scales. Diurnal and seasonal effects are obvious but patterns also occur at intermediate frequencies (e.g. Craddock 1957; Dumont & Boyce 1974) which means that even after accounting for seasonal variation, day to day variation may not automatically be treated as a series of independent events.

2. The weather is largely the result of changes in the heat balance at the earth's surface. This itself is a dynamic system and it is well understood that distinct measurable attributes of the weather may be correlated. What is perhaps less frequently mentioned is that as a result of the working of the dynamic system, e.g. radiation increase → heat gain → increase in v.p.d., such correlations may take place with a time lag, and, because we may also

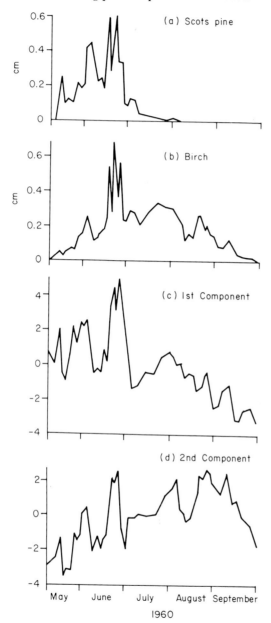

FIG. 19.6. Within-season variation of height growth of two species of trees showing correlation with variation in weather. (a) and (b) Mean height increment for each measurement period (1–10 days) for Scots pine and birch respectively, expressed as height increment per day. (c) and (d) First and second components respectively of a principle component analysis of 20 meteorological variables (White 1972).

anticipate time lags in growth responses as the environment changes, due allowance must be made in the analysis.

3. There may be time varying interactions between weather variables, plant community structure and the interaction of the latter on its own microclimate. Thus conditions favourable to growth for a field crop, e.g. intense solar radiation and high temperature, are more likely to produce a more rapid depletion of soil moisture for similar conditions late in the season when a full leaf canopy has developed than during an early phase of growth.

4. The receptiveness of a plant to environmental variation may be influenced by its developmental stage. Peacock (1975) transferred *Lolium perenne* from the field into controlled environments in both spring and autumn and measured leaf extension in relation to air temperature. Rates of extension at the same temperature were higher for spring than autumn plants, hence confirming field observations. Attempts have been made to build this type of variation into models. In a model of shoot elongation for *Picea abies*, Hari (1976) specified increment in terms of functions of temperature and the physiological stage of development. The latter function used three linear components to represent a curvilinear seasonal trend. However, such trends may not be independent of trends in environmental variation (Fig. 19.6) and their detailed representation requires justification.

5. Plants may acclimatize to differences in environmental conditions: Denne (1976) grew seedlings of *Picea sitchensis* in a nursery and transferred samples at different times to a range of different light intensities in controlled environments. All transferred plants showed an initial decrease in the rate of cell production by the vascular cambium compared to plants remaining in the nursery. The decrease was larger, i.e. to 65%, and lasted for longer, up to 35 days, in the low light treatment. Following the decrease plants in all treatments at least regained rates of cell production comparable to those of nursery plants. Denne (1976) suggested that this mechanism could buffer the plant against environmental change. However its effectiveness would depend upon the magnitude and duration of change.

6. In communities of large plants with strongly developed microclimates distinct spatial variation can occur in growth rates both between individuals and between different plant parts, e.g. shoots at different heights in a tree canopy. Cochrane & Ford (1978) measured a change in the relationship between leader and branch growth at the apex of plantation grown *Picea sitchensis*. After the lower branches of neighbouring young trees came to touch, when the crop was about six years old, there was apparent competition for resources between the leader and branches of the terminal whorl—the greater number of branches the shorter the extension of the leader. Subsequently, when the canopy had developed further, i.e. from year 11, and competition between trees took place large trees produced more branches and had greater branch growth than their smaller neighbours.

Thus to summarize, any mathematical methods used to elucidate the relationship between changes in weather and plant growth as a stochastic, dynamic system must be sufficiently powerful to allow these situations listed above to be accounted for. In the next section a technique, time series analysis, of considerable power is introduced and a simple application is shown for illustrative purposes.

TIME SERIES ANALYSIS

When it is possible to record changes in plant response concurrently with changes in the weather the techniques of time series analysis (Kendall 1973; Box & Jenkins 1970) can be used to identify influences of environment on plant growth. The technique requires a large number of measurements, i.e. usually greater than 30, of both the environmental variable and plant response, to be taken at equally spaced times and inevitably this means that the technique is most suitable where non-destructive measurements can be made automatically.

The basic model (Fig. 19.7) assumes (i) an input, e.g. a weather variable, (ii) an output, i.e. an attribute of plant function which responds to variation in the input and (iii) a variable, called the 'noise series', which measures all the factors other than the input that influence the estimate of the relationship between input and output. Through time series analysis a 'transfer function' relating output to input as a dynamic system can be estimated. The method does not assume that the fitted transfer function can match plant function to

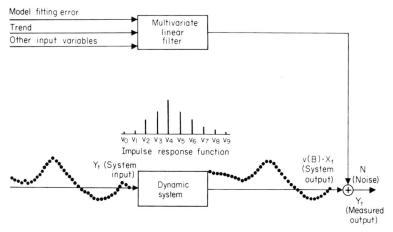

Fig. 19.7. Input and output time series in relation to a dynamic system. The effect of a given impulse response function, v (B) and an input, X_t, is shown and the addition of a multicomponent noise series, N_t, is included to illustrate how time series analysis treats a measured variable Yt.

environment exactly but its failure to do so is estimated as the values of the noise series. Obviously the magnitudes of the values in the noise series are an important criteria of fit, the variance of the noise series should be less than the variance of the original plant response series but perhaps more important is whether the noise series is a random sequence or has a distinct temporal pattern. If the latter, then it is an indication that an additional input variable is influencing the measured plant response and that influence remains in the values of the noise series. By further analysis of the noise series using the same techniques again, it may be possible to quantify this further effect of environment on plant growth.

The technique can be generalized to be multivariate but for illustrative purposes, as below, or to improve the success of application, it is best restricted to describing the relationship between two variables close in the chain of environmental and physiological mechanisms. Thus the following describes its application to the quantification of the relationship between shoot water potential and trunk radius of *Picea sitchensis* in a plantation forest. Values of shoot water potential, P_t, and trunk radius (relative to an arbitrary origin), R_t, at hourly intervals over a period of seven days are presented in Fig. 19.8.

On examination of the time series of stem radii it is noticed (Fig. 19.8) that the stem shrank as well as expanded during the seven days. Net expansion was about 140 μm and shrinkage was about 60 μm day^{-1}. It is well known that two processes are operating together to give this pattern of measurements. First, the expansion and contraction of the wood cells due to variation in their water status, and second, the irreversible growth process, i.e. production and expansion of new cells.

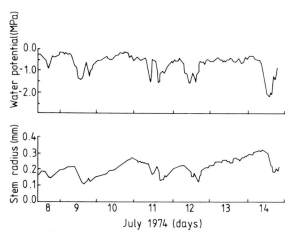

FIG. 19.8.   Shoot water potential (P_t) and stem radius (R_t) measured at hourly intervals for 12 year old Sitka spruce growing in a plantation at Greskine Forest, Scotland.

Earlier work on modelling the effect of shoot water potential on stem radius usually involved linear regression of these two variables and the use of the fitted equation to correct stem radius measurements if required (Ahti 1973; Worrall 1966). The dynamic nature of the processes however was not considered and this is a serious omission as the measurements show a 'hysteresis', with evening and morning diameters being equal but potentials for that size being different. Time series analysis, however, implicitly includes the dynamic effects in its estimate from the measured data of the relationship between shoot water potential and stem radius.

Now the measurements shown in Fig. 19.8 were taken at equally spaced times and the appropriate form of dynamic relationship for such sampled data is the discrete-difference equation. Typical orders of such equations which time series analysis can use are

'1st order' discrete-difference equation

$$(1 + \zeta_1 \nabla)R_t = gP_t \tag{19.8}$$

'2nd order' discrete difference equation

$$(1 + \zeta_1 \nabla + \zeta_2 \nabla^2)R_t = gP_t \tag{19.9}$$

where ∇ is the Difference Operator

i.e.

$$\nabla R_t = R_t - R_{t-1} \tag{19.10}$$

$$\nabla^2 R_t = R_t - 2R_{t-1} + R_{t-2} \tag{19.11}$$

and ζ_1, ζ_2 and g are constants.
These equations have equivalent differential equations for the continuous measures of the same variables

'1st order' differential equation

$$(1 + T_1 D)R(t) = GP(t) \tag{19.12}$$

'2nd order' differential equation

$$(1 + T_1 D + T_2 D^2)R(t) = GP(t) \tag{19.13}$$

where D is the differential operator d/dt and T_1, T_2 and G are constants.
If required the discrete-difference form of model obtained by time-series analysis can subsequently be converted to the differential form. The discrete-difference equations can also be written in other forms

'2nd order' in B notation

$$\{1 - \delta_1 B - \delta_2 B^2\}R_t = \omega_o P_t \tag{19.14}$$

where

$$(1 - \nabla)R_t = BR_t = R_{t-1} \tag{19.15}$$

δ_1, δ_2 and ω_o are constants and general form in Impulse Function notation

$$R_{t_0} = \sum_{t=0}^{\infty} v_t P_{t_0-t} \qquad (19.16)$$

where v_t's are constants and called the Impulse Function.

The latter form is of particular interest as it shows that present values of radius are a weighted sum of earlier values (see Fig. 19.7).

There are many different possible orders of equation which might fit the data, the first part of the technique described here is a method of choosing a suitable order using the measured data. This identification involves estimation of the serial cross-correlation function between the output (R_t) and the input (P_t)

i.e.

$$r_k = C_k/C_o \qquad (19.17a)$$

where

$$C_k = 1/N \sum_{t=0}^{N-k} (R_t - \bar{R})(P_{t-k} - \bar{P}) \qquad (19.17b)$$

for $k = 0, 1, 2, \ldots K$ (lag values) and $\bar{R}$, and $\bar{P}$ are mean values.

The shape of this function is then compared with the various theoretically possible shapes of the family of difference equations (Box & Jenkins 1970). The appropriate order of equation can usually be readily chosen, the number of parameters being normally limited to two or three. The theoretically possible cross-correlation shapes are those of the appropriate Impulse Function since the latter relates present outputs to earlier inputs as does the Impulse Function. However there is a condition that applies to this technique: it can only be used directly if the input to the system is completely random. The reason for this is that any persistence relations, i.e. serial dependence of data values, in the input will appear in the cross-correlation function as well as the effect of the system alone, a situation which does not occur with a completely random input variable. Now this condition cannot normally be met in plant systems in the field as was discussed earlier. The input measurements must then be 'prewhitened' before estimation of the cross-correlation function. The non-randomness of a variable can be seen from its auto-correlation function:

$$a_k = b_k/b_o \qquad (19.18a)$$

where

$$b_k = 1/N \sum_{t=0}^{N-k} (P_t - \bar{P})(P_{t-k} - \bar{P}) \qquad (19.18b)$$

for $k = 0, 1, 2, \ldots K$

This function can show two aspects of non-randomness, (i) nonstationarity, and (ii) persistence or serial correlation of adjacent values. Nonstationarity can usually be removed by differencing (equation 19.10) of the raw data revealing the persistence pattern in the autocorrelation function of the differenced data series. A simple 1 or 2 parameter model of this persistence can usually then be proposed and with the differencing fitted to the raw input variable.

Thus the autocorrelation function of the shoot water potential data and of its first differences are shown in Fig. 19.9. It can be seen that the actual values are highly autocorrelated (i.e. persistence is present) or perhaps the data is nonstationary (non-zero mean). The first difference however shows little if any serial correlation. It was decided therefore to remove non-stationarity by differencing and to model the persistence by the one parameter 1st order autoregressive or Markov relationship. Thus the following equation was iteratively fitted to the input data

$$(1 - \delta B)\nabla P_t = a_t \tag{19.19}$$

where a_t is value of completely random variable at time t and δ is a constant.

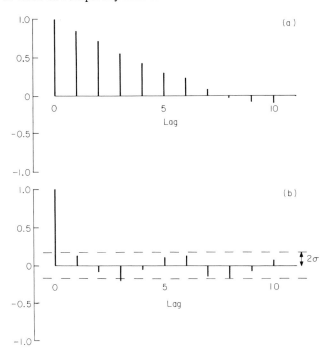

FIG. 19.9. Autocorrelation functions of (a) shoot water potential (P_t) and (b) 1st differences of shoot water potential (∇P_t). Note the high serial correlation over several adjacent values of P_t indicating non-stationarity but negligible such correlations for ∇P_t.

δ was found to have a value of 0·1211 (actually non-significant suggesting that ∇P_t would be more simply modelled as random) and σ_a^2 was 902·1. Once the input variable model is available it is the principle of 'prewhitening' to use that model to transform the input measured data to a completely random noise and the same model form and parameter values to weight the measured output data. The persistence structure of the input is thus equally corrected for in input and output. The cross-correlation function of the corrected data thus shows only the effect of the system Impulse Function. The potential and radius data were thus weighted according to the models

$$\alpha_t = (1 - 0·1211B)\,\nabla P_t \qquad\qquad (19.20a)$$

$$\beta_t = (1 - 0·1211B)\,\nabla R_t \qquad\qquad (19.20b)$$

where α_t and β_t are the weighted or 'prewhitened' input and output respectively.

On comparison with the various possible shapes the cross-correlation function of α_t and β_t showed most similarity with that expected from a second order difference equation model (Fig. 19.10). Note that there also was a time delay in the correlation function of ≤ 1 hour, although this was probably an artefact of the relative timing of the sampling of water potential and radius in the field, for completeness it was not discarded.

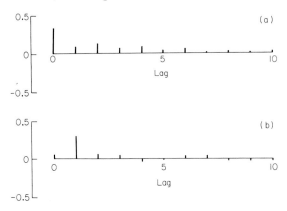

FIG. 19.10. (a) Impulse Response of typical '2nd order system'. (b) Estimated impulse response of stem radius (R_t) versus shoot water potential (P_t). Note the similarity (except for a shift of 1 lag unit) of decay of the values for the theoretical and estimated responses.

It is therefore proposed that shoot water potential and stem radius are related by

$$(1 - \delta_1 B - \delta_2 B^2)R_t = \omega_o P_{t-1} + n_t \qquad\qquad (19.21)$$

where n_t is an interfering noise series.

Iterative least squares fitting then gave optimal parameter values of

$$\delta_1 = 0\cdot339$$
$$\delta_2 = 0\cdot264$$
$$\omega_o = 0\cdot314$$

From equation 19.21 it can be seen that the variation in R_t can be split into two components where

$$R_t = Q_t + N_t \tag{19.22a}$$

and

$$Q_t = f(B)\cdot P_{t-1} \tag{19.22b}$$

Q_t is the influence of shoot water potential on stem radius while N_t is a residual noise series. Fig. 19.11 illustrates these relationships as a block diagram. The N_t's, being that part of stem radius not due to water potential changes, are a measure of irreversible growth, the effect of other weather variables, and the fitting error of the water system model; of these, growth must be the major component. These residuals, N_t, which can be considered as stem radii corrected for the effect of water potential, are shown in Fig. 19.12 along with the original radii R_t and the values of Q_t, which are the effects of shrinkage. It can be seen that the large effect of shrinkage has been eliminated leaving a series whose largest component must indicate the slow variation of irreversible growth.

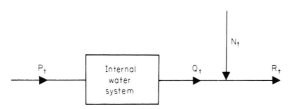

FIG. 19.11. Block diagram of model relating stem radius measurements (R_t) to shoot water potential (P_t). Q_t is the output of the internal water system and N_t is an interfering noise series.

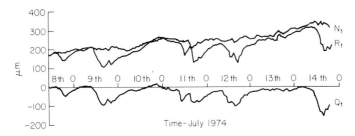

FIG. 19.12   Measured stem diameter (R_t) and estimated shrinkage effect (Q_t) due to changes in shoot water potential. N_t is the residual 'noise' series after removing the effect of Q_t from R_t and hence must indicate true growth of the stem as well as the effect of errors in fitting and interference from unmeasured variables.

In summary this example illustrates how time series analysis can help to elucidate relationships in plant growth analysis. The method offers two particular advantages: (i) a structured system for the rigorous examination of error which also allows estimation of residual variation time series; (ii) specification of relationships in an implicitly dynamic form which may be discrete difference or continuous differential equations.

CONCLUSIONS

The ecologist who chooses to ask questions of the type 'Which environmental factors or properties of my plant limit its growth?', can only ever obtain a partial answer. Both weather and plant growth are dynamic systems varying over time in such a way that component processes are constantly changing in their rate of activity. However, a full description of weather and plants as dynamic systems in a particular situation may not only be impossible for the ecologist to obtain; it may be an entirely inappropriate objective. He has to be prepared to place a limit on his description which can be used to judge the usefulness of his model as his purposes change. To this end we urge that where a modelling system is to be used then a specification of error is essential. For models of dynamic systems this must be considered in terms of both absolute value and the patterns of error in time.

Over the past decade, since the last B.E.S. symposium on the topic (Wadsworth 1958), considerable advances have been made in the understanding of environment to plant relationships but generally only under well controlled conditions. There has also been a recognition that attributes hitherto considered as parameters which could be simply estimated must be considered as variables over longer time periods and under different baseline conditions.

A similar difficulty has emerged in attempts to link the 'fast' physiological processes with the 'slow' process of growth (Cooper & Thornley 1976). One conclusion from these difficulties is that the ecologist must continue with the process of systems identification and not assume that his task is simply the quantification of models already written by the physicist and physiologist. We have shown in this paper that the parsimonious incorporation of parameters into time series models in a manner similar to that proposed by Box & Jenkins (1970) offers considerable advantages in quantifying that stochastic dynamic system which is the growing plant.

REFERENCES

Ahti E. (1973) Correcting stem girth measures for variations induced by soil moisture changes. *Communicationes Instituti Forestalis Fenniae*, **78**, 4.

Austin R.B., Nelder J.A. & Berry G. (1964) The use of a mathematical model for the analysis of manurial and weather effects on the growth of carrots. *Annals of Botany*, **28**, 153–162.

Boyd W.J.R., Goodchild N.A., Waterhouse W.K. & Singh B.B. (1976) An analysis of climatic environments for plant-breeding purposes. *Australian Journal of Agricultural Research*, **27**, 19–33.

Box G.E.P. & Jenkins G.M. (1970) *Time Series Analysis: Forecasting and Control*. Holden-Day, San Francisco.

Cochrane L.A. & Ford E.D. (1978) Growth of Sitka spruce plantation. Analysis and stochastic description of the development of the branching structure. *Journal of Applied Ecology*, **15**, 227–244.

Cooper A.J. & Thornley J.H.M. (1976) Response of dry matter partitioning, growth, and carbon and nitrogen lengths in the tomato plant to changes in root temperature: Experiment and theory. *Annals of Botany*, **40**, 1139–1152.

Cormack R.M. (1971) A review of classification (with discussion). *Journal of the Royal Statistical Society*, Series A, **134**, 321–367.

Craddock J.M. (1957) An analysis of the slower temperature variations at Kew Observatory by means of mutually exclusive band pass filters. *Journal of the Royal Statistical Society*, Series A, **120**, 387–397.

Deans J.D. (1979) Fluctuations of the soil environment and fine root growth in a young Sitka spruce plantation. *Plant & Soil*, **52**, 195–208.

Denne M.D. (1976) Effects of environmental change on wood production and wood structure in *Picea sitchensis* seedlings. *Annals of Botany*, **40**, 1017–1028.

Dumont A.G. & Boyce D.S. (1974) The probabilistic simulation of weather variables. *Journal of Agricultural Engineering Research*, **19**, 131–145.

Erickson R.O. (1976) Modelling of plant growth. *Annual Review of Plant Physiology*, **27**, 407–434.

Evans G.C. (1972) *The Quantitative Analysis of Plant Growth*. Blackwell, Oxford.

Evans G.C. (1976) A sack of uncut diamonds. The study of ecosystems and the future resources of mankind. *Journal of Applied Ecology*, **13**, 1–39.

Fisher R.A. (1924) The influence of rainfall on the yield of wheat at Rothamstead. *Philosophical Transactions of the Royal Society, London*, Series B, **213**, 89–142.

Ford E.D., Robards A.W. & Piney M.D. (1978) Influence of environmental factors on cell production and differentiation in the early wood of *Picea sitchensis*. *Annals of Botany*, **42**, 683–692.

Fritts H.C. (1976) *Tree Rings and Climate*. Academic Press, New York.

Gittins R. (1969) The application of ordination techniques. *Ecological Aspects of the Mineral Nutrition of Plants*, Symposia of the British Ecological Society 9 (Ed. I.H. Rorison), pp. 37–66. Blackwell, Oxford.

Gold H.J. (1977) *Mathematical Modelling of Biological Systems—An Introductory Guide*. Wiley–Interscience, New York.

Goodchild N.A. & Boyd W.J.R. (1975) Regional and temporate variations in wheat yield in Western Australia and their implications in plant breeding. *Australian Journal of Agricultural Research*, **26**, 209–217.

Hari P. (1976) An approach to the use of differential and integral calculus in plant autecology. *Research Note, 13*. Department of Silviculture, University of Helsinki.

Hunt R. (1979) Plant growth analysis: the rationale behind the use of the fitted mathematical function. *Annals of Botany*, **43**, 245–249.

Hurd R.G. & Thornley J.H.M. (1974) An analysis of the growth of young tomato plants in water culture at different light integrals and CO_2 concentrations. I. Physiological aspects. *Annals of Botany*, **38**, 375–388.

Julian P.R. & Fritts H.C. (1968) On the possibility of quantitatively extending climatic records by means of dendroclimatalogical analysis. *Proceedings of the First Statistical Meteorology Conference*, pp. 76–82. American Meteorological Society, Hartford, Connecticut.

Kendall M.G. (1973) *Time Series*. Griffin, London.

LaMarche V.C. (1974a) Paleoclimatic inferences from long tree-ring records. *Science*, **183**, 1043–1048.

LaMarche V.C. (1974b) Frequency-dependent relationships between tree-rings series along an ecological gradient and some dentroclimatic implications. *Tree-Ring Bulletin*, **34**, 1–20.

LaMarche V.C. & Fritts H.C. (1971) Tree rings, glacial advance, and climate in the Alps. *Zeitschrift für Gletscherkunde und Glazialgeologie*, **7**, 125–131.

Malcolm D.C. (1970) Site factors and the growth of Sitka spruce. Ph.D. Thesis, University of Edinburgh.

Mikola D. (1962) Temperature and tree growth near the northern timber line. *Tree Growth* (Ed. T.T. Kozlowski), pp. 442. Ronald Press, New York.

Miller H.G. & Cooper J.M. (1976) Tree growth and climatic cycles in the rain shadow of the Grampian mountains. *Nature, London*, **260**, 697–698.

Milsum J.H. (1966) *Biological Control Systems Analysis*. McGraw-Hill, New York.

Nelder J.A., Austin R.B., Bleasdale J.K.A. & Slater P.J. (1960) An approach to the study of yearly and other variation in crop yields. *Journal of Horticultural Science*, **35**, 73–82.

Patterson J.G., Goodchild N.A. & Boyd W.J.R. (1978) Classifying environments for sampling purposes using a principal component analysis of climatic data. *Agricultural Meteorology*, **19**, 349–362.

Peacock J.M. (1975) Temperature and leaf growth in *Lolium perenne*, III Factors affecting seasonal differences. *Journal of Applied Ecology*, **12**, 685–697.

Ralston C.W. (1964) Evaluation of forest site productivity. *International Review of Forest Research*, **1**, 171–201.

Richards F.J. (1959) A flexible growth function for empirical use. *Journal of Experimental Botany*, **10**, 290–300.

Thornley J.H.M. & Hurd R.G. (1974) An analysis of the growth of young tomato plants in water culture at different light integrals and CO_2 concentrations. II A mathematical model. *Annals of Botany*, **38**, 389–400.

Waring R.D. & Cleary B.D. (1967) Plant moisture stress: evaluation by pressure bomb. *Science*, **155**, 1248–1254.

Wadsworth R.M. (Ed.) (1968) *The Measurement of Environmental Factors in Terrestrial Ecology*. Symposia of the British Ecological Society 8. Blackwell, Oxford.

White E.J. (1972) Orthogonalized regressions of height increments on meteorological variables. *Research Papers in Forest Meteorology: an Aberystwyth Symposium* (Ed. J.A. Taylor), pp. 109–125. Cambrian News, Aberystwyth.

White E.J. (1974) Multivariate analysis of tree height increment on meteorological variables, near the altitudinal tree limit in northern England. *International Journal of Biometeorology*, **18**, 199–210.

Worrall J. (1966) A method of correcting dendrometer measures of tree diameters for variations induced by moisture stress change. *Forest Science*, **12**, 427–429.

20. THERMAL TIME AND TEA

G.R. SQUIRE

Department of Physiology and Environmental Studies, University of Nottingham, Sutton Bonington, Loughborough, Leicestershire

SUMMARY

In well managed tea plantations, yield is not normally limited by shortage of assimilate, but by the effects of atmospheric environment on the growth of shoots. Field observations and glasshouse experiments in Central Africa have shown air temperature to be the most important factor determining the rate at which shoots extend, except during the seasonal drought when extension is limited by dry air even when the soil is irrigated. Preliminary studies suggest that the expression of shoot extension in thermal time (day-degrees) may be useful in comparing the growth of shoots at different times of the year and in different parts of the world.

INTRODUCTION

Tea (*Camellia sinensis* L.), if left to grow naturally, forms a large shrub or medium sized tree, but in commercial plantations is cultivated to form flat topped bushes about one metre high. Vegetative shoots, which originate from the axils of leaves on the tops of bushes, are harvested when 10–12 cm long, and then axillary buds below the remaining stumps begin to grow and so continue the cycle of shoot production. In many parts of the world the yield of shoots from plantations is not evenly distributed throughout the year. The workforce and factories employed in harvesting and manufacturing tea can be overstretched during months of heavy yield, but redundant when yield is light or negligible. The need for a solution to the social and economic problems caused by poor distribution of yield has been the major stimulus for research into the effect of the atmospheric environment on the tea plant. In Malawi the Ministry of Overseas Development has financed a series of projects over the last twelve years at the Tea Research Foundation of Central Africa with the aim of understanding the physiology of shoot growth and the environmental factors which affect it. The climate, the weather and the nature of the problem have been described by Fordham (1970), Ellis (1971) and Green (1971). Most of the shoots are harvested in only five months and then yield remains small throughout the cool season (May to August) and the dry season (September to November). The seasonal pattern of yield is determined largely by changes in the rate of growth of shoots, which require six or seven weeks to reach harvestable size in the main cropping season, and approximately double this time in the cool season.

Four stages in the transfer of energy from sunlight to the harvested shoot have been examined: (1) reception and (2) interception of radiation, (3) efficiency of conversion of radiant energy to carbohydrate, and (4) partitioning of assimilate between the shoots and the rest of the bush. For plantations in Malawi, the first three steps can be eliminated as causes of seasonality in shoot growth. In May, when yield first becomes reduced, the level of mean daily irradiance is similar to values for the previous five months, and well managed plantations having a leaf area index of around six intercept all this available energy. Photosynthesis shows no marked reduction in the cool season (Squire 1977), which suggests that the production of shoots is not closely coupled to current assimilation. Because shoots are removed when they are small, their absolute growth rate is very slow for a large part of their life, so the production of shoots under current management practice is not usually limited by shortage of assimilate (Tanton 1979). Seasonality of shoot growth is caused by factors which affect the partitioning of assimilate. The search for the important environmental factors has involved statistical correlations between meteorological records and monthly yields. However, relating the *yield* of shoots to *current* weather has been misleading, since yields of tea can be depressed for several months as a result of particular combinations of management and weather (T. W. Tanton, unpublished). The matter has also been confused by the undue emphasis placed in India and Sri Lanka on factors such as radiation which affect assimilation directly (Hadfield 1974; Devanathan 1975). The critical factors must be ones which affect the rate of growth of shoots on plants where the supply of assimilate is not limiting, and so far photoperiod and air temperature have received most attention.

From the observation that seasonality is hardly detectable on the equator and becomes more pronounced in plantations further from the equator photoperiod has gained favour as the principal cause (Laycock 1969; Barua 1969; Carr 1972). The main evidence to substantiate this view came from experiments in which daylength was extended or long nights interrupted in the winter season by means of artificial lighting above the canopy (Barua 1969; Fordham 1970; Herd & Squire 1976). However, the size of the effect obtained in these experiments was insufficient to account for seasonality.

The importance of air temperature on shoot extension and leaf expansion in Malawi was suggested by the statistical correlations of Fordham (1970) and Green (1971), but no decisive result emerged, possibly because of the effects of other environmental variables. In a later study in Malawi, Squire (1979) examined weekly rates of shoot extension measured with a ruler in relation to mean air temperatures and to vapour pressure deficits recorded by mercury-in-glass thermometers in a Stevenson screen. Measurements made

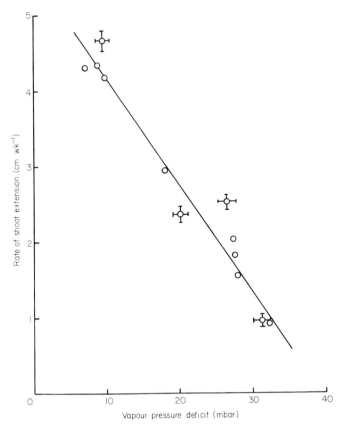

FIG. 20.1. Weekly rates of shoot extension of irrigated tea in relation to mean vapour pressure deficit (measured at 14.00 h each day). The data shown are restricted to periods when mean air temperature was 20–22 °C, out of a total of thirty-six measurements made during all seasons of the year. (Reprinted from *Experimental Agriculture.*)

in all seasons revealed an important effect of vapour pressure deficit on shoot extension even when the soil was irrigated during the dry season (Fig. 20.1). Shoot water potential fell approximately linearly with rise in vapour pressure deficit which was a major factor driving evaporation (Green 1971; Squire 1979). Although the relevant physiological measurements have not been made, the fall in potential caused by the rise in vapour pressure deficit would probably have been accompanied by a fall in turgor sufficient to reduce the rates of cell division and expansion.

This influence of dry air on shoot growth probably obscured the true response to temperature in previous statistical analyses, but if the data were confined to a narrow range of vapour pressure deficit (10 mbar), shoot extension was related linearly to mean air temperature within the range 16–24 °C above an apparent base temperature of about 13 °C (Squire 1979). More detailed measurements by T.W. Tanton (unpublished) confirmed the linearity in response and defined more precisely a base temperature of 12·5 °C. These measurements refer only to air temperature recorded in a Stevenson screen, but the tissue temperature of small, rapidly transpiring leaves and their stems would be within 2–3 °C of air temperature for a large part of the year. The field observations made over the last decade by Fordham, Green, Squire and Tanton all point to mean air temperature as being the most important environmental factor affecting the rate of shoot growth and the seasonality of yield in Malawi, but the type of analysis used could give misleading results if other factors correlated with temperature also changed from season to season. However, the importance of changes in mean air temperature was demonstrated experimentally in a glasshouse built over established bushes (Squire 1978). When mean air temperature was raised over weekly periods during the cooler part of the year when yields are low, shoots grew at rates similar to those observed in the main cropping season.

Using a base temperature (T_o) of 12·5 °C, the growth of a population of shoots is expressed in Fig. 20.2 in terms of thermal time (accumulated day-degrees). Provided that the minimum temperature is above the base temperature, the number of day-degrees for each day is calculated from $\bar{T} - T_o$, where $\bar{T}$ is the mean of the daily maximum and minimum temperatures.

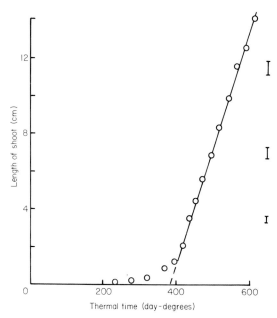

Fig. 20.2. Growth in length of a population of 100 shoots in relation to thermal time. (Reprinted from *Experimental Agriculture*.)

When the minimum falls below the base, day-degrees can be calculated using a pencil and graph paper (and plenty of spare time) or a computer (Gallagher 1976). The number of day-degrees required by shoots for their development and growth to harvestable size has been found by T.W. Tanton (unpublished) to be similar during both the main cropping season and the following cool season, confirming that seasonal changes in mean temperature are the principle cause of seasonality of shoot growth. The relation between growth and thermal time broke down only in the dry season between September and November when saturation deficits were large.

In having a linear response to temperature above a clearly defined base temperature the growth of tea shoots is similar to many other processes of growth and development in plants (Arnold 1959; Iwata 1975; Monteith 1977). Large differences in base temperature have been found between species, ranging from 0 °C for leaf development and extension of winter wheat (Biscoe & Gallagher 1978) to about 20 °C for leaf expansion of cowpea in the humid tropics (Littleton *et al.* 1979). The only analysis of this kind for tea has been made in Malawi. It remains to be seen whether the base temperature of 12·5 °C is universal, or whether tea grown in other parts of the world has adapted to local climate.

REFERENCES

Arnold C.Y. (1959) The determination and significance of the base temperature in a linear heat unit system. *Proceedings of the American Society of Horticultural Science*, **74**, 430–445.

Barua D.N. (1969) Seasonal dormancy in tea. *Nature, London*, **224**, 514.

Biscoe P.V. & Gallagher J.N. (1978) A physiological analysis of cereal yield. 1. Production of dry matter. *Agricultural Progress*, **53**, 34–50.

Carr M.K.V. (1972) The climatic requirements of the tea plant: A review. *Experimental Agriculture*, **8**, 1–14.

Devanathan M.A.V. (1975) Weather and the yield of a crop. *Experimental Agriculture*, **11**, 183–186.

Ellis R.T. (1971) The general background to plant water relation investigations on tea in Central Africa and some findings to date. *Water and the Tea Plant* (Ed. by M.K.V. Carr & S. Carr), pp. 49–57. Tea Research Institute, Kericho, Kenya.

Fordham R. (1970) *Factors affecting tea yields in Malawi.* Ph.D. thesis, University of Bristol.

Gallagher J.N. (1976) *The growth of cereals in relation to weather.* Ph.D. thesis, University of Nottingham.

Green R.M. (1971) *Weather and the seasonal growth of tea in Malawi.* M.Phil. thesis, University of Nottingham.

Hadfield W. (1974) Shade in North-East Indian tea plantations II. Foliar illumination and canopy characteristics. *Journal of Applied Ecology*, **11**, 179–199.

Herd E.M. & Squire G.R. (1976) Observations on the winter dormancy of tea (*Camellia sinensis* L.) in Malawi. *Journal of Horticultural Science*, **51**, 267–279.

Iwata F. (1975) Heat unit concept of crop maturity. *Physiological Aspects of Dryland Farming* (Ed. by U.S. Gupta), pp. 351–370. I.B.H. Publishing Company, New Delhi and Oxford.

Laycock D.H. (1969) Latitude, daylength, and crop distribution. *Proceedings of the Tocklai Experiment Station, Jorhat, Assam, India for 1969.*

Littleton E.J., Dennett M.D., Elston J. & Monteith J.L. (1979). The growth and development of cowpea (*Vigna unguiculata*) under tropical field conditions. 1. Leaf area. *Journal of Agricultural Science, Cambridge,* **93**, 291–307.

Monteith J.L. (1977) Climate. *Ecophysiology of Tropical Crops* (Ed. by P. de T. Alvim and T.T. Kozlowski), pp. 1–27. Academic Press, New York.

Squire G.R. (1977) Seasonal changes in photosynthesis of tea (*Camellia sinensis* L.). *Journal of Applied Ecology,* **14**, 303–316.

Squire G.R. (1978) A simple temperature-controlled glasshouse for field experimentation in the tropics. *Experimental Agriculture,* **14**, 7–12.

Squire G.R. (1979) Weather, physiology and seasonality of tea (*Camellia sinensis* L.) yields in Malawi. *Experimental Agriculture,* **15**, 321–330.

Tanton T.W. (1979) Some factors limiting yields of tea (*Camellia sinensis* L.) *Experimental Agriculture,* **15**, 187–191.

21. THE USE OF MODELS IN INTERPRETING PLANT RESPONSE TO WEATHER

J.J. LANDSBERG

*Long Ashton Research Station, Long Ashton, Bristol BS*18 9*AF*

SUMMARY

One of the more important reasons for using models is that they constitute precise, testable statements of our hypotheses about the workings of a particular system, and its responses to stimuli. As such they should incorporate existing knowledge, explain observations and identify areas where knowledge is most deficient. Statistical analyses of data—say of the multivariate type—not based on a mechanistic model, provide a description of the data but tell us nothing that is not contained in them. When applied to plant responses to weather this is a particularly serious limitation because of the correlations between weather factors and the almost infinite possible permutations and combinations of weather factors. However, statistical analyses may be useful aids in the formulation of hypotheses and have a vital role to play in comparing alternative models, and in deriving the constants and coefficients for model equations.

An outline of a generalized model of plant growth in relation to weather is presented and used to show how such a model, incorporating many currently accepted theories but also including speculative mechanisms, may be used as a framework for studies on plant responses to weather and as an aid in interpreting the significance of those responses. The model consists of a family of sub-models; the form which those dealing with radiation interception, stomatal conductance, transpiration, plant water status, net photosynthesis, respiration, assimilate partitioning, root growth and soil moisture status might take is outlined. The effect of weather on plant development is considered briefly.

INTRODUCTION

The proposition put forward in this paper is that models are not only useful, but are essential tools for interpreting plant response to weather. A mathematical model provides a representation of a system. It should therefore resemble the system and, if the system is dynamic, should be capable of simulating its movements (Thornley, 1976). We may also define a model as a formal statement, or a set of statements, constituting a precise description of a hypothesis about the workings of a particular system and its responses to stimuli. It follows from this definition that models should be testable; i.e. it should be possible to devise experiments which may invalidate them—or parts of them. In strict logic a hypothesis (or model) can never be proved to be true; it can only be 'not invalidated', in which case it stands as an accept-

369

able description of the system. Questions about whether differences between observation and hypotheses are large enough to invalidate the hypothesis, and the extent to which it is justifiable to explain experimental results which do not conform with prediction by proposing subsidiary hypotheses (sub-models), are basic to the philosophy of scientific research (see, for example, Popper (1972)—particularly his first chapter) but space precludes their discussion here.

The formulation of a model should therefore precede *any* experimental work, at *any* level. The procedure will clarify the questions asked, determine the type of experiments which should be done or observations to be made and indicate those areas where data and information are most deficient. Thornley (1976) has given a clear statement of these and other advantages of models.

It is often possible to formulate alternative models which may appear to provide equally valid descriptions of a particular system and, at least in the biological sciences, it will usually be necessary to use statistical methods to decide which is the best. Statistical techniques are also extremely valuable for fitting the equations of a model to experimental data, testing goodness of fit and obtaining numerical values for constants and coefficients. It is some-times possible to fit more than one function to a set of data, and not be able to distinguish, on statistical grounds, which is best. In this case a choice of function may have to be made on mechanistic grounds (is it possible to attribute biological meaning to the constants and coefficients?), or by applying Occam's razor (choosing the simplest form). However, statistical analysis of data, to determine the inter-relationships between a dependent variable and one or more independent variables, does not constitute the testing of a hypothesis. It provides a non-generalizable description of the particular data set and tells us nothing about the mechanisms responsible for the interactions. Because of the correlations between weather factors, and the almost infinite possible permutations and combinations of weather factors, the usefulness of such analyses is particularly questionable when they are applied to plant responses to weather. However, multivariate analysis may be a useful aid in hypothesis formulation (Ford, this volume).

Models may be written at a number of levels but progress in understanding the operation of a system, and in our ability to predict its responses to a wide range of stimuli, is arguably most rapid when the system is described in terms of its sub-systems; i.e. if possible a model should be written in terms of the mechanisms one level of organization lower than that with which we are concerned. For example dry matter production by a plant community might be described in terms of the photosynthetic and respiratory characteristics of single leaves, together with a model of light interception by the canopy and the distribution of light within the canopy. A simpler model in this case might be the postulate that dry matter production over a given time interval is proportional to total radiant energy. This may be adequate in many cases but it would be inadequate in some situations, leading to the need to develop a more complex and powerful model. It is clear that there will be a feed-back system: model → experiment → revised model → experiment. At each step previously unforeseen problems are likely to become apparent and decisions will be required as to whether these require solution before further progress can be made, or whether they can be bypassed.

In formulating models we naturally rely heavily on information already available about the system under study, but areas where knowledge is lacking rapidly become apparent and we may then (legitimately) make assumptions based on guesswork or intuition. Furthermore the formulation of a model, at any level, inevitably involves simplifications deriving from decisions about which processes or effects are 'first-order' and must be included, and which can be ignored. Without such simplifications models would become as complex as the systems under study. Should a model be shown, by experiment, not to represent adequately the system or its behaviour, the simplifications made will be among the factors to consider on re-examining it.

It is essential, when developing a model, that the purpose for which it is required is clearly defined. In the context of plant responses to weather, we must decide whether we are concerned primarily with understanding the effects of weather at the process level (e.g. on photosynthesis, respiration, the development of water stress in particular tissues, cell division, bud morphogenesis); on whole plants (e.g. dry matter production, average time to flower), or on plant communities (e.g. community productivity, species distribution and survival). The final objective in all cases must be the successful simulation of the response of the system to different weather conditions. Ideally we would aim at the development of a hierarchy of models, those at the lower (process) levels providing the means for testing the consequences of the simplifications which will be made in higher level, longer time-scale, models.

It is axiomatic that the time scale across which the weather factors considered to affect a given physiological process are averaged should be of the same order as the response time of the process. For example if plant water potential is to be modelled as a 'steady-state' function of transpiration rate, and a change in transpiration rate is expected to result in a new 'equilibrium' value of water potential within about 10 minutes, there is little purpose in attempting to analyse water potentials in relation to measurements from which instantaneous values of leaf energy balance, and hence transpiration, are calculated. The appropriate averaging time for field measurements would be of the order of half an hour to an hour. Similarly if a change in the morphogenetic state of a plant, attributed to temperature, can only be unequivocally identified across time intervals of about a week, then such changes would be evaluated in terms of mean weekly temperature, or perhaps integration of temperature in terms of time over the week.

To illustrate, for the purposes of this paper, the use of models in interpreting plant response to weather, I present an outline of a generalized model describing the effects of weather on the growth of plants, over a period of months. Developmental processes are treated briefly in a separate section. Such a model would include sub-models describing the interactions, at the physical level, between plants and the weather, and sub-models describing the effects of environmental factors on physiological processes. The time scale of the sub-models may be of the order of hours or days. The complete model provides a framework for integrating and evaluating the accuracy of our knowledge about plant response to weather over a whole season. The sub-models, and hence the whole model, should conform to the requirements stated earlier, that is, they will consist of a series of precise, testable, mathematical statements describing current hypotheses.

The models proposed here are not, of course, the only possible ones; they may be challenged experimentally or on the ground of their assumptions, the mechanisms invoked, their logical development or their sheer inadequacy. However, to be constructive, any challenger should put forward alternative hypotheses which not only successfully explain as much of the behaviour of the system as the originals, but also account for at least some of the anomalies or rectify some of the inadequacies.

A GENERALIZED MODEL OF THE EFFECT OF WEATHER ON PLANT GROWTH

As a basis for modelling plant responses to weather, we require information on the effects of individual factors on plant growth and development. It is common, in experimental work, to consider the effects of factors such as radiant energy, temperature and plant water status separately, acknowledging only in passing that there is considerable interaction at both the physical and physiological level. However, the interactions cannot be avoided in any model purporting to deal with whole plants, and indeed one of the major contributions which models can make is to provide a means of exploring these interactions without the need for long, difficult and expensive multi-factorial experiments in a wide range of environments.

The physical interactions can, in general, be modelled with greater confidence and accuracy than the physiological; for example interactions between radiant energy and the temperature of plant organs can be completely described in terms of energy balance equations, and the feed-back between transpiration and depletion of soil water is relatively straightforward. However, the mechanisms involved in interactions between the physiological processes are often not so clear; carbohydrate accumulation, in leaves or other organs, may reduce photosynthesis and temperature and

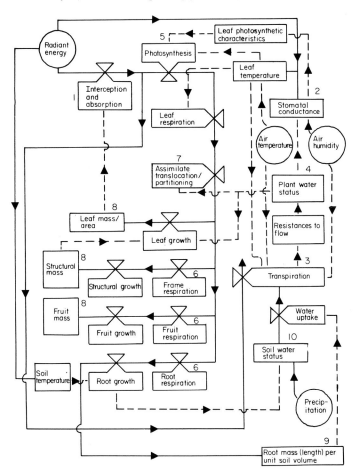

FIG. 21.1.   Schematic presentation of a model of plant growth in relation to weather. Solid lines represent flows of material, dotted lines flows of 'information'. Valves (⋈) represent rates, boxes represent states and circles enclose the environmental driving variables. Numbers indicate the sequence of calculations for programming purposes, i.e. calculate (1) energy interception, (2) stomatal conductance – feed forward to net photosynthesis, transpiration, (3) transpiration, (4) leaf water status – feed back to stomatal conductance (iterate), xylem water potential – feed forward to partitioning equations, (5) net photosynthesis, (6) organ respiration, (7) partition accumulated assimilates, (8) update mass of component parts (convert assimilates to leaf area, root length etc.), (9) volume of soil exploited by roots, (10) update soil status in root zone – feed back to transpiration.

plant water status may alter dry matter allocation patterns, but attempts to simulate these phenomena are likely to be based either on empirical relation-ships or on arbitrary feed-back functions, pools, switches etc., which may be somewhat divorced from reality. Some of these problems are considered in the following discussion of the generalized model of plant growth; this consists of a carbon balance model, incorporating sub-models for photosynthesis, respiration, plant water status and assimilate partitioning. It is outlined as a flow diagram in Fig. 21.1, where the number in some of the 'rate' and 'state' boxes correspond to the sequence of calculations which might be used should such a model be programmed onto a computer; the same sequence is used in the discussions of the form of the main sub-models, which follows.

The sub-models

Radiation interception

There are a great many models of light interception and absorption by plant communities; the principles are well treated by Ross (1975) and have been reviewed by Monteith (1969) and Lemeur & Blad (1974). Theoretically, if the objective is to calculate photosynthesis, the irradiance of each leaf element should be calculated, across a time interval selected as suitable in terms of the response time of photosynthesis. Most mathematical treatments assume that stands are horizontally uniform, i.e. that the characteristics of the stand, in terms of leaf angle and distribution, do not vary within a horizontal layer, and depend only on height (z). If the irradiance on a canopy surface is $Q_s(h)$ the average flux density at any level in the stand ($Q_s(z)$) depends on the amount of energy intercepted by the leaves between z and $z = h$, i.e.

$$\frac{dQ_s}{dA} = -kQ_s(h) \tag{21.1}$$

where k is an extinction coefficient and $A = \int_z^h a(z)$, where $a(z)$ is the leaf area density (surface area of leaves, one side only, per unit volume of canopy space) at height z. When the integration is between $z = 0$ and $z = h$ then $A = A_h$, the leaf area index. Integration of 21.1 gives an expression

$$Q_s(z) = Q_s(h) \exp(-kA) \tag{21.2}$$

which has formed the basis of a number of empirical and theoretical studies of radiation in plant communities. Some of the theoretical treatments are impracticably complex and for most purposes a relatively simple model will suffice. Equation 21.2 is useful as it stands although it is only strictly applicable when foliage is randomly distributed; values for k can be established empirically and these may provide estimates of $Q_s(z)$ which are relatively more accurate than the information on leaf photosynthesis, and its variation with leaf age and condition, which must be combined with the equation to estimate canopy photosynthesis. A less simple, but generally adequate—and easily manageable—model is that given by Monteith (1965a). Models of energy interception by isolated plants (Charles-Edwards & Thornley 1973), widely-spaced row crops (Allen 1974) and hedgerows (Charles-Edwards & Thorpe 1976) are also available.

Radiant energy is a major driving variable for transpiration, but it is not usually necessary to calculate the distribution of net radiation within a canopy; many workers have shown net radiation to be linearly related to short-wave radiation flux density on the canopy top (e.g. Monteith & Szeicz 1961; Fritschen 1967) and estimates of canopy net radiation derived in this way are usually adequate for transpiration calculations. Thorpe (1978) has demonstrated that the model of short-wave radiation absorption by hedgerows (Charles-Edwards & Thorpe 1976) can be used to calculate net radiation per unit leaf area, and hence, via the energy balance equations, transpiration rates for hedgerow apple orchards.

Stomatal conductance

There are a number of extant models of stomatal functioning and response to factors of the external environment (Penning de Vries 1972; Takakura, Goudriaan & Louwerse 1975; Jarvis 1976; Thorpe, Warrit & Landsberg 1980). The models by Penning de Vries and by Takakura *et al.* were shown to mimic stomatal behaviour satisfactorily, but contain a large number of parameters and embody untested assumptions about underlying processes governing stomatal behaviour. The model suggested by Jarvis incorporates empirically derived responses to light, temperature, humidity and leaf water potential, assuming that there is no interaction between these component sub-models. The model by Thorpe *et al.* is similar to that of Jarvis, but simpler, including only responses to light and air humidity. In their experimental work with apples, Warrit, Landsberg & Thorpe (1980) observed the usual 'critical point' response to leaf water potential, but they omitted it from the model on the grounds that it occurred outside the range of water potentials normally observed in the field. It is of interest, in this respect, that Schulze & Küppers (1979) have recently demonstrated that short-term (hours–days) periods of water stress had no effect on the stomatal conductance of hazel leaves, but long-term stress caused progressive reductions in stomatal conductance.

In view of the wide range of stomatal responses to external factors, the enormous leaf to leaf variation (Warrit *et al.* 1980), changing stomatal behaviour with leaf age (discussed by Burrows & Milthorpe 1976) and the present inadequate state of our knowledge about stomatal functioning in general, relatively simple empirical models probably represent the best approach to simulating the stomatal behaviour of plants in the field. These may be tested at the level of their output by measurements of stomatal conductance values in known environmental conditions or in terms of transpiration rates calculated with equations requiring stomatal conductance values (see below).

Transpiration

In the context of modelling plant responses to weather, estimates of transpiration rate and amount are required to calculate plant water potentials and
for soil water balance calculations. One of the most successful models in the
field of plant-weather interactions is the combination (energy balance—mass
transfer) equation for calculating transpiration. The best known and currently
most widely used version is the equation which has become known as the
Penman-Monteith equation; the derivation is given by Monteith (1965b,
1973).

$$\lambda \mathbf{E} = \frac{s\mathbf{Q}_n + \rho c_p \, \partial e g_a}{s + \gamma(1 + g_a/g_s)} \tag{21.3}$$

where $\lambda \mathbf{E}$ is latent heat flux, $\mathbf{Q}_n$ in net radiation, ∂e is the vapour pressure
deficit of the air, s, ρ, c_p and γ represent the slope of the saturation vapour
pressure/temperature curve, air density and specific heat and the psychrometric constant, respectively; g_s and g_a are stomatal and boundary layer
conductances, respectively.

One of the great values of equation 21.3 is that all the weather data
needed as inputs can be obtained from standard weather measurements. The
equation may be applied to single leaves, canopies of all types and individual
trees. When used to calculate transpiration from single leaves it is necessary
to take into account the ratio of the conductances of the two leaf surfaces
(Burrows & Milthorpe 1976). The boundary layer conductance of single
leaves is a function of wind speed and leaf dimensions, but allowance should
be made for mutual interference between leaves (Landsberg & Thom 1971;
Landsberg & Powell 1973). When applied to canopies the boundary layer
conductance is the aerodynamic conductance of the canopy, which depends
on wind speed and physical characteristics such as plant height and spacing.
The stomatal conductance in this case is usually called canopy conductance
(g_c) and can be estimated as $g_c = \Sigma (A_i \cdot g_{s,i})$, i.e., by summation of the
product of leaf area in layer i times stomatal conductance in that layer.
When there is good reason to believe that g_s varies with depth into the
canopy, either because of leaf age distribution or adaptation to shading, the
best estimate of g_c will probably be obtained as the weighted average of the
stomatal conductance values considered appropriate to the different leaf
layers. Equation 21.3 works well for crops with dry leaf surfaces; when leaf
surfaces are partially wet the situation becomes more complex and the
accuracy of the calculation is open to doubt (Shuttleworth 1976).

In principle, plant water potential can be calculated from the flux of water
through the plants and knowledge of the frictional resistances in the flow
pathways. Calculations of transpiration from canopies provide flux density

of water vapour; to obtain flux per stem it will be necessary to divide by the number of stems per unit area.

Plant water status

Simple resistance analogues have been widely used for the analysis of variations in plant water potential in relation to flow rate, but the recurrence in the literature of non-linear relationships between leaf water potential and transpiration rates has effectively invalidated them as adequate general models and it has become clear that the capacitance of the system must, in many cases, be taken into account if we are to predict plant water potential accurately. An attempt to do this was made by Powell & Thorpe (1977) for apple trees. Wallace (1978) recently showed that water storage capacity of wheat is significant and developed a model which accounts for the commonly observed hysteresis between leaf water potential (ψ_l) and flux through the plant (q) but which does not require knowledge of the soil hydraulic properties or the exact location of the hydraulic capacitances within the system. His main result is given by the equation:

$$\Psi_l(t) = \overline{\Psi}_s - R\{q(t_1) + mt - mRC + (k/RC)\exp(-t/RC\} \quad (21.4)$$

where t is time, $m = \Delta q/\Delta t$, R is hydraulic resistance, C the hydraulic capacitance, $\overline{\Psi}_s$ the effective water potential of the soil and k is a constant of integration, determined by the initial conditions at time t_1. Note that RC has the dimensions of a time constant.

More empirically, but based on a similar hypothesis, *viz.* that the flow of water through the soil-plant system can be described by a resistance–capacitance network, we (D.B. Powell and J.J. Landsberg) have fitted the equation

$$\Psi_l(t) = \overline{\Psi}_s - bE + c\,\mathrm{d}\mathbf{E}/\mathrm{d}t \quad (21.5)$$

to diurnal changes in the water potential of apple leaves. $\mathbf{E}$ is the transpiration rate and the expression accounts for a high proportion of observed variance. The parameter b has the dimensions of (axial) resistance; c has the dimensions of resistance $\times$ time and accounts for the movement of water out of and into storage as the transpiration rate increases or decreases.

Equations 21.4 and 21.5 reflect the fact that plant water status is dynamic, continually changing with soil water status and aerial environmental conditions. Successful modelling of plant water status *per se* is a necessary condition for analysing, and modelling, the effects of plant water status on plant growth. The problem is how to model these effects: is growth reduced by any reduction in turgor, and are the effects of water potential on growth and assimilate partitioning linear, so that they may be analysed in terms of the integral of water potential over time? There are few answers to such questions, and space precludes detailed discussion here, but the reader is referred to Hsiao *et al.* (1976) for an excellent treatment.

Net photosynthesis

The variation of leaf net photosynthesis (P_n) with light intensity (photon flux density, Q_p) can generally be represented by some form of rectangular hyperbola; the equations used have ranged from the very simple (e.g. $P_n = P_{max}/(1 + c/Q_p)$ where P_{max} is maximum observed P_n and c is a constant) to highly complex formulations (see Acock *et al.* 1976 for a summary). We (Watson, Landsberg & Thorpe 1978; Thorpe *et al.* 1978) have found the simplified form of a model by Charles-Edwards & Ludwig (1974) to be very useful. It encompasses virtually all the information which can be obtained from conventional gas exchange studies and includes explicitly boundary layer, stomatal and mesophyll conductances. The equation is

$$P_n = \frac{\alpha Q_p C_i g_m}{\alpha Q_p + g_m C_i} - R_d \qquad (21.6)$$

This model involves the assumption that $P_{max} \simeq g_m C_i$. The initial slope (α) gives the quantum efficiency, C_i is inter-cellular CO_2 concentration, obtained from the relation $C_i = C_a - P_n/G$, where C_a is ambient CO_2 concentration and $G = g_a g_s/(g_a + g_s)$, R_d is dark respiration and g_m is mesophyll conductance. To solve the model C_i is eliminated, leading to a quadratic. Combining the equations for P_{max} and C_i we obtain

$$P_{max} = \frac{g_m C_a - R_d}{1 + g_m/G} \qquad (21.7)$$

from which the relative importance of the conductances can be evaluated.

It is clear that for accurate simulation of leaf net photosynthesis the parameter values (α, g_m) appropriate to leaves of different age or condition should be known. This information is seldom available but if some indication of the range of values can be obtained the consequences of using the same parameter values for all leaves can be evaluated by sensitivity analysis. Available evidence suggests that variations in the photosynthetic characteristics of leaves has little influence on estimates of canopy photosynthesis. Acock *et al.* (1976) concluded that a simple rectangular hyperbola, with a term for dark respiration, gives an adequate summary of crop responses to light and CO_2 and Thorpe *et al.* (1978) used equation 21.6, with the same parameter values for all leaf types and ages, and successfully simulated net photosynthesis by an apple tree. However, ignoring stomatal response to environmental factors may cause serious errors in estimates of canopy net photosynthesis. In their study Thorpe *et al.* used a sub-model describing the effects of light and humidity on stomatal conductance. Acock *et al.* did not and while this apparently introduced no errors in their analysis of a green pepper crop grown in a glasshouse, Acock, Charles-Edwards & Hand (1976) in another study had no satisfactory explanation for the effects of humidity on tomato canopy photosynthesis, which they analysed using a model which did not take account of the effects of humidity on stomatal conductance.

The purpose of models of leaf net photosynthesis, combined with models of radiant energy interception and distribution, is to provide estimates of plant or canopy productivity. For this they must be combined with models of respiration, considered in the following section.

Respiration

Respiration reflects the activity of plants as a whole and their component organs. It is generally found to be dependent on the mass of living tissue (Fukai & Silsbury 1977) and on temperature. McCree (1970) analysed whole plant growth in terms of two components: growth, or synthesis, and maintenance respiration. He used the equation

$$\mathbf{R} = c_1 \mathbf{P} + c_2 W \tag{21.8}$$

where $\mathbf{R}$ is the 24 h total of dark respiration, $\mathbf{P}$ is the 12 h total of gross photosynthesis and W is the CO_2 equivalent of the dry weight of the plant. The coefficients c_1 and c_2 define the rates of growth and maintenance, respectively. McCree (1974) found that the growth coefficient appeared to be independent of temperature but maintenance was strongly temperature dependent. This approach has been extended and developed by McCree himself (McCree 1974) and by Thornley and others (see Thornley 1976) and has been successfully used in crop growth models and to analyse field measurements of CO_2 exchange (Biscoe, Scott & Monteith 1975; Jones *et al.* 1978). Thornley (1977) has recently re-examined this view of respiration and proposed a model with three dependent variables: storage dry weight, the degradable component of structural weight and the non-degradable component of structural weight. This appears to overcome some of the problems of the 'growth' and 'maintenance' concepts and is more attractive from several points of view. The main problem with Thornley's new approach is that it introduces five parameters, and for non-steady state growth numerical solutions to the differential equations will be necessary. It seems unlikely to supersede equation 21.8 as a basis for estimating plant or stand respiration, at least in the immediate future.

The respiration rates of individual organs are important because they reflect the sink activity ('mobilizing ability', Wareing & Patrick 1975) of a particular organ. To illustrate this we may adopt an approach similar to that used by Thornley (1976) in his analysis of respiration. Assume that growth is steady state, so that an amount of substrate Δs supplied to an organ in time Δt is completely utilized. Some (Δs_g) will be used for growth (defined as dry matter increase, ΔW, which may consist of stored carbohydrate, protein, structural material etc.) and some will be used to satisfy the requirement for maintenance respiration (Δs_m), i.e.

$$\Delta s = \Delta s_g + \Delta s_m \tag{21.9}$$

An amount of substrate Δs_r is respired to meet the requirements of growth, or synthesis, hence

$$\Delta s_g = \Delta s_r + \Delta W \tag{21.10}$$

or

$$\Delta W = Y_G \, \Delta s_g \tag{21.11}$$

where

$$Y_G = \frac{\Delta W}{\Delta W + \Delta s_r}$$

Clearly Δs_r will vary according to the transformation taking place, being lowest when sugars are being converted to high molecular weight carbohydrates, higher when proteins are being synthesized and higher still for fats. If there is no growth in an organ $(\Delta s_g = 0)$ then it cannot utilize more substrate than is required to satisfy its maintenance requirements and the flux of substrate to the organ can only be $\Delta s/\Delta t = \Delta s_m/\Delta t$. If $\Delta s_g > 0$ then, for any period or phase of growth during which Δs_r can be taken as constant, flux will be proportional to respiration rate $(\Delta s_r + \Delta s_m)$. If, therefore, we can estimate the relative respiration rates of tissues we can use this information to model partitioning.

Maintenance respiration rates can be taken to be proportional to tissue mass and dependent on temperature, it is probably reasonable to assume, at least initially, that they are the same for all organs and can be estimated from knowledge of the maintenance respiration rate of the plant as a whole. However, estimating Δs_r is more of a problem. A simple model which may provide a means of doing this is suggested below.

Assimilate partitioning

Estimates of total dry matter production alone are often of limited value; it is necessary to know how the dry matter is distributed between the component parts of plants. Since the mechanisms governing assimilate partitioning are not well understood, the basis of some mechanistic models of this process (e.g. Thornley 1972a, b) is open to doubt, and it is often difficult to obtain the necessary parameter values. Furthermore since we are concerned with plant response to weather, any model must make allowance for changes in assimilate partitioning ratios caused by factors such as temperature (Davidson 1969; Hofstra 1972) or water stress (Fisher & Turner 1978).

An obvious initial premise is that assimilates will go to the regions of greatest activity. Charles-Edwards (1976) has modelled partitioning in terms of the specific activities of carbon and nitrogen and Barnes (1979) invoked the concept of competitive sinks to model the partitioning of assimilates between shoots and storage roots. A similar approach to the problem, outlined here, offers the possibility of analysing the growth patterns of plants with any number of component parts in terms of the effects of environmental factors on the sink activity of the different organs.

We define the specific activity of organ *i* as

$$\mu_i = \frac{1}{Y_g} \frac{1}{W_i} \frac{dW_i}{dt} \qquad (21.12)$$

where W_i is the weight of organ *i* and Y_g $(= \Delta W/\Delta s)$ defines the efficiency with which labile carbohydrates are converted to plant material (dry weight): μ_i therefore has the dimensions $\Delta s_g/\Delta t \cdot W_i$. Assuming that μ_i is dependent upon the concentration of carbon substrate $\{C\}$ available from the phloem, according to Michaelis-Menten kinetics, it is given by

$$\mu_i = \frac{k_i\{C\}}{K_i + \{C\}} \qquad (21.13)$$

where k_i, K_i are activity constants.

From equations 21.12 and 21.13

$$\frac{dW_i}{dt} = k_i W_i Y_g \left[\frac{\{C\}}{K_i + \{C\}} \right] \qquad (21.14)$$

If we take $K_i = K$ for all parts of the plant and if the activity constant k_i is approximately constant over time interval 0, *t*, we can obtain a solution for equation 21.14 as follows

$$\ln W_i(t) = \frac{k_i}{k_j} \ln W_j(t) + [\ln W_i(0) - \frac{k_i}{k_j} \ln W_j(0)] \qquad (21.15)$$

i.e.

$$\frac{W_i(t)}{W_i(0)} = \left[\frac{W_j(t)}{W_j(0)} \right]^{k_i/k_j}$$

Equation 21.15 can be fitted to dry weight increments of plant organs across any time interval during which the assumptions are likely to hold, giving a standard allometric-type equation with the term in square brackets given by the intercept, and yielding a value for the ratio k_i/k_j. Since $k_i = \mu_i$ when $\{C\}$ is non-limiting (eqn 21.13), this is the ratio of the specific activities of the plant parts *i* and *j* or (*cf.* eqn 21.4) the ratio of their specific growth respiration rates. This might be tested by direct measurement of respiration. The ratios would be expected to change with time, as tissues mature or the plant goes from the vegetative to the reproductive phase or if environmental factors affect the activity of particular organs.

Size of component parts

If we can calculate total dry matter productivity and allocate the assimilate to component parts of the plants, conversion to leaf area, root length etc. simply requires the definition of appropriate conversion factors. These may of course, vary with environmental conditions.

Volume of soil exploited by roots

A preliminary model describing the volume of soil exploited by roots might embody the assumptions that root length per unit volume of soil (L_V) decreases exponentially with depth (z) (Gerwitz & Page 1974). For a single plant in uniform soil if the radius (r) of the volume occupied by the root system decreases exponentially with depth (i.e. $r = r_o \exp(-\beta z)$), the total volume of soil exploited by the roots can be shown to be $\pi r_o^2/2\beta$, where r_o is the radius at the surface and β would have to be measured. If L_V also decreases exponentially with z, and the total length of roots (L) is known, then the appropriate integrations lead to

$$L = \frac{\pi r_o^2 L_V(0)}{(2\beta + \beta')} \tag{21.16}$$

where $L_V(0)$ is the value of L_V at $z = 0$ and β' is the constant defining the rate of change of L_V with depth.

A model of this type would almost certainly only be applicable to uniform soils, but it provides a means of allocating root material in the soil and, combined with information on soil moisture characteristics and models of water uptake and movement through plant roots (Landsberg & Fowkes 1978; Rowse, Stone & Gerwitz 1978; Seaton & Landsberg 1978) it could provide the basis for calculating plant water uptake and the soil water balance in the root zone.

Soil moisture status in the root zone

The amount of water in the root zone (θ) depends on the water holding capacity of the soil, the rooting depth and the balance between precipitation (PR) and run-off (RO), and water loss by evapotranspiration (ET) and drainage (DR) across a given period. It can be written

$$\Delta\theta = \sum_{t_0}^{t} (PR - ET - DR - RO)\, \Delta t \tag{21.17}$$

where ΔW is the change in water content. Calculations of W usually start from some known value. Information on infiltration rates in relation to rainfall intensity and soil dryness may be needed, unless it is considered that these (and possibly run-off) can be ignored. In more detail, particularly where we are concerned to calculate resistances to uptake by and movement of water through plants, with a view to calculating plant water potential, it will be necessary to calculate water uptake from different levels in the soil, which will depend on root density and soil moisture potential at these levels. This necessitates calculation of water movement from layer to layer, for which information on soil hydraulic conductivity is required.

Resistance to water uptake by plants is a consequence of resistance to water movement through the soil to the roots (R_s, see Gardner 1960; Cowan 1965) and resistance to water movement into and through roots, where the major component appears to be radial resistance (Landsberg & Fowkes 1978). Assuming that total root resistance (R_T) in any layer can be estimated as the summation in parallel of radial resistance per unit root surface area, R_T can be estimated as R_R/a_R, where a_R is total root surface area in the layer (estimated from L_V). We may therefore, write, for any layer of thickness Δz, centred on the depth z,

$$q(z) = (\psi_s - \psi_o)/(R_s + R_T) \tag{21.18}$$

where q is the volume flux of water into the roots and ψ_o is the water potential at the junction of roots and stem base. Seaton & Landsberg (1978) provide a solution for a similar model for three soil layers. Rowse et al. (1978), who also used this form of model, calculated R_s by using a method first put forward by Gardner (1960). By analogy with steady state diffusion between concentric cylinders

$$R_s = \frac{\ln (r_1^2/r_2^2)}{4\pi k} \tag{21.19}$$

where r_1 is the radius of the outer cylinder, assumed to be equal to half the mean distance between roots (calculated as $r_1 = 1/(\pi L_V)^{1/2}$), r_2 is the radius of the inner cylinder, assumed to be equal to root radius, and k is the mean of the hydraulic conductivity in the bulk soil and at the root surface. (The estimation of $\bar{k}$ causes problems.)

It will be clear, from this discussion, that the total uptake from all layers (i.e. $\Sigma_{z=0}^{z} q(z) \Delta z$) must equal transpiration rate, unless there is significant capacitance in the plants, in which case uptake rates may be less than transpiration rate when water is being removed from store, and more when it is being replaced. The interactions and feed-backs between the aerial environ-mental conditions which determine potential transpiration rates, resistance in the flow pathways causing reduction in potential, and hence possibly stomatal closure, and soil moisture content in the root zone, should also be apparent.

PLANT DEVELOPMENT

In any attempt to model the growth of a plant through a season it is essential to consider developmental processes as well as dry matter production. I have discussed the question of modelling plant development in relation to weather elsewhere (Landsberg 1977) and will not elaborate to any great extent here. Many will argue that attempts to model developmental processes are likely to be abortive because we do not know enough about the mechanisms involved in, say, the switch from vegetative to reproductive growth (conifers provide an excellent example of this) to formulate precise hypotheses. This argument has as its premise the idea that we need to know a great deal about a system at the lower levels of organization (i.e. that we should have well validated sub-models) before we attempt to formulate system (whole plant) models. I would argue that we can use quite imprecise ('black box') models to complete a whole plant model, accepting that for the purposes of simulation these constitute a (possibly serious) weakness. However, these models would be tested and refined, so that we work downwards towards greater understanding of the mechanism involved rather than upwards from knowledge of the mechanism to incorporation of that knowledge in a whole plant model.

The definition of the point(s) at which morphogenetic changes occur is a central problem in modelling development. Kirby (1974) has shown a very marked change in the rate of initiation of primordia in the shoot apex of wheat after the change-over from leaf to spikelet primordia production. He suggests that the transition may be dependent upon the number of leaf primordia and the rate at which they are formed. It was possible to analyse his data using a model which allowed estimation of the rate of leaf primordia production, the rate of spikelet primordia production and the total number of primordia formed. The model used (by P.M. Lerman) is unpublished but it seems likely that the values of some of its parameters are dependent upon (say) temperature.

The volume of the apical meristem may be critical in the induction of flowering. Charles-Edwards *et al.* (1979) have modelled flowering in *Chrysanthemum* assuming that primordial initiation is regulated by the presence of an inhibitor in the apical dome. The analysis embodies the 'dual requirement' hypothesis; that is, flowering depends upon both the level of assimilate supplied to the dome through its specific growth rate and is also dependent on another factor, the presence of the inhibitor. Changes in the supply of assimilate, through manipulation of, say, the plant light environment, affect both the numbers of vegetative primordia and floral part primordia.

In some perennial plants (e.g. apples) buds become dormant after the production of reproductive parts, and their subsequent rate of development (in spring) depends not only on the temperatures in spring but also on the degree of winter chilling to which they have been subjected. Landsberg (1974) modelled this process by postulating the existence of an inhibitor, responsible for dormancy, which is broken down by low temperatures. The model, which combines summation of temperature reciprocals and straightforward temperature summation (both with limits) to simulate breakdown of the inhibitor, restoration of the level of the promoter and morphological development, has been remarkably successful in its predictions. It has not been tested at the level of the assumptions about mechanisms (changes in inhibitor levels) but this is feasible.

DISCUSSION

The position which I have adopted with regard to the need for models is often challenged on the grounds that all we need are qualitative hypotheses, on the basis of which we carry out experiments. We then analyse the data to 'see what they tell us' and explain the results in terms of qualitative descriptions (which cannot be rigorously tested) of the mechanisms thought to be involved. This approach leads to the accumulation of information but not to much gain of knowledge. Further, in the absence of quantitative hypotheses it is not possible to evaluate the consequences of changes in the rate of a process or level of a substance in relation to other processes taking place at the same time. To return to a point made earlier, I accept that we may often have to base our preliminary models on empirical relationships arising from vague hypothesis of the type '*a* is related to *b*', but to demonstrate causality we need to describe in mathematical terms the mechanism(s) which we consider will explain the association. The formulation should predict the observed relationship and allow us to explore, algebraically or numerically, the form it may take under other conditions.

Models of the type outlined in Fig. 21.1 and in the sub-models constitute a summary of knowledge, at a particular level, about the subject area. The objective here has been to demonstrate that, in the present state of knowledge, quite precise deterministic hypotheses about the way weather affects plants can be formulated, and that such hypotheses provide a useful framework for experimentation. It may be argued that deterministic models are often inappropriate, especially in the ecological context, since the variability associated with plant response to weather is often such that a model may give results which are correct for one set of plants and conditions and apparently incorrect for another. A partial answer to this argument lies in the computation of confidence limits and probability statistics for the various parameters in the model, based on assessment of the variability in plant response to particular factors. However, the question of whether to use deterministic or stochastic models is a large one, inadequately explored.

Much of the information currently available about plant responses to weather comes from work on agriculturally important plants, and it is, perhaps, relatively easy to see the application of whole plant/plant community models to monocultures, where conditions are much less variable than those in natural plant communities. But the ecologist might ask the question: 'of what value are these models to me in dealing with the problems posed by complex communities in a range of conditions?' The answer to this must be that such models are of tremendous value. Studies on the physiology of plants, and their responses to environmental factors, lead to parameter values for the models. The models may then be used to explore the consequences, for the plants in question, of particular environmental conditions, and they provide a means of extrapolating observed responses to conditions other than those in which the observations were made. Field observations in turn provide tests for models. The essential argument remains the same—models provide a framework within which present khowledge is summarized, a guide to the experiments to be done or the observations to be made, and a means of exploring, in quantitative terms, the consequences of the hypotheses embodied in the models.

In conclusion, we must accept that plant responses to weather are extremely complex, and it is unlikely that a model which will simulate them completely accurately can ever be developed—in fact it is questionable whether this would be a useful objective. However, the process of model formulation forces rigorous and critical thought and the results provide an excellent framework for experimentation. A good model should contain within it a summary of our knowledge of the way the system works, allowing us to evaluate the consequences of manipulating it by altering either the environment or the plant.

ACKNOWLEDGEMENTS

I am grateful to Dr K. Cockshull (Glasshouse Crops Research Institute, Littlehampton) for a copy of the paper by Charles-Edwards, Cockshull, Horridge & Thornley (1979) which at time of writing had not yet appeared in the press.

REFERENCES

Acock B., Hand D.W., Thornley J.H.M. & Warren-Wilson J. (1976) Photosynthesis in stands of green peppers. An application of empirical and mechanistic models to controlled-environment data. *Annals of Botany*, **40**, 1293–1307.

Acock B., Charles-Edwards C.A. & Hand D.W. (1976) An analysis of some effects of humidity on photosynthesis by a tomato canopy under winter light conditions and a range of carbon dioxide concentrations. *Journal of Experimental Botany*, **27**, 933–941.

Allen L.H. (1974) Model of light penetration into a wide-row crop. *Agronomy Journal*, **66**, 41–47.

Barnes A. (1979) Vegetable plant part relationships. II. A quantitative hypothesis for shoot-storage root development. *Annals of Botany*, **43**, 487–499.

Biscoe P.V., Scott R.K. & Monteith J.L. (1975) Barley and its environment. III. Carbon budget of the stand. *Journal of Applied Ecology*, **12**, 269–293.

Burrows F.J. & Milthorpe F.L. (1976) Stomatal conductance in the control of gas exchange. *Water Deficits and Plant Growth*. Vol. IV (Ed. T.T. Kozlowski), pp. 103–152. Academic Press, New York.

Charles-Edwards D.A. (1976) Shoot and root activities during steady-state plant growth. *Annals of Botany*, **40**, 767–772.

Charles-Edwards D.A. & Thornley J.H.M. (1973) Light interception by an isolated plant. A simple model. *Annals of Botany*, **37**, 919–928.

Charles-Edwards D.A. & Ludwig J. (1974) A model for leaf photosynthesis by C_3 plant species. *Annals of Botany*, **38**, 921–930.

Charles-Edwards D.A. & Thorpe M.R. (1976) Interception of diffuse and direct-beam radiation by a hedgerow orchard. *Annals of Botany*, **40**, 603–613.

Charles-Edwards D.A., Cockshull K.E., Horridge J.S. & Thornley J.H.M. (1979) A model of flowering in chrysanthemums. *Annals of Botany*, **43**.

Cowan I.R. (1965) Transport of water in the soil-plant atmosphere system. *Journal of Applied Ecology*, **2**, 221–239.

Davidson R.L. (1969) Effects of root-leaf temperature differentials on root/shoot ratios on some pasture grasses and clover. *Annals of Botany*, **33**, 561–569.

Fischer R.A. & Turner N.C. (1978) Plant productivity in the arid and semi-arid zones. *Annual Review of Plant Physiology*, **29**, 277–317.

Fritschen L.J. (1967) Net and solar radiation relations over irrigated field crops. *Agricultural Meteorology*, **4**, 55–62.

Fukai S. & Silsbury J.N. (1977) Responses of subterranean clover communities to temperature. II. Effects of temperature on dark respiration rate. *Australian Journal of Plant Physiology*, **4**, 159–167.

Gardner W.R. (1960) Dynamic aspects of water availability to plants. *Soil Science*, **89**, 63–73.

Gerwitz A. & Page E.R. (1974) An empirical mathematical model to describe plant root systems. *Journal of Applied Ecology*, **11**, 773–782.

Hofstra G. (1972) Response of soybeans to temperature under high light intensities. *Canadian Journal of Plant Science*, **52**, 535–544.

Hsiao T.C., Fereres E., Acevedo E. & Henderson D.W. (1976) Water stress and dynamics of growth and yield of crop plants. *Water and Plant Life: Problems and Modern Approaches* (Eds O.L. Lange, L. Kappen and E.-D. Schulze), pp. 281–305. Springer-Verlag, Berlin.

Jarvis P.G. (1976) The interpretation of the variations in leaf water potential and stomatal conductance found in canopies in the field. *Philosophical Transactions of the Royal Society of London*, **B**. **273**, 593–610.

Jones M.B., Leafe E.L., Stiles W. & Collett B. (1978) Pattern of respiration of a perennial ryegrass crop in the field. *Annals of Botany*, **42**, 693–703.

Kirby E.J.M. (1974) Ear development of spring wheat. *Journal of Agricultural Science, Cambridge*, **82**, 437–447.

Landsberg J.J. (1974) Apple fruit bud development and growth: Analysis and an empirical model. *Annals of Botany*, **38**, 1013–1032.

Landsberg J.J. (1977) Effects of weather on plant development. *Environmental Effects on Crop Physiology* (Ed. by J.J. Landsberg and C.V. Cutting), pp. 289–307. Academic Press, London.

Landsberg J.J. & Thom, A.S. (1973) Aerodynamic properties of a plant of complex structure. *Quarterly Journal of the Royal Meteorological Society*, **97**, 565–570.

Landsberg J.J. & Powell D.B.B. (1973) Surface exchange characteristics of leaves subject to mutual interference. *Agricultural Meteorology*, **12**, 169–184.

Landsberg J.J. & Fowkes N.D. (1978) Water movement through plant roots. *Annals of Botany*, **42**, 493–508.

Lemeur R. & Blad B.L. (1974) A critical review of major types of light models for estimating the short-wave regime of plant canopies. *Mededelingen Fakulteit Landbouw-wetenschappen, Gent*, **4**, 1535–1585.

McCree K.J. (1970) An equation for the rate of respiration of white clover plants grown under controlled conditions. *Prediction and Measurement of Photosynthetic Productivity* (Ed. Z. Setlik), pp. 332–339. Wageningen.

McCree K.J. (1974) Equations for the rate of dark respiration of white clover and grain sorghum, as a function of dry weight, photosynthetic rate and temperature. *Crop Science*, **14**, 509–514.

Monteith J.L. (1965a) Light distribution and photosynthesis in field crops. *Annals of Botany*, **29**, 17–37.

Monteith J.L. (1965b) Evaporation and environment. Symposium of the Society for Experimental Biology 19. pp. 205–234. Cambridge University Press.

Monteith J.L. (1969) Light interception and radiation exchange in crop stands. *Physiological Aspects of Crop Yield* (Eds. J.O. Eastin, F.A. Haskins, O.Y. Sullivan and C.H.M. Van Bavel), pp. 81–111, American Society of Agronomy, Madison.

Monteith J.L. (1973) *Principles of Environmental Physics*. Edward Arnold, London.

Monteith J.L. & Sceicz C. (1961) The radiation balance of bare soil and vegetation. *Quarterly Journal of the Royal Meteorological Society*, **87**, 159–170.

Penning de Vries F.W.T. (1972) A model for simulation of transpiration of leaves with special attention to stomatal functioning. *Journal of Applied Ecology*, **9**, 57–78.

Popper K.R. (1972) *Objective Knowledge: An Evolutionary Approach*. Oxford University Press.

Powell D.B.B. & Thorpe M.R. (1977) Dynamic aspects of plant-water relations. *Environmental Effects on Crop Physiology* (Eds J.J. Landsberg and C.V. Cutting), pp. 259–279. Academic Press, London.

Ross J. (1975) Radiative transfer in plant communities. *Vegetation and the Atmosphere*, Vol. I (Ed. J.L. Monteith), pp. 13–55. Academic Press, London.

Rowse H.R., Stone D.A. & Gerwitz A. (1978) Simulation of the water distribution in soils. II. The model for cropped soil and its comparison with experiment. *Plant and Soil*, **49**, 533–550.

Schulze E.-D. & Küppers M. (1979) Short-term and long-term effects of plant water deficits on stomatal response to humidity in *Corylus avellana* L. *Planta*, **146**, 319–326.

Seaton K.A. & Landsberg J.J. (1978) Resistances to water movement through wheat root systems. *Australian Journal of Agricultural Research*, **29**, 913–924.

Shuttleworth J. (1976) Experimental evidence for the failure of the Penman-Monteith equation in partially wet conditions. *Boundary-Layer Meteorology*, **10**, 91–94.

Takakura T., Goudriaan J. & Louwerse W. (1975) A behavioural model to simulate stomatal resistance. *Agricultural Meteorology*, **15**, 393–404.

Thornley J.H.M. (1972a) A model to describe the partitioning of photosynthate during vegetative plant growth. *Annals of Botany*, **36**, 419–430.

Thornley J.H.M. (1972b) A balanced quantitative model for root: shoot ratios in vegetative. *Annals of Botany*, **36**, 431–441.

Thornley J.H.M. (1976) *Mathematical Models in Plant Physiology.* Academic Press, London.

Thornley J.H.M. (1977) Growth, maintenance and respiration: a re-interpretation. *Annals of Botany*, **41**, 1191–1203.

Thorpe, M.R. (1978) Net radiation and transpiration of apple trees in rows. *Agricultural Meteorology*, **19**, 41–57.

Thorpe M.R., Saugier B., Auger S., Berger A. & Methy M. (1978) Photosynthesis and transpiration of an isolated tree: model and validation. *Plant, Cell and Environment*, **1**, 269–277.

Thorpe M.R., Warrit B. & Landsberg J.J. (1980) Responses of apple leaf stomata: a model for single leaves and a whole tree. *Plant, Cell and Environment*, **3**, 23–27.

Wallace J.S. (1978) Water transport and leaf-water relations in winter wheat crops. Ph.D. thesis, University of Nottingham.

Wareing, P.F. & Patrick J. (1975) Source-sink relations and the partition of assimilates in the plant. *Photosynthesis and Productivity in Different Environments* (Ed. J.P. Cooper), pp. 481–507. Cambridge University Press.

Warrit B., Landsberg J.J. & Thorpe M.R. (1980) Responses of apple leaf stomata to environmental factors. *Plant, Cell and Environment*, **3**, 13–22.

Watson R.L., Landsberg J.J. & Thorpe M.R. (1978) Photosynthetic characteristics of the leaves of Golden Delicious apple trees. *Plant, Cell and Environment*, **1**, 51–58.

22. CONTROLLED ENVIRONMENT AGRICULTURE FOR HOT DESERT REGIONS

J. GALE

Desert Research Institute, Sde Boqer, Ben Gurion University of the Negev

SUMMARY

Controlled or closed system agriculture offers a number of potential advantages for crop production in arid regions. These include: reduction of irrigation water requirement, use of solar energy for heating in winter months, and for desalinating brackish water, and maximizing plant use of solar energy by means of carbon dioxide fertilization. How to build such a system at an acceptable cost is the main problem to be solved and includes questions of engineering, meteorology and environmental plant physiology. A number of possible solutions are discussed.

INTRODUCTION

Lack of fresh water and high potential evapotranspiration are, of course, the main disadvantages of hot desert regions for cultivation of plants. To this may be added large seasonal and diurnal fluctuations in temperature and lack of fertile soil. However the same arid land characteristics also offer some benefits. The most notable is a high level of annual solar radiation, which in many areas exceeds $8300 \text{ MJ m}^{-2} \text{ yr}^{-1}$ in comparison with about $4000 \text{ MJ m}^{-2} \text{ yr}^{-1}$ for the UK (Sellers 1965). Some other characteristics of desert regions which may be turned to advantage are the relatively high winter temperatures, the availability of land and the lack of pollution. Small quantities of brackish water (1000 to 4000 mg l^{-1} total salts) are often available and may be used for irrigation after desalination, or directly, if special techniques are employed (Issar 1975).

Under these desert conditions, closed or semi-closed, artificial environments are very attractive for growing plants. In this context a 'closed' environment is a greenhouse in which energy and water are partly conserved and there is only limited and controlled exchange of the gaseous environment of the crop with the outside air.

Such closed systems offer a number of potential advantages. Among these are low water consumption and the possibility of using water of higher than

391

usual salt content, carbon dioxide fertilization, protection from wind and dust abrasion, conservation of energy (lower fuel requirements for heating in the winter months) and compatability with solar systems for desalinating water.

One of the main subjects of this type of study is engineering. There are a large number of possible engineering solutions (de Bivort, Taylor & Fontes 1978). Systems must be devised which are sufficiently inexpensive so as to be economically acceptable, with respect to the value of the crops grown. The grower may have to compromise with respect to the environment. For example, it may be less expensive to produce a hot, humid climate than a cooler, drier one. Information as to plant response to such conditions is required, which may not be immediately available. Furthermore optimal conditions will certainly vary for different crops and it may be more advantageous to choose crops for the given environment than the reverse.

ADVANTAGES AND PROBLEMS OF CLOSED ENVIRONMENTS

Water use

Under field or greenhouse conditions, once the soil is completely covered by a crop, transpiration accounts for almost the entire water loss. The rate of transpiration is a function of the leaf to air vapour density gradient and the resistances in the diffusion pathway (Gaastra 1959), the most significant of the latter being those of the stomata and boundary layers. When the rate of water loss exceeds that of uptake and transport to the leaves, stomata close. This reduces vapour loss but also the uptake of carbon dioxide. Under the conditions of a closed system, air humidity will probably be high. If leaf temperatures can be controlled by air movement within the system, the leaf to air vapour gradient and hence transpiration will be low. Even in partly open greenhouses water use is only about one third that of open field agriculture. We estimate that in an almost closed system, in an arid region, water use will be only about one-fifth to one-tenth of that of a crop growing in an open field. The latter may be as high as 10–15 mm day^{-1} depending upon meteorological conditions. In desert regions small plots of irrigated plants are especially subjected to the 'oasis effect'. Under the latter conditions the plants receive an input of advective energy, from hot dry winds, in addition to the immediate solar radiation load. This much increases their transpiration. Furthermore, with such a system, the farmer may need only about 1/20 of the area to make his living from a high cash crop, compared with irrigated, open-field agriculture. Thus the water use per farming family may be lowered by a factor of 100–200.

Plant growth under conditions of low transpiration presents other problems especially with respect to leaf temperature, mineral nutrition and plant disease. The leaf energy balance may be described by the equation:

$$S_t + L_d = L_e \pm C + \lambda E \dots \qquad (22.1)$$

where S_t and L_d are the absorbed short and long wave radiations respectively and L_e, C and λE the dissipation terms. L_e is the long wave radiation from the leaf, C the sensible heat exchange and λ the latent heat lost in E, transpiration (Raschke 1956). L_e, C and λE are all functions of temperature. If λE is small L_e and C must increase if leaf temperature is to be maintained. However, L_e can only be increased at a cost of elevated leaf temperature. Consequently excessive leaf temperature under conditions of low transpiration in the closed system must be solved by either increasing sensible heat exchange C by increased air turbulence around the leaf, or by reducing S_t or L_d. The latter may be achieved by reducing the incoming short wave radiant flux density, in the non-photosynthetic wavebands and the incoming long wave radiation, by lowering the temperature of the roof (see below).

There is also the possibility that the uptake and transport of certain mineral nutrients may be deficient when the soil to-leaf mass flow is very low (Rudd-Jones & Winsor 1978). This warrants further investigation.

Problems of plant disease may arise in an atmosphere of high humidity. Whereas high humidity is generally desirable, condensation on the leaf or drops falling from the roof, in which pathogenic fungal spores can germinate, must be avoided.

As noted above small quantities of brackish water are often available in desert regions. These 'fossil' waters have accumulated underground in previous geological epochs and, although a limited resource, could be used in significant quantities for many years (Issar 1975). In an environment of high humidity and moderate temperatures the deleterious effect of saline water on the growth of plants may be alleviated in many species (Gale 1975). Consequently the available brackish water may be an important factor for closed systems, especially if diluted with solar-distilled water (see below).

Conservation of energy

At present, the most economically worthwhile plants, for cultivation in controlled environments, are out-of-season (usually winter) high cash crops. In arid regions, such as North Africa and the Middle East, solar energy may be sufficient to provide all the heating which, in more northerly regions, is accomplished in greenhouses with increasingly costly fossil fuel. However, this requires capture and storage of surplus daytime solar energy and its release to the greenhouse at night. The same heat storage system should be used, in the summer months, for dissipating surplus energy. The solution of this heat transfer problem, at reasonable cost, would enable the system to remain closed throughout most hours of the day. This would bring about the water saving discussed above and also enable fertilization with carbon dioxide. With a heat store as part of the system, it may also be possible to choose between heating the air or heating the roots. Heating the roots may often be more advantageous to plant growth than heating the plant tops, especially in winter and spring.

Carbon dioxide fertilization and productivity

Carbon dioxide fertilization has been recognized for many years as of considerable potential for raising the rate of photosynthesis and yields of many greenhouse crops. In northerly latitudes it may be profitable even with the prevailing relatively low quantum flux densities (Hand & Cockshull 1976, 1977). However, the largest response is found at higher quantum fluxes (Enoch, Rylski & Samish 1970; Rudd-Jones, Calvert & Slack 1978).

In arid zones, excessive heat during the day makes ventilation of greenhouses necessary, unless there is some internal means of heat dissipation such as by a heat exchanger (de Bivort *et al.* 1978). This precludes the application of CO_2, during the midday hours when it would be most advantageous. Nearly all greenhouse plants have 'C_3' photosynthesis and at high temperatures have elevated levels of photo-oxidation. However, by raising the level of ambient CO_2 this photo-oxidation can be partly inhibited and hence net-photosynthesis and growth increased (Zelitch 1971).

Kimball & Mitchell (1978) investigated the potential of CO_2 supplementation for growing crops in closed systems, in arid climates. They used a closed greenhouse in Arizona cooled by conventional refrigeration. Tomato plants were grown in this unventilated greenhouse, at a CO_2 concentration of 1000 μl l^{-1} during the daylight hours, as compared with ~ 320 μl l^{-1} for normal air. In this simulated closed system, yields were increased more than 60% by the CO_2 treatment.

Bassham (1977) has analysed the potential for plant growth in closed systems in arid zones, using CO_2 supplementation. He estimates that with a continuously harvested crop of alfalfa *Medicago sativa* a theoretical maximum of 200 tonnes per hectare could be produced annually. Alfalfa is a high protein fodder crop. He suggests that with appropriate closed system and protein extraction technology this may prove to be a means for producing protein for cattle, poultry and humans in desert regions.

Extremely high levels of starch production from sweet potatoes *Ipomoea batatas* grown from single leaf and node cuttings have been reported by

Yabuki & Uewada (in press). They grew these cuttings in a closed system with CO_2 concentration up to 2400 μl l^{-1}.

Such high levels of productivity as predicted by Bassham (1977), de Bivort *et al.* (1978) and Yabuki & Uewada (in press) are based on a 12-month growing season of controlled temperatures, optimal root nutrition and water balance, high levels of carbon dioxide and high levels of solar radiation. However, as noted by Yabuki & Uewada (in press) and found in greenhouse practice, not all plants respond well to CO_2 supplementation.

PLANTS, WATER AND CARBON DIOXIDE FOR CLOSED SYSTEMS

Choice of plants

The greenhouse industry today concentrates on the production of high cash, intensive, generally out-of-season crops; the main growth period being autumn to spring. Vegetable crops and ornamentals are the main staples. Two operating closed systems in the arid zone of the Persian Gulf and one in the U.S. arid region have specialized in vegetable crops (de Bivort *et al.* 1978).

It is somewhat paradoxical that in arid regions a year-round tropical environment is produced in a closed system at smaller expense than a drier more temperate one. To exploit such conditions to the maximum one may have to look towards a crop with an all round cropping and growth cycle such as alfalfa (Bassham 1977), or to new tropical crops. An example of the latter is the high protein winged bean *Psophocarpus tetragonolobus* (N.A.S. 1975).

Crop species may have to be bred for the special conditions of the system. Special attention will have to be given to growth throughout the year, response to CO_2 and resistance to disease in the high humidity environment.

CO_2 supply

A cheap source of carbon dioxide is required in the operation of closed systems. Unlike ordinary ventilated greenhouses, CO_2 supplementation in a closed system is a necessity, not just a possible benefit. Without it the CO_2 level will quickly fall during the daylight hours. The quantity of CO_2 required is large and may be as high as 850 kg CO_2 ha^{-1} day^{-1} (Table 22.1).

Industrial waste gas such as from cement factories is one of the most attractive sources of CO_2 but transportation and the necessity for scrubbing are serious problems. In Israel the Negev desert contains considerable

TABLE 22.1. Estimate of CO_2 requirement for closed system agriculture in arid regions.

For photosynthesis	at normal levels of CO_2 (320 $\mu l\,l^{-1}$)	300 kg CO_2 ha^{-1} day^{-1}
	at elevated CO_2 (1000 $\mu l\,l^{-1}$)	500 kg CO_2 ha^{-1} day^{-1}
CO_2 lost from system, with one air exchange per hour (CO_2 level maintained at 1000 $\mu l\,l^{-1}$)		350 kg CO_2 ha^{-1} day^{-1}
Estimated total requirement of closed system		850 kg CO_2 ha^{-1} day^{-1}

deposits of bituminous shale rocks. Much of this is of low quality for heat generation, as it contains up to 80% carbonates. We are presently investigating the possibility of burning this shale and thus releasing CO_2 from the carbonates and organic material. This would be carried out in small village kilns. The kilns could provide CO_2 by direct piping to greenhouses and, at the same time, small quantities of heat for domestic use.

Solar desalination of brackish water

Even in a completely closed system there would be a small requirement for water for growth—approximately equivalent to the fresh weight of the crop. However in reality the system will never be completely closed. At midday, during hot days, exchange of air with the outside may be required. A small rate of transpiration will be inevitable. As noted above, an advantage of such transpiration would be the facilitation of mineral transport from root to tops. A low rate of transpiration could be realized by maintaining some part of the closed system at a temperature below that of the leaves. In the same way condensation on the leaves would be prevented.

One possible way to supply the fresh water requirements would be by solar distillation of the available brackish, fossil, water. The stills could be either an integral part of the closed system (Trombe & Foex 1961; Bettaque 1977) or a separate unit (Howe & Tliemat 1974). Because of the following considerations it seems that solar distillation may be an ideal source of fresh water for closed systems in desert areas:

(a) High solar irradiance > 8000 MJ m^{-2} yr^{-1}.
(b) Low cost of land.
(c) Availability of brackish water.
(d) Medium volume requirements.
(e) No storage required. Fresh water can be fed directly to the plants in the closed system.
(f) Linkage of rate of production and demand. Both depend upon solar radiation.

Furthermore, as noted above, under closed system conditions (high humidity and CO_2, moderate temperatures) plants are expected to have

TABLE 22.2. Estimate of area of solar-still required for desalinating brackish water for closed system agriculture in desert regions.

	Worst case	Best case
	mm day^{-1}	
Solar still water production	4	6
Water available after 1 : 1 dilution with brackish water	8	12
Plant water requirement in closed system	2	1
Required ratio of area of still to area of plant growth	1 : 4	1 : 12

Estimate based on use of brackish water containing 2000 mg l^{-1} total salts.

Under closed system conditions many plants are expected not to respond adversely to water containing 1000 mg l^{-1} total salts (see text).

increased resistance to salinity. Consequently, the effective quantity of 'fresh' water can be increased. An estimate of the area of still required for a closed system is given in Table 22.2.

ENGINEERING ALTERNATIVES

Liquid optical filter systems

De Bivort *et al.* (1978) have pointed out that there are an extremely large number of possible engineering approaches to the solution of the main problem of closed systems—namely the capture, storage and release at night of surplus energy from the daytime period. The systems we are studying at the Desert Research Institute are based on the optical filter principle (Morris *et al.* 1958; Canham 1962; Spencer *et al.* 1974; Chiapale *et al.* 1976). Only about 48% of the solar radiation used by plants in photosynthesis is in the 400–700 nm waveband (Szeicz 1974). The remaining ultra-violet (< 400 nm) and near infra-red wavebands (700–3000 nm) are not directly used by plants in photosynthesis. Consequently if a liquid which absorbs the non-photo-synthetic radiation can be passed over, or through, the greenhouse roof, the heat load can be considerably reduced and part of the incoming radia-tion energy conserved. The principle of this type of system is illustrated in Fig. 22.1a.

The greenhouse is built of a hollow (honeycomb), 6–12 mm thick, transparent, plastic roof. Water is circulated through this roof to a tank, which serves for heat storage, either directly or via a heat exchanger. A material is added to the water which absorbs the ultra-violet and near infra-red and transmits the 400–700 nm waveband. In this way about 50% of the solar energy is absorbed before entering the plant growing space. However,

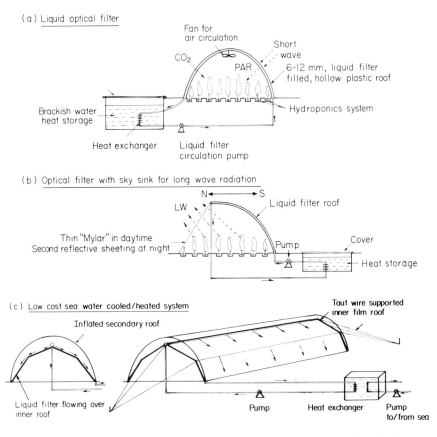

Fig. 22.1. Liquid optical filter systems for controlled environment agriculture in hot desert regions. For description of operation see text.

most of the radiation in the 400–700 nm waveband is also transformed to heat within the greenhouse, apart from the few per cent which is changed to chemical energy in the process of photosynthesis. This heat must also be removed. As the roof is held at a temperature a few degrees lower than that of the plants, heat is transferred to the roof by convection, conduction and to a small extent by long wave radiation. A further important contribution to dissipation of heat is by way of condensation on the roof of water transpired from the plants. The latter runs off, and may be returned to the plant roots. In this way excessively high temperatures are prevented during the day. At night, as a result of the low sky temperatures common in arid regions and consequent net loss of long wave radiation, the greenhouse tends to cool rapidly. This is prevented by recycling the stored warm water through the roof. This night-time recycling also serves to cool the water, as the liquid filter loses long wave radiation to the sky.

Our initial calculations have indicated that in the hottest hours of the day, during the summer season, some exchange of air with the outside will be unavoidable to enable cooling of the plants by transpiration at the expense of increased water use.

An alternative scheme based on ideas of Professor Went (University of Nevada) is shown in Fig. 22.1b. As the sky in the desert has a low temperature it may be used even during daylight hours as a heat sink. A thin 'Mylar' film, transparent to long wave radiation (~ 10 μm) is placed on the north-facing side of the structure. It would allow heat loss to the sky during the day. A second heat-reflecting plastic sheet is put in place at night. About 70% of the energy of the diffuse sky radiation is in the 400–700 nm photosynthetically active waveband (Szeicz 1974) hence the effectiveness of the optical filter roof would in any case be diminished on the north-facing side of the greenhouse. The latter has been recognized by Chiapale *et al.* (1976) who have a liquid filled roof section only on the south-facing side of their greenhouse.

There are a number of problems with this type of system which are presently being investigated at the Desert Research Institute and elsewhere. These include development of a non-phytotoxic, stable solution, which transmits all the 400–700 nm waveband. $CuCl_2$ has sometimes been used but it also absorbs a considerable fraction of the photosynthetic light and, if a leak occurs, is very toxic to plants.

Heat storage is also a problem. If no more than $10\,°C$ is allowed for day to night temperature differential of the fluid roof, then we estimate that a 4000–6000 m^3 water tank is required per hectare greenhouse for heat storage. This heat storage could be of brackish or sea water but it is relatively expensive to build.

A possible low cost alternative system suitable for desert sea coasts, designed at the Desert Research Institute, is shown in Fig. 22.1c. A taut wire plastic structure is covered by a second, inflated, ultra-violet resistant plastic roof. The liquid filter is circulated over the inner roof. It is then cooled by sea water at $15\,°C$ brought up from a depth of about 100 m. Air temperatures are expected to be higher in summer than in the systems described in Fig. 22.1a and b as the absorption of the near infra-red solar radiation would be necessarily small in the thin film of water streaming down the inner surface of the roof (Morris *et al.* 1958).

Solar stills for closed systems

Possible combinations of closed systems and solar stills are shown in Fig. 22.2. The principle of the Bettaque type 'salt-water' greenhouse, which has been operating for the past two years at the Desert Research Institute (Bettaque 1977) is depicted in Fig. 22.2a. The system produces all the fresh

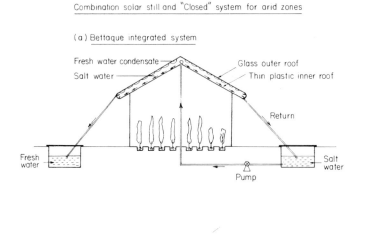

Combination solar still and "Closed" system for arid zones

(a) Bettaque integrated system

Fresh water condensate —
Salt water —
Glass outer roof
Thin plastic inner roof

Return

Fresh water
Salt water
Pump

(b) Semi-integrated system

Liquid optical filter system

Solar still

Fresh water

Heat exchanger
Pump

FIG. 22.2. Combination solar stills and closed systems for hot desert controlled environment agriculture. For description of operation see text.

water required for growing plants from water of a salinity equivalent to sea water. The main disadvantage of this type of system is that in order to obtain such quantities of fresh water the inner roof must have a high absorbance for solar radiation. This results in a reduction of photosynthetic light, and an increase of air temperature within the greenhouse. A possible mode of operation is to completely blacken the inner roof towards the end of the spring season. During the four hottest summer months the system would be used only as a solar still and the fresh water produced stored until the autumn to spring growing period, when the blackening is removed. A possible combination of optical filter greenhouse and conventional solar still is depicted in Fig. 22.2b. This system is presently in the design stage.

Whatever type of system is involved, its thermodynamic analysis requires a fairly complicated dynamic simulation model. This is because of the many sources and sinks of energy and their diurnal and seasonal variations. Secondary simulation models must describe the growth response of plants to the conditions created in the system and fluctuating economic factors. First attempts to model this type of system have been reported by Van Bavel & Demagnez (in press).

ACKNOWLEDGEMENTS

The present review is the outcome of many discussions with scientists of the Haifa Technion, The Hebrew and Tel-Aviv Universities, the I.B.M. Co-Israel Science Center and colleagues from Germany, who are cooperating within the Closed System Agriculture Project of the Desert Research Institute, Sde Boqer.

Part of this work has been supported by a grant from the BMFT-Germany, and part from a grant of the E.D. Bergman fund.

REFERENCES

Bassham J.A. (1977) Increasing crop production through more controlled photosynthesis. *Science* **197**, 630–638.

Bettaque R. (1977) Agua para tierras sedientas. *El Campo* **61**, 16–26.

de Bivort L.H., Taylor T.B. & Fontes M. (1978) An assessment of controlled environment agriculture technology. *Natural Technical Information Service U.S. Department of Commerce*, PB-279-211.

Canham A.E. (1962) Shading glasshouses with liquid films. *British Electrical & Allied Industries Research Association.* Report w/T40.

Chiapale J.P., Demagnez J., Denis P. & Jourdan P. (1976) La serre solaire. 12e *Colloque National des Plastiques en Agriculture*, pp. 87–90.

Enoch H., Rylski I. & Samish Y. (1970) CO_2 enrichment to cucumber, lettuce and sweet pepper plants grown in low plastic tunnels in a subtropical climate. *Israel Journal of Agricultural Research*, **20**, 63–69.

Gaastra P. (1959) Photosynthesis of crop plants as influenced by light, carbon dioxide, temperature and stomatal diffusion resistance. *Mededelingen van de Landbouwhogeschool te Wageningen, Nederland*, **59**, 1–68.

Gale J. (1975) The combined effect of environmental factors and salinity on plant growth. *Plants in Saline Environments.* (Ed. by A. Poljakoff-Mayber and J. Gale), pp. 186–192. Springer Verlag, Berlin.

Hand D.W. & Cockshull K.E. (1976) The effects of CO_2 concentration on the canopy photosynthesis and winter bloom production of the glasshouse rose 'Sonia' (syn. Sweet Promise). *Acta Horticultura*, **51**, 243–252.

Hand D.W. & Cockshull K.E. (1977) Roses: the effects of controlled levels of CO_2 enrichment on winter bloom production. *Experimental Horticulture*, **29**, 72–79.

Howe E.D. & Tliemat B.W. (1974) Twenty years of work of solar desalination at the University of California. *Solar Energy*, **16**, 97–105.

Issar A. (Ed.) (1975) *Proceedings of the International Symposium on Brackish Water as a Factor in Development.* Desert Research Institute Ben Gurion University.

Kimball B.A. & Mitchell S.T. (1978) CO_2 enrichment of tomatoes in unventilated greenhouses in an arid climate. *Abstracts of the International Symposium Potential Productivity in Protected Cultivation.* Kyoto, Japan.

Morris L.G., Trickett E.S., Vanstone F.H. & Wells D.A. (1958) The limitation of maximum temperature in a glasshouse by the use of a water film on the roof. *Journal of Agricultural Engineering Research*, **3**, 121–130.

N.A.S. (1975) *The Winged Bean: A High Protein Crop for the Tropics*. National Academy of Sciences, Washington D.C.

Raschke K. (1956) Über die physikalischen Beziehungen zwischen Wärmeübergangszahl, Strahlung, Austausch, Temperatur und Transpiration eines Blattes. *Planta*, **48**, 200–237.

Rudd-Jones D., Clavert A. & Slack G. (1978) CO_2 enrichment and light dependent temperature control in glasshouse tomato production. *Abstracts of the International Symposium Potential Productivity in Protected Cultivation*. Kyoto, Japan.

Rudd-Jones D. & Winsor G.W. (1978) Environmental control in the root zone: nutrient film culture. *Abstracts of the International Symposium Potential Productivity in Protected Cultivation Climatology*. Kyoto, Japan.

Sellers W.D. (1965) *Physical Climatology*. University of Chicago Press, Chicago & London.

Spencer D.L., Daunicht H.J. & Smith T.F. (1974) Variable shading for greenhouses. *ASME Paper No. 74-WA/Sol-12*. Winter Annual Meeting American Society of Mechanical Engineers Solar Energy Division New York, N.Y.

Szeicz G. (1974) Field measurements of energy in the 0·4–0·7 micron range. *Light as an Ecological Factor: II*. Symposia of the British Ecological Society 16. (Eds. G.C. Evans, R. Bainbridge & O. Rackman), pp. 513–519. Blackwell, Oxford.

Trombe F. & Foex M. (1961) Utilization of still solar energy for simultaneous distillation of brackish water and air conditioning greenhouses in arid regions. *United Nations Conference on New Sources of Energy*. Rome. Paper 35/S/64.

Van Bavel C.H.M. & Demagnez J. A simulation model for energy storage and savings of a fluid-roof solar greenhouse. International Society Horticultural Science. Symposium. *More Profitable Use of Energy under Protected Cultivation*. In press.

Yabuki K. & Uewada T. High efficient starch production with single leaf and node cutting culture of sweet potato. *Acta Horticulture*, **87**. In press.

Zelitch I. (1971) *Photosynthesis, Photorespiration and Plant Productivity*. Academic Press, London.

23. CONCLUDING REMARKS

A.J. RUTTER

*Department of Botany, Imperial College of Science and Technology,
Prince Consort Road, London SW7 2BB*

It was originally suggested to me that this concluding paper might discuss the contribution of environmental physiology to plant ecology. It does not appear under this title in the programme for the field seemed too great to be reviewed thoroughly at the end of the last afternoon of the symposium. Nevertheless, I propose to make some remarks on this subject in the light of the symposium before commenting on a few points that have occurred to me as I listened to the papers or their discussion.

Given that the first role of ecologists is to describe certain natural phenomena, and that there is still descriptive work to be done and good descriptive work being done, the following questions, capable of analysis in terms of environment and physiology, then arise and have been of perennial interest to plant ecologists:

1. How are the observed distributions of plants to be explained?

2. What are the mechanisms which suit plants to their environments?

3. What are the physiological bases of the inter-relations of individuals and species in populations and communities?

These are not all the questions that plant ecologists may ask of environmental physiology, nor are they entirely distinct or mutually exclusive. If the first is given its widest possible meaning it embraces the other two. Furthermore, we cannot attempt complete answers to any of these questions in relation to plants with roots or rhizoids without considering responses to soil as well as climate, and the interactions between climatic and edaphic effects and between shoot and root systems. I will, however, confine myself, as the context of the symposium demands, to climatic effects as far as possible.

SPECIES DISTRIBUTIONS

Plant species have characteristic distributions, evidently in part related to climate, on a range of scales that grade into one another but which can be roughly distinguished as:

1. Geographical location as defined by latitude and longitude; distance from an ocean might also be included in this largest scale.

2. Topographic situation, characterized by altitude, aspect, slope and degree of exposure or shelter.

3. Plant community, and associated characteristic microclimate, in which the species occurs.

In the first of these ranges climate is undoubtedly a major determinant of plant distribution although the evolution of the earth's surface and of the distribution of its continents, concurrently with the evolution of plant life and later the evolution of man and of anthropogenic effects, constitute an interaction from which the selective effects of climate have to be disentangled. In considering topographic and other more local variations in distribution one has to be careful to distinguish edaphic and biotic from climatic influences and clearly the performance of a plant in a particular community is dependent on interactions between species in the soil as well as in the atmosphere.

In relation to topographic variation Dr Cernusca gave us an account of the contrasted energy balance and microclimate over a range of sites differing in altitude and I found the analysis of mechanisms which produce almost a uniformity of temperature within the vegetation (albeit in summer only) at three locations of markedly different altitude extremely interesting. It was very valuable to have Dr Körner's parallel investigations of the effects of these microclimates on stomatal behaviour, especially since the circumstances allow comparisons both between species at given altitude and, for some species, within species but at different altitudes.

In Great Britain we have a rich source of examples of contrasted plant distributions in the *Atlas of the British Flora* (Perring & Walters 1962). On a scale of presence in 10 km squares, many show either predominantly northern or southern distribution, or restriction to varying altitudinal limits (Pigott 1974). Professor Pigott and his students (Pigott 1974; Woodward 1975; Woodward & Pigott 1975; Prince 1976) and also Davison (1977) have analysed some of the effects of climate in Great Britain in physiological terms. From these analyses a number of points emerge.

Firstly, the critical factors determining limits of distribution or, in the case of barley, satisfactory yield (Prince 1976) very frequently act through effects on the physiology of reproduction, e.g. time of flowering, success of fertilization, seed and embryo growth, and germination vigour; a field of physiological response somewhat neglected in this symposium. Secondly, explanations of plant distribution can evidently be made at that physiological level that deals with the description of growth rates of plants and plant parts in relation to environment. This statement is not intended to decry the further analysis in terms of 'process' physiology and I have appreciated the ecological significance of contributions to the symposium on cell water relations and carbon metabolism (Tyree & Karamanos; Björkmann; Dunn, Long & Thomas, this volume). Thirdly, the contrasted altitudinal distributions of *Sedum telephium* and *S. rosea* (Woodward & Pigott 1975), though dependent on differential temperature responses, can only be understood by reference also to competition, a factor which must modify many plant responses to climate.

The investigations on the distributions of British plants that I have quoted have tended to treat time of flowering as something to be recorded and have largely ignored, as we have too, the complexity of the influences of climate on flower production. It may well be that the effects of climate on flowering are sufficiently well understood and that filling in the details for individual species is now almost a routine, but the symposium should not pass without reference to the extraordinarily delicate way in which flowering and other processes are geared to the annual cycles of temperature and day length and their interaction; effects whose physiological mechanisms are still imperfectly understood.

Finally in this section, I note the complexity of simulating the natural environment in controlled experiments and distinguishing its significant features that was brought out in Dr Rorison's contribution.

ADAPTATION TO ENVIRONMENT

Educated as I was before the last war, I spent a considerable time as an undergraduate (or so it seems now) in cutting sections of plant parts and interpreting or receiving interpretations of the way in which their anatomy adapted them to their environment. Even now, on my first year field course, I can rely on students telling me that the rolling of the leaves of *Ammophila arenaria* is a response that restricts transpiration. In a lecture Bennet-Clark (1945) once stated that the rolling of *A. arenaria* leaves hardly affects their transpiration and he suggested that the effect of this movement is to maintain humid air around the stomata and so to delay stomatal closure and the restriction of photosynthesis. I have never been able to trace the detailed evidence on which he based his remarks, nor do I know which is the more correct interpretation, but this example illustrates a paradox in the interpretation of several of the xeromorphic characters of leaves, namely that, without experimental evidence, they could as credibly be expected to keep stomata open as to restrict transpiration. Another example of the folklore surrounding xeromorphy is given by Parkhurst (1978): 'As an undergraduate I was taught that hypostomatous leaves were adapted to dry conditions and the present research grew from that prejudice'. In fact his search of the literature for evidence of correlation of the hypostomatous condition with more xeric habitats showed that this evidence was so thin as to be almost non-existent.

Small leaves are often said to be an adaptation characteristic of xerophytes and Parkhurst & Loucks (1972) attempted a theoretical analysis of the significance of leaf size in the relation between photosynthesis and transpiration. They concluded that natural selection favours the reduction of leaf size in atmospheric conditions conducive to high transpiration rates and especially in high radiation intensity, and that selection for large leaves should operate only in warm or hot environments with low radiation intensity. They discussed fairly credibly some obvious exceptions to this prediction. Although the assumptions of their model can be criticized, it is firmly based in the theory of the exchanges of radiant energy, heat and gas between the plant and its environment, that coupling of the plant to the atmosphere which Professor Monteith so lucidly expounded in our first session. Leaf xeromorphy may be variably associated with restricting transpiration, maintaining carbon assimilation or ensuring survival in dry conditions, and the very imperfect correlation between xeromorphy and xerophytism has for long puzzled botanists. I suggest, however, that xeromorphy is a subject of great interest which is ripe for renewed experimentation in the light of modern understanding of the principles of energy and gas exchanges in leaves.

One of the weaknesses—acknowledged by the authors—of Parkhurst and Louck's model is its neglect of the temperature relations of photosynthesis. These are not easy to handle in a model in view of the phenomenon of acclimation which was discussed by Professor Björkmann, but effects of temperature may be significant in the interpretation of leaf structure. In a comparison of different populations of *Sesleria caerulea*, Lloyd & Woolhouse (1978) found considerable differences in leaf diffusion resistance attributable to differences in maximum stomatal pore width. Since this plant does not grow in water deficient habitats, high stomatal resistance can hardly be a xerophytic adaptation but the authors comment that it may act to raise leaf temperature, and that an increase of 1–2 °C above ambient temperature in the range 10 °C to 20 °C would increase photosynthesis by 10 to 15 % in this species. It has a northerly distribution in Britain and the observed differences between populations might be significant in their distribution between sites differing in mean temperature.

Parkhurst and Louck's basic idea is that the maximization of photosynthesis is of first importance to the plant while water remains available. This idea has been used also by Jones (1976) in a model which relates not just to the leaf but to the whole stand or crop. It takes account of such things as canopy architecture, light penetration and the distribution of sources and sinks in the exchanges of mass and energy with the atmosphere. He, too, makes the point that the most effective combination of physiological and structural characters changes in complex ways with the environment and he concludes, 'the greatest scope for maximizing assimilation (in relation to transpiration) may be offered by selection for stomatal behaviour appropriate to the expected environment', a conclusion in line with the growing recognition that stomatal behaviour is an extremely delicate mechanism of adjustment to a continually varying environment and reinforcing the importance of investigations of stomatal physiology.

So far in this section I have discussed some *models* of adaptation and acclimation. How far have eco-physiological studies enabled us to analyse the adaptation and acclimation of any particular species? This question is too large to review comprehensively, so since our meeting is in Edinburgh and organized by Professor Jarvis, I will simply refer to the admirable study of Sitka spruce made by him and his associates during the last ten years. This work has been concerned mainly with photosynthesis, stomatal behaviour and water relations. Consequently questions about responses and environmental limits of reproduction cannot be asked of the data. However, questions that one would like to ask, with fair confidence that reasonable answers could be given in terms of the net photosynthesis of forest stands, are: What are the physiological and structural characters of this plant, evolved in the west of North America, that suit it for photosynthesis (and high productivity) in parts of Britain? In the knowledge that productivity is less in the drier (or warmer?) parts of Britain, can a physiological explanation of this variation be advanced? What are the characters which should be selected or bred from the available range of provenances (ecotypes) to maximize net photosynthesis of Sitka spruce stands in the range of climates in which we would wish to grow them? I have tacitly assumed that there is some relatively simple relation between net photosynthesis and production; otherwise to the practical forester the questions are hardly worth asking. If my questions are naive or impertinent I am sure Professor Jarvis can ignore them; given the detailed and exhaustive nature of the data they may only serve to show how difficult it is to obtain fundamental insights into physiological adaptation. I am fairly optimistic that answers could be given but I know that Professor Jarvis could claim that I am merely demonstrating that too little resource is available for investigation of one of our most important and productive crops.

INTER-SPECIFIC RELATIONS IN PLANT COMMUNITIES

If ecology were simply the investigation of the responses of organisms to their physical and chemical environments, it would be no more than an extension of physiology. However, almost all organisms have to live in association with other organisms; the distinguishing concepts of ecology are those of population and community and, so far as plants are concerned, the main effects of neighbours are to modify the climatic and soil environments. The mechanisms of response to influences mediated by population and community are therefore of prime interest to plant ecologists.

There have been numerous sophisticated studies of the effects of canopy structure on light attenuation in mono-specific communities and of the resultant effects on community photosynthesis. Norman & Jarvis (1974, 1975) and Watts, Neilson & Jarvis (1976) provide an example of such a study in Sitka spruce. However, our understanding at the physiological level of all the relationships between individuals in mono-specific communities or between species in mixed communities is still rather rudimentary, as a perusal of the chapter on 'Mechanisms of Interaction between Species' in Harper (1977) will show. There are undoubtedly forbidding difficulties. In the first place, the spatial relationships between the leaves and roots of species in mixtures change in time as a result of their concurrent growth and interaction. Haizel (1972), for instance, has shown that the canopy architecture of barley is profoundly modified by the presence of *Sinapis alba*. Secondly, the competitive situation cannot be analysed without reference to the interaction of shoot and root or of microclimate and soil, e.g. shade restricts root development and reduced nutrient uptake in turn affects the development of leaves. Finally, while a certain amount of attention has been given to the attenuation of light in mixed communities, e.g. by Stern & Donald (1962), Harper & Clatworthy (1963), the effects of the gradients of vapour pressure and leaf temperature have hardly been considered. Perhaps the simplest entry to the study of the physiology of species in layered communities in relation to the gradients of microclimate is initially through investigation of the physiological ecology of the woodland ground flora. Here again, most studies to date have concentrated on the effects of reduced light intensity and modified spectral composition, and so it is encouraging to find Dr Woodward including a consideration of water relations in his investigation of the woodland plant *Circaea lutetiana*.

The relationships between species populations in mixed communities remain one of the greatest challenges to plant ecologists.

THEORETICAL AND TECHNICAL DEVELOPMENTS

I will now make some comments that have occurred or been suggested to me during the symposium.

During the last decade the theoretical analysis of micro-meteorological observations to give estimates of the exchanges of matter and energy between vegetation and atmosphere has proved to be a very powerful tool. There are problems, as Dr Unsworth has explained, in dealing with situations where free convection is important and the possibility that the micro-meteorological technique could be elaborated to discriminate the photosynthesis and transpiration of separate layers of the leafy canopy has been frustrated by both technical and theoretical difficulties. The method, however, is finding important new applications in the study of the behaviour of gaseous pollutants in relation to vegetation, as the contributions of Dr Unsworth and Dr Fowler have shown, and we have heard of an entirely new but cognate development in Dr Chamberlain and Dr Little's work on the behaviour of particles. Embraced in this development is the behaviour of spores and also, in work being initiated by Dr Unsworth but not discussed in the symposium, the behaviour of mist and fog droplets. Unsworth is primarily interested in such droplets as the vehicle of one of the inputs of sulphur to vegetation, but droplet capture is also a largely unknown quantity in the water input to forests. In some situations, where fog or low cloud blow continually, droplet capture is visibly important and causes trees to drip when there is no precipitation in the open, but beyond this there is little reliable quantitative information, nor any model of general application.

It is interesting to note how our views on the significance of laminar boundary layers have been modified. In outdoor situations they are of rather small importance in relation to photosynthesis and transpiration, but we see now (Chamberlain & Little) that they are of critical importance in particle capture. It has been interesting, too, to see the concepts of diffusion resistance applied to the ecology of bryophytes and intertidal algae in the contributions of Dr Proctor and Dr Jones, respectively.

Because micro-meteorological techniques cannot at present reveal gas exchanges in different layers of a canopy, and because we now have very convenient porometers for investigating stomatal behaviour and cuvettes for measuring photosynthesis on individual leaves or shoots, both Dr Unsworth and Professor Jarvis have urged us to make greater use of these. They can give us a detailed picture of gas exchange over the plant surface and may assist in the problem that I discussed earlier, the investigation of the physiology of plants in multi-species communities. However, if these methods are to be used to build up integrated estimates of ecosystem function, e.g. photosynthesis or transpiration per unit land area, I should like to reiterate the note of caution that Professor Jarvis sounded when he stressed the importance of designing and testing sampling programmes before putting them into effect. Twenty years ago I estimated the annual course of transpiration in pine forest by making large numbers of measurements of the transpiration (Rutter 1966) and diffusion resistance (Rutter 1967) of individual leaves in different layers of the canopy. As I was finishing this work, micrometeorological methods were beginning to be used in forests and, although I did not think I had wasted my time, I did not think either that anyone would attempt to use this approach again. Now the wheel has circled and the advantages of this approach are being stressed, albeit with more convenient techniques now available. A point that I want to make is that, although I did design my sampling systems to reduce sampling error and eliminate bias as far as possible, I remained sceptical of my estimates of forest transpiration and checked them against a soil water balance. Field methods of measuring ecosystem function are so subject to systematic as well as sampling error that I think it is always desirable to check estimates by two independent methods.

SCOPE AND SUCCESS OF THE SYMPOSIUM

Many ecologists have to concern themselves with three levels of organization. If their main interest is in the relation of organisms to their environment (i.e. they are working at the organism level), they will wish to interpret their results at the level of process or tissue physiology and to extrapolate them to the level of population or community organization. They may well also be interested in comparisons, at the same level of organization, between different species or different environments. In yet a third dimension, they need to understand the physics and chemistry that underlie the exchanges between the plant and its environment. Virtually all these interactions between levels of biological organization and between the biological and physical aspects of environmental study have been represented in our symposium. Its success has derived from this and from the very high quality of the contributions. I feel sure that all participants have enjoyed it as much as I have and on their behalf I would like sincerely to thank the organizers.

REFERENCES

Bennet-Clark T.A. (1945) Adaptation to drought. *Scientific Journal of the Royal College of Science*, **15**, 99–102.

Davison A.W. (1977) The ecology of *Hordeum murinum* L. III. Some effects of adverse climate. *Journal of Ecology*, **65**, 523–530.

Haizel K.A. (1972) The canopy relationship of pure and mixed populations of barley (*Hordeum vulgare* L.), white mustard (*Sinapis alba* L.) and wild oats (*Avena fatua* L.). *Journal of Applied Ecology*, **9**, 589–608.

Harper J.L. (1977) *Population Biology of Plants*. Academic Press, London.

Harper J.L. & Clatworthy J.N. (1963) The comparative biology of closely related species living in the same area. VI. Analysis of the growth of *Trifolium repens* and *T. fragiferum* in pure and mixed populations. *Journal of Experimental Botany*, **14**, 172–190.

Jones H.G. (1976) Crop characteristics and the ratio between assimilation and transpiration. *Journal of Applied Ecology*, **13**, 605–622.

Lloyd N.D.H. & Woolhouse H.W. (1978) Leaf resistances in different populations of *Sesleria caerulea* (L.) Ard. *New Phytologist*, **80**, 79–85.

Norman J.M. & Jarvis P.G. (1974) Photosynthesis in Sitka spruce (*Picea sitchensis* (Bong.) Carr). III. Measurements of canopy structure and interception of radiation. *Journal of Applied Ecology*, **11**, 375–398.

Norman J.M. & Jarvis P.G. (1975) Photosynthesis in Sitka spruce (*Picea sitchensis* (Bong.) Carr). V. Radiation penetration theory and a test case. *Journal of Applied Ecology*, **12**, 839–878.

Parkhurst D.F. (1978) The adaptive significance of stomatal occurrence on one or both surfaces of leaves. *Journal of Ecology*, **66**, 367–383.

Parkhurst D.F. & Loucks O.L. (1972) Optimal leaf size in relation to environment. *Journal of Ecology*, **60**, 505–537.

Perring F.H. & Walters S.M. (1962) *Atlas of the British Flora.* Botanical Society of the British Isles and Thomas Nelson, London and Edinburgh.

Pigott C.D. (1974) The responses of plants to climate and climatic changes. *The Flora of a Changing Britain* (Ed. by F.H. Perring). Classey Ltd., Faringdon, Berks.

Prince S.D. (1976) The effect of climate on grain development in barley at an upland site. *New Phytologist*, **76**, 377–389.

Rutter A.J. (1966) Studies on the water relations of *Pinus sylvestris* in plantation conditions. IV. Direct observations on the rates of transpiration, evaporation of intercepted water, and evaporation from the soil surface. *Journal of Applied Ecology*, **3**, 393–405.

Rutter A.J. (1967) An analysis of evaporation from a stand of Scots pine. *Forest Hydrology* (Ed. by W.E. Sopper and H.W. Lull). Pergamon Press, Oxford.

Stern W.R. & Donald C.M. (1962) Light relationships in grass—clover swards. *Australian Journal of Agricultural Research*, **13**, 599–614.

Watts W.R., Neilson R.E. & Jarvis P.G. (1976) Photosynthesis in Sitka spruce (*Picea sitchensis* (Bong.) Carr). VII. Measurements of stomatal conductance and $^{14}CO_2$ uptake in a forest canopy. *Journal of Applied Ecology*, **13**, 623–638.

Woodward F.I. (1975) The climatic control of the altitudinal distribution of *Sedum rosea* (L.) Scop. and *S. telephium* L. II. The analysis of plant growth in controlled environments. *New Phytologist*, **74**, 335–348.

Woodward F.I. & Pigott C.D. (1975) The climatic control of the altitudinal distribution of *Sedum rosea* (L.) Scop. and *S. telephium* L. I. Field observations. *New Phytologist*, **74**, 323–334.

INDEX

413